CLIMATE-RESILIENT AGRICULTURE FOR SUSTAINABLE CROP PRODUCTION

New Adaptation and Mitigation Strategies

CLIMATE-RESILIENT AGRICULTURE FOR SUSTAINABLE CROP PRODUCTION

New Adaptation and Mitigation Strategies

Sanjay Kumar, PhD
Shivani Ranjan, PhD
Sumit Sow, PhD
Smruti Ranjan Padhan, PhD
Dhirendra Kumar Roy, PhD
Editors

APPLE ACADEMIC PRESS

First edition published 2026

Apple Academic Press Inc.
1265 Goldenrod Circle, NE,
Palm Bay, FL 32905 USA

760 Laurentian Drive, Unit 19,
Burlington, ON L7N 0A4, Canada

CRC Press
2385 NW Executive Center Drive,
Suite 320, Boca Raton FL 33431

4 Park Square, Milton Park,
Abingdon, Oxon, OX14 4RN, UK

Library and Archives Canada Cataloguing in Publication

CIP data on file with Canada Library and Archives

Library of Congress Cataloging-in-Publication Data

CIP data on file with US Library of Congress

ISBN: 978-1-998511-14-3 (hbk)
ISBN: 978-1-998511-15-0 (ebk)
DOI: 10.1201/9781998511150

About the Editors

Sanjay Kumar, PhD
Professor-cum-Chief Scientist, Department of Agronomy,
Bihar Agricultural University, Sabour, Bhaglapur, Bihar, India

Sanjay Kumar, PhD, is currently working as a University Professor-*cum*-Chief Scientist in the Department of Agronomy at Bihar Agricultural University (BAU), Sabour, Bhagalpur, Bihar, India. His research focuses on sustainability, crop productivity, and climate-smart agricultural practices. He has supervised six postgraduate and PhD students and has 12 years of research and teaching experience at the undergraduate, postgraduate, and PhD levels. He has held various administrative and research management positions, including Coordinator of Conservation Agriculture, Nodal Officer of the Climate Resilient Agriculture Programme, Nodal Officer of the Right to Information (RTI), Principal Investigator for the All India Coordinated Research Project (AICRP) on Integrated Farming Systems (IFS), Chairman of the Department of Agronomy at BAU, Sabour, and Scientist at Krishi Vigyan Kendra (KVK), Giridih (2004–2011).

His research achievements include scaling out zero tillage under the IRRI project. He promoted Conservation Agriculture-based Sustainable Intensification (CASI) technologies through the SRFSI project. He also established innovation platforms and business models, such as the ARANYAK and DEHAAT models, under the same project. Additionally, he developed five new technologies at BAU, Sabour.

He has published more than 40 research articles and 30 popular articles. Additionally, he has served as an editor for several books and as a guest editor for special issues in reputed journals.

Shivani Ranjan, PhD
Assistant Professor, Department of Agronomy, Tirhut College of Agriculture, Dr. Rajendra Prasad Central Agricultural University, Pusa, Bihar, India

Shivani Ranjan, PhD, is an Assistant Professor in the Department of Agronomy at Tirhut College of Agriculture, Dr. Rajendra Prasad Central Agricultural University, Pusa, Bihar, India. She completed her B.Sc. (Ag.) from Dr. Rajendra Prasad Central Agricultural University (Dr. RPCAU), Pusa, Bihar; M.Sc. in Agronomy from Bihar Agricultural University, Sabour, Bihar, India; and PhD in Agronomy from Dr. RPCAU, Pusa, Bihar, India. She has qualified the Indian Council of Agricultural Research National Eligibility Test (ICAR-NET) the University Grants Commission National Eligibility Test (UGC-NET), and is a recipient of the UGC fellowship. She has published 20 research and review articles, 33 popular articles, and 20 book chapters. Dr. Ranjan is a life member of the Indian Society of Weed Science and Indian Society of Agronomy. She has received awards for best poster presentation, best article and best oral presentation. Her research interests include conservation agriculture, carbon footprint management, integrated nutrient management, greenhouse gas emissions, and energy management in the rice-wheat cropping system.

Sumit Sow, PhD
Senior Research Fellow, Soil Health Project funded by Tata Cornell Institute, Cornell University, Ithaca, NY, USA, Department of Soil Science, Dr. Rajendra Prasad Central Agricultural University, Pusa, Bihar, India

Sumit Sow, PhD, is currently working as a Senior Research Fellow, Soil Health Project funded by Tata Cornell Institute, Cornell University, Ithaca, NY, USA, Department of Soil Science, Dr. Rajendra Prasad Central Agricultural University, Pusa, Bihar, India. He completed his graduation from Visva-Bharati, West Bengal, India, and his M.Sc. in Agronomy with an ICAR-Junior Research Fellowship (ICAR-JRF) from Bihar Agricultural University (BAU), Sabour, Bihar, India. He has qualified ICAR-NET and UGC-NET. Dr. Sow has published 10 research articles, 8 review articles, 34 popular articles, and 22 book chapters. He is the author of two books and has served as an editor for four edited books. His research interests include site-specific nutrient management using crop canopy sensors in rice-wheat cropping systems, integrated nutrient management, standardization of planting materials in sugarcane, and weed management in rice-based cropping systems.

Smruti Ranjan Padhan, PhD
Subject Matter Specialist (Agronomy), ICAR-Krishi Vigyan Kendra, East Sikkim, Sikkim, India

Smruti Ranjan Padhan, PhD, is currently working as a Subject Matter Specialist (Agronomy) at ICAR-Krishi Vigyan Kendra, East Sikkim, Sikkim, India. He completed his undergraduate studies at Odisha University of Agriculture and Technology (OUAT), Bhubaneswar, Odisha, India. By securing All India Rank 03 (AIR-03) in Agronomy in the ICAR-AIEEA-PG (JRF), he successfully completed his Master's degree as a topper from the Division of Agronomy at the ICAR-IARI, New Delhi, India. He also qualified for the ICAR-AIEEA-PhD (SRF) entrance examination during his PhD program. He has published several peer-reviewed research and review papers in reputed national and international journals, in addition to contributing many book chapters and popular articles. His current research focuses on various aspects of weed management and conservation agriculture-based nutrient management options for cropping systems under the maize sequence.

Dhirendra Kumar Roy, PhD
Professor-cum-Chief Scientist and Director, Seeds, Dr. Rajendra Prasad Central Agricultural University, Pusa, Bihar, India

Dhirendra Kumar Roy, PhD, is a Professor-*cum*-Chief Scientist and Director, Seeds at Dr. Rajendra Prasad Central Agricultural University (Dr. RPCAU), Pusa, Bihar, India. He has also held esteemed positions such as Deputy Director (Pusa Farm) and head of various regional research stations. With more than 16 years of teaching and research experience, he has guided 23 MSc and 5 PhD students.

In addition, he has successfully completed two prestigious externally funded projects and six consulting projects. Dr. Roy has published 65 research papers, 32 popular articles, 10 technical bulletins, 4 extension bulletins, 6 pamphlets, 3 booklets, 5 folders, 10 book chapters, 3 books, and 1 practical manual. He has also actively participated in the program committees of several national and international conferences and serves on the editorial boards of eminent international journals.

Contents

Contributors

Basit Abrar
Division of Basic Sciences and Humanities, Faculty of Agriculture, Sher-e-Kashmir University of Agricultural Sciences and Technology of Kashmir, Jammu and Kashmir, India

Ashwaq M. Alnemari
Biology Department, College of Science and Humanities, Prince Sattam bin Abdulaziz University, Al-Kharj, Saudi Arabia

Aiman Ashraf
Division of Basic Sciences and Humanities, Faculty of Agriculture, Sher-e-Kashmir University of Agricultural Sciences and Technology of Kashmir, Jammu and Kashmir, India

Malik Asif Aziz
Division of Basic Sciences and Humanities, Faculty of Agriculture, Sher-e-Kashmir University of Agricultural Sciences and Technology of Kashmir, Jammu and Kashmir, India

Purna Kanta Barua
Department of Plant Breeding and Genetics, Faculty of Agriculture, Assam Agricultural University, Jorhat, Assam, India

Adarsha Gopalakrishna Bhat
Division of Agricultural Engineering, ICAR–Indian Agricultural Research Institute, New Delhi, India

Rajarshi Roy Burman
Division of Agricultural Extension, Indian Council of Agricultural Research, New Delhi, India

Luciana Raphael Chappa
Mwalimu Nyerere University of Agriculture and Technology (MNUAT); Department of Agricultural sciences. Butiama-Mara Tanzania

Sharmila Dutta Deka
AICRP on Crops, Department of Plant Breeding and Genetics, Faculty of Agriculture, Assam Agricultural University, Jorhat, Assam, India

Dharminder
Department of Agronomy, Dr. Rajendra Prasad Central Agricultural University, Pusa, Samastipur, Bihar, India

Jerry C. Dlamini
Unit for Environmental Sciences and Management, Faculty of Natural and Agricultural Sciences, North-West University, Potchefstroom, South Africa

Shripati Dwivedi
Department of Agricultural Economics, Dr. Rajendra Prasad Central Agricultural University, Pusa, Samastipur, Bihar, India

Mukul Kumar Gandhi
Department of Agricultural Biotechnology and Molecular Biology, Dr. Rajendra Prasad Central Agricultural University, Pusa, Bihar, India

Gousia Gani
Division of Basic Sciences and Humanities, Faculty of Agriculture, Sher-e-Kashmir University of Agricultural Sciences and Technology of Kashmir, Jammu and Kashmir, India

Mainak Ghosh
Department of Agronomy, Bihar Agricultural University, Sabour, Bhagalpur, Bihar, India

Harun Ireri Gitari
Department of Agricultural Sciences, Mwalimu Julius Nyerere University of Agriculture and Technology, Butiama-Mara, Tanzania

Ratnesh Kumar Jha
Department of Agronomy, Center for Advanced Studies on Climate Change (CASCC), Dr. Rajendra Prasad Central Agricultural University, Pusa, Samastipur, Bihar, India

Abhishek Kumar
Department of Genetics and Plant Breeding, Dr. Rajendra Prasad Central Agricultural University, Pusa, Bihar, India

Navnit Kumar
Sugarcane Research Institute, Dr. Rajendra Prasad Central Agricultural University, Pusa, Bihar, India

Raj Kumar
Department of Agricultural Biotechnology and Molecular Biology, Dr. Rajendra Prasad Central Agricultural University, Pusa, Bihar, India

Rajan Kumar
Department of Agronomy, Dr. Rajendra Prasad Central Agricultural University, Pusa, Samastipur, Bihar, India

Sanjay Kumar
Department of Agronomy, Bihar Agricultural University, Sabour, Bhagalpur, Bihar, India

Shambhavi Kumari
Department of Genetics and Plant Breeding, Dr. Rajendra Prasad Central Agricultural University, Pusa, Bihar, India

Sudhanand Prasad Lal
Department of Agricultural Extension Education (PGCA), Dr. Rajendra Prasad Central Agricultural University, Pusa, Samastipur, Bihar, India

Yamikani Harry Makwinja
Department of Forestry, Ministry of Natural Resources and Climate Change, Blantyre, Malawi

Imtiyaz Ahmad Mir
Division of Plant Pathology, Faculty of Agriculture, Sher-e-Kashmir University of Agricultural Sciences and Technology of Kashmir, Jammu and Kashmir, India

Yasir Hanif Mir
Division of Soil Science and Agricultural Chemistry, Faculty of Agriculture, Sher-e-Kashmir University of Agricultural Sciences and Technology of Kashmir, Jammu and Kashmir, India

Divyanshi Mishra
Department of Agricultural Economics, College of Agriculture, Acharya Narendra Deva University of Agriculture and Technology, Kumarganj, Ayodhya, Uttar Pradesh, India

Harshit Mishra
Department of Agricultural Economics, College of Agriculture, Acharya Narendra Deva University of Agriculture and Technology, Kumarganj, Ayodhya, Uttar Pradesh, India

Dhiman Mukherjee
Regional Research Station (Red and Laterite Zone), Bidhan Chandra Krishi Viswavidyalaya, Jhargram, West Bengal, India

Sweety Mukherjee
Division of Agricultural Extension, ICAR–Indian Agricultural Research Institute, New Delhi, India

Riziki Mwadalu
Central Highlands Eco-Region Research Program, Kenya Forestry Research Institute, Nairobi, Kenya

Yogesh Dashrath Naik
Department of Agricultural Biotechnology and Molecular Biology, Dr. Rajendra Prasad Central Agricultural University, Pusa, Bihar, India

Seema Nargis
Division of Basic Sciences and Humanities, Faculty of Agriculture, Sher-e-Kashmir University of Agricultural Sciences and Technology of Kashmir, Jammu and Kashmir, India

Dibyajyoti Nath
Department of Soil Science, Dr. Rajendra Prasad Central Agricultural University, Pusa, Samastipur, Bihar, India

Vedasto Victor Ngaiza
School of Life Sciences and Bioengineering, The Nelson Mandela Institution of Sciences and Technology, Arusha, Tanzania

Emmanuely Zephaline Nungula
Department of Environment and Sustainable Development, Mzumbe University, Morogoro, Tanzania

Rabindra Nath Padaria
Division of Agricultural Extension, ICAR–Indian Agricultural Research Institute, New Delhi, India

Smruti Ranjan Padhan
Division of Agronomy, ICAR–Indian Agricultural Research Institute, New Delhi, India

Abhishek Raj
Pandit Deendayal Upadhyay College of Horticulture and Forestry, Dr. Rajendra Prasad Central Agriculture University, Samastipur, Bihar, India

Imran Ramzan
Division of Basic Sciences and Humanities, Faculty of Agriculture, Sher-e-Kashmir University of Agricultural Sciences and Technology of Kashmir, Jammu and Kashmir, India

Lalita Rana
Department of Agronomy, Dr. Rajendra Prasad Central Agricultural University, Pusa, Samastipur, Bihar, India

Shivani Ranjan
Department of Agronomy, Dr. Rajendra Prasad Central Agricultural University (RPCAU), Pusa, Samastipur, Bihar, India

Dhirendra Kumar Roy
Department of Agronomy, Dr. Rajendra Prasad Central Agricultural University, Pusa, Samastipur, Bihar, India

Monish Roy
Department of Seed Science and Technology, Faculty of Agriculture, Uttar Banga Krishi Vishwavidyalaya, Pundibari, Cooch Behar, West Bengal, India

Souvik Sadhu
Department of Soil Science and Agricultural Chemistry, Bihar Agricultural University, Sabour, Bihar, India

Shazia Sadiq
Division of Soil Science and Agricultural Chemistry, Faculty of Agriculture, Sher-e-Kashmir University of Agricultural Sciences and Technology of Kashmir, Jammu and Kashmir, India

Sushmita Saini
Division of Agricultural Extension, ICAR–Indian Agricultural Research Institute, New Delhi, India

Mahmoud F. Seleiman
Plant Production Department, College of Food and Agriculture Sciences, King Saud University, Riyadh, Saudi Arabia

Kishore Kumar Sharma
Department of Plant Breeding and Genetics, Faculty of Agriculture, Assam Agricultural University, Jorhat, Assam, India

Priyanka Sharma
AICRP on Seed (Crops), Department of Plant Breeding of Genetics, Faculty of Agriculture, Assam Agricultural University, Jorhat, Assam, India

Gyan Shukla
Department of Extension Education, Institute of Agricultural Sciences, Banaras Hindu University, Varanasi, Uttar Pradesh, India

Sumit Sow
Department of Agronomy, Dr. Rajendra Prasad Central Agricultural University, Pusa, Samastipur, Bihar, India

Diksha Srivastava
Department of Extension Education and Communication Management, Dr. Rajendra Prasad Central Agricultural University, Pusa, Samastipur, Bihar, India

D. Dinesh Varma
Department of Genetics and Plant Breeding, Dr. Rajendra Prasad Central Agricultural University, Pusa, Bihar, India

M. Vennela
Department of Genetics and Plant Breeding, Dr. Rajendra Prasad Central Agricultural University, Pusa, Bihar, India

Abbreviations

ACC	1-aminocyclopropane-1-carboxylic acid
3D CNNs	3D convolutional neural networks
ABA	abscisic acid
ANFIS	adaptive neuro-fuzzy inference system
APEDA	Agricultural and Processed Food Products Export Development Authority
AI	artificial intelligence
ANN	artificial neural networks
BYDV	barley yellow dwarf virus
BCAs	biological control agents
CO_2	carbon dioxide
CC	climate change
CSA	climate-smart agriculture
CBA	community-based adaptation
CBOs	community-based organizations
CSA	community-supported agriculture
CAGR	compound annual growth rate
CNNs	convolutional neural networks
CBA	cost-benefit analysis
DTR	decision tree regression
DT	decision trees
DL	deep learning
DNNs	deep neural networks
DQN	deep recurrent Q-network
DRR	disaster risk reduction
ENSO	El Niño-Southern Oscillation
EU	European Union
ELM	extreme learning machine
FFS	farmer field schools
FFNN	feedforward neural network
FAO	Food and Agriculture Organization
GMOs	genetically modified organisms
GWAS	genome-wide association studies
GS	genomic selection
GBS	genotyping-by-sequencing
GHG	greenhouse gas

GDP	gross domestic product
HTP	high-throughput phenotyping
HR	homologous recombination
IMD	India Meteorological Department
IAA	indoleacetic acid
ICM	integrated crop management
IEWM	integrated expert water management
IPM	integrated pest management
IPCC	Intergovernmental Panel on Climate Change
IFOAM	International Federation of Organic Agriculture Movements
IoT	Internet of Things
ILs	introgression lines
JA	jasmonic acid
KNN	k-nearest neighbors
KVK	Krishi Vigyan Kendra
LR	linear regression
LSTM	long short-term memory
LEISA	low external input sustainable agriculture
ML	machine learning
MJO	Madden–Julian Oscillation
CH_4	methane
MTL	multi-task learning
MLP	multilayer perceptron
MLR	multiple linear regression
NB	Naive Bayes
NBC	Naive Bayes classifier
NAAS	National Academy of Agricultural Sciences
NICRA	National Innovations on Climate Resilient Agriculture
NF	natural farming
NLP	natural language processing
NGS	next-generation sequencing
N_2O	nitrous oxide
NWP	numerical weather prediction
OF	organic farming
PSO	particle swarm optimization
PLFA	phospholipid fatty acids
PPPs	plant protection products
PCA	principal component analysis
PAM	protospacer adjacent motif
QTL	quantitative trait locus
RF	random forest
RFR	random forest regression

ROS	reactive oxygen species
RILs	recombinant inbred lines
RH	relative humidity
ROI	return on investment
RNPs	ribonucleoproteins
SHGs	self-help groups
SSNM	site-specific nutrient management
SOC	soil organic carbon
SOM	soil organic matter
S2S	sub-seasonal-to-seasonal
SVM	support vector machine
SVR	support vector regression
T-RFLP	terminal-restriction fragment length polymorphism
UN	United Nations
WGRS	whole genome resequencing
WSNs	wireless sensor networks
WFP	World Food Program
ZBNF	zero budget natural farming

Foreword

The impacts of climate change on agriculture are increasingly evident, posing significant challenges to global food systems. Shifts in temperature and precipitation patterns disrupt traditional growing seasons and affect crop development, leading to altered yields and reduced nutritional content. Extreme weather events, such as hurricanes, droughts, and floods, cause substantial crop losses, impacting the livelihoods of farmers and threatening food security. Pests and diseases, previously constrained by specific climatic zones, are expanding their ranges, introducing new threats to crops and livestock.

The changing climate also exacerbates water scarcity, with altered rainfall patterns and increased evaporation reducing irrigation availability and further stressing agricultural systems. Beyond immediate production concerns, climate change has broader socioeconomic implications. Smallholder farmers, especially in developing countries, often lack the resources to adapt to rapidly changing conditions, heightening their vulnerability. The interconnected nature of global food supply chains means that disruptions in one region can have far-reaching consequences, potentially leading to food price spikes and increased market volatility.

Addressing the impacts of climate change on agriculture requires a multifaceted approach. This includes implementing resilient farming practices, improving water management, and fostering international cooperation to build adaptive capacities and ensure the long-term sustainability of food production systems.

The edited book, *Climate-Resilient Agriculture for Sustainable Crop Production: New Adaptation and Mitigation Strategies*, offers a comprehensive exploration of innovative strategies designed to revitalize agriculture in the face of a changing climate. The interdisciplinary approach embraced in this book integrates scientific advancements, technological solutions, and sustainable practices to provide a holistic understanding of climate-resilient agriculture.

From site-specific nutrient management and advanced breeding approaches to extension services, the book delves into a spectrum of adaptive measures. Furthermore, it underscores the importance of community engagement, policy frameworks, and international collaborations in adopting resilience at both local and global scales.

I hope that this book will be very useful for individuals and institutions committed to navigating the nexus of agriculture and climate resilience, leading to a harmonious coexistence between humanity and the environment.

I appreciate the efforts of the editors to focus on the emerging technologies to deal with climate change and bring out this book.

—Dr. D. R. Singh
Vice Chancellor,
Bihar Agricultural University Sabour, Bhagalpur, Bihar, India

Preface

Climate change stands as one of the defining challenges of our time, casting a long and ominous shadow over global food security. As we witness the intensification of extreme weather events, shifts in precipitation patterns, and the inexorable rise in temperatures, the impact on agriculture becomes increasingly evident. In the face of this existential threat, the imperative to develop and implement climate-resilient agricultural practices has never been more urgent.

The edited book, *Climate-Resilient Agriculture for Sustainable Crop Production: New Adaptation and Mitigation Strategies*, represents a collective effort to address the intricate relationship between climate dynamics and the complexities of global agriculture. Bringing together insights from eminent scholars, researchers, and dedicated practitioners, this book explores innovative solutions and fosters a deeper understanding of how to cultivate resilience in the face of a changing climate.

The genesis of the idea for this edited book lies in the recognition that the impacts of climate change are not confined to scientific discourse or policy debates but have tangible consequences for the farmers toiling in the fields, the communities reliant on agriculture, and the delicate balance of our ecosystems. This book seeks to transcend disciplinary boundaries, offering a holistic perspective that integrates diverse fields such as agronomy, climatology, genetics, and socioeconomics. By doing so, we aim to provide a comprehensive resource that can guide both researchers and practitioners in navigating the complex challenges posed by climate change to global agriculture.

The chapters encompass the latest advancements in crop science, the deployment of cutting-edge technologies, and the development of sustainable farming practices. From the micro-level intricacies of crop genetics to the macro-scale considerations of global food systems, the contributors illuminate the multifaceted dimensions of climate-resilient agriculture. Importantly, the edited book emphasizes not only the adaptation strategies required to withstand climate impacts but also the mitigation measures essential for creating a more sustainable agricultural future.

As editors, we acknowledge the collaborative nature of this endeavor. The diverse perspectives presented within these pages reflect the collective wisdom of experts who have dedicated their careers to understanding and addressing the challenges posed by climate change in agriculture. We extend our deepest gratitude to the contributors, whose passion and expertise have enriched this volume, making it a valuable resource for scholars, students, policymakers, and practitioners alike.

In the spirit of fostering dialogue and inspiring action, we invite readers to engage critically with the content, reflect on the urgency of the issues at hand, and consider how each of us, in our respective roles, can contribute to a more resilient and sustainable future for global agriculture. This edited book aims to serve as a catalyst for transformative change, guiding us toward a shared vision of climate-resilient agriculture that nourishes both the planet and its people.

—Editors

CHAPTER 1

Artificial Intelligence–Driven Innovations for Climate-Resilient Agriculture

SWEETY MUKHERJEE,[1,*] ADARSHA GOPALAKRISHNA BHAT,[2] RABINDRA NATH PADARIA,[1] SUSHMITA SAINI,[1] and RAJARSHI ROY BURMAN[3]

[1]*Division of Agricultural Extension, ICAR–Indian Agricultural Research Institute, New Delhi, India*

[2]*Division of Agricultural Engineering, ICAR–Indian Agricultural Research Institute, New Delhi, India*

[3]*Division of Agricultural Extension, Indian Council of Agricultural Research, New Delhi, India*

*Corresponding author

ABSTRACT

Climate change continues to endanger global food security by modifying weather patterns, escalating the frequency of catastrophic events, and placing greater environmental strain on agriculture. In this context, incorporating artificial intelligence (AI) innovations into agriculture can be considered a potential strategy to mitigate climate change effects while simultaneously boosting crop output and resource efficiency. With AI-driven precision agriculture, farms can be monitored and managed with unparalleled accuracy using remote sensing, satellite imaging, and Internet of Things devices. These advances reduce water wastage, limit the use of synthetic chemicals, and ultimately reduce agriculture's carbon

Climate-Resilient Agriculture for Sustainable Crop Production: New Adaptation and Mitigation Strategies. Sanjay Kumar, Shivani Ranjan, Sumit Sow, Smruti Ranjan Padhan, and Dhirendra Kumar Roy (Eds.)

DOI: 10.1201/9781998511150-1

footprint. AI-based predictive models also help farmers adapt to changing climatic conditions by enabling proactive decision-making through crop yield predictions, disease outbreak forecasts, and weather pattern analysis. AI-guided smart farming systems encourage agroforestry, cover cropping, and crop rotation techniques that improve soil health and carbon sequestration. Despite these benefits, India remains far from achieving widespread AI adoption due to highly unstructured agricultural data. Overcoming challenges related to accessibility, privacy, and information dissemination is essential. Collaboration among governments, researchers, and the agricultural industry is crucial to ensure fair and responsible AI deployment. In an era of environmental unpredictability, AI holds significant potential for fostering sustainable and climate-resilient food systems.

1.1 INTRODUCTION

Global food production is projected to be directly impacted by climate change. A rise in the average seasonal temperature can reduce the growing season for many crops, lowering their final yield. In places where temperatures are already close to the physiological maximum for crops, warming will immediately affect yields. By 2080, it is anticipated that global agricultural productivity will decrease anywhere between 3% and 16%. In developing nations, where the majority of average temperatures are already close to or beyond crop tolerance levels, agricultural productivity is expected to fall by 10%–25% on average by the year 2080 (Raghuvanshi et al., 2018). Countries with lower average temperatures will experience less of a negative impact (up to 6%) on their agricultural productivity. In fact, some of these countries may even see a slight increase (up to 8%) in productivity. On the other hand, developing countries, which often have higher average temperatures, are predicted to experience a more significant decline in agricultural productivity due to climate change. For example, India may see a drop of 30%–40% in agricultural productivity (Mahato, 2014). Since the 19th century, anthropogenic activities have caused an increase of 0.9°C in the average global temperature; by 2050, this increase is predicted to reach 1.5°C (Eftekhari, 2022).

Climate change is projected to have significant impacts on agriculture through direct and indirect effects on crops, soils, livestock, and pests (Hatfield et al., 2020). Despite the fact that climate change is a gradual process that involves only small changes in temperature and precipitation over extended periods of time, these changes nonetheless have an impact on a number of soil processes, notably those that are connected to soil fertility. According to Qiu et al. (2023), changes in soil moisture conditions, and increases in soil temperature and CO_2 levels as a consequence, are projected to be the principal way that climate change would affect soil. It is a challenging endeavor to increase food production in order to fulfill the increasing needs brought on by the growing population, especially in light of the threats posed by climate change (Bibi and Rahman, 2023). A recent study found that these biotic and abiotic factors cause widespread reductions of 30%–50% in agricultural productivity (Rajput et al., 2021). It can be challenging to determine the extent to which climate change is affecting the physical attributes of soil. The most frequent and important direct effects of climate change on altering soil structure include the potential of raindrops to cause damage, water and surface runoff filtration, and intense rain events (Liu et al., 2023).

Recent research has demonstrated that increased soil microbial activity as a result of rising atmospheric carbon dioxide levels causes a significant reduction in soil organic matter. Thus, it causes a greater turnover of carbon into the atmosphere, resulting in a self-reinforcing cycle. As more carbon is released into the atmosphere, it further contributes to climate change, potentially leading to more warming (Wessolek et al., 2022). According to Nievola et al. (2017), the majority of crops have a rather limited range of survival over high temperatures (40°C–45°C), and heat stress and high temperatures are linked to a number of physiochemical pathways in plants, including cellular damage, membrane lipid peroxidation, and oxidative stress. According to Byers et al. (2018), the yield of crops could drop by as much as 5%–10% in the future for every 1°C increase in temperature. Higher temperatures force plants to complete their growth cycle more quickly. As a result, plants have less time to reproduce, which ultimately reduces production significantly (Taylor et al., 2022). Therefore, to mitigate the adverse effects of climate change and simultaneously enhance agricultural productivity in a sustainable manner, the adoption of climate-resilient practices within the agricultural sector is imperative. However, these challenges are multifaceted and may result in decreased crop yields, compromised crop quality, and elevated risks for farmers. In light of the aforementioned challenges, it is decisive to embrace a range of climate-resilient agricultural practices.

1.2 CLIMATE-RESILIENT PRACTICES IN AGRICULTURE

The Intergovernmental Panel on Climate Change (IPCC) describes climate-resilient agriculture as a sustainable strategy for ensuring food security that combines adaptation and mitigation tactics (IPCC, 2014). These adaptation strategies are climate-resilient if they can achieve resource management (Sain et al., 2017), reduce greenhouse gas (GHG) emissions, and improve carbon sequestration (Campbell et al., 2014). A significant issue still exists in how to increase agricultural productivity while taking into account the effects of climate change and without endangering social as well as environmental resources (Islam et al., 2022). According to IPCC (2014) estimates, there is a possibility of a 25% loss in yield by 2050 due to a rise in temperature. According to the IPCC (2014), the threefold objectives of an agricultural system to be climate-resilient are sustainability in production, climate change adaptation, and climate change mitigation. Alvar-Beltrán et al. (2021) put forward a range of climate-resilient strategies within agriculture, livestock management, agroforestry, and the governance of climate-smart agricultural practices. They also recommended leveraging communication technologies that are fitting, pioneering, cost-effective, and easily accessible to end-users, such as artificial intelligence (AI).

1.3 NECESSITY OF AI AMID CLIMATE CHANGE

The integration of AI into climate resilience efforts is crucial for its predictive abilities, adaptability, and data-driven decision-making. It equips farmers and other stakeholders in agriculture with the tools to better understand, respond to, and mitigate the challenges posed by the changing climate. The essentiality of AI in the climate resilience of agriculture is explained in Table 1.1.

AI tools such as natural language processing (NLP), machine learning (ML), and predictive analytics can bridge the gaps in the existing climate change advisories. Through the utilization of AI, organizations such as the Indian Meteorological Department (IMD) are furnishing farmers and communities with more precise, tailored, and prompt information. The Mausam app by IMD provides data on four essential weather parameters, namely temperature, relative humidity (RH), wind speed, and wind

TABLE 1.1 Necessities of AI in Climate-Resilient Agriculture

Sl. No.	Challenges Arising from Climate Change	Features of AI	Examples
1.	Delay and inaccuracy of climatic data	Real-time and accurate climatic information	Kisan Suvidha (India), Carrot Weather (worldwide)
2.	Generalized, static, and fixed packages of practices	Personalized and context-specific information	Agrivi, Granular, Trimble, FarmERP, FarmLogs, Agworld, AgriWebb, Conservis
3.	Limited accessibility and non-native language barriers	Vernacular language support and accessible climate knowledge	SweetEat FarmBot and Bot for Papaya Natural Farm (chatbots), AI-DISC (India)
4.	Lack of comprehensive insights and practical recommendations	Decision support systems based on historical data	Climate FieldView platform
5.	Delayed and overdue notification systems	Early warning systems	The Weather Company, IBM
6.	Unidirectional flow of information	Feedback loops	FarmBeats (Microsoft)

direction. This information is updated every three hours and predicts weather conditions for the upcoming seven days, thus augmenting climate resilience in Indian agriculture and other domains (Rutenberg et al., 2021).

1.4 APPLICATION OF EMERGING TECHNOLOGIES IN AGRICULTURE

The term AI describes the simulation of human intelligence in machines that have been designed to think, learn, and solve problems in a manner similar to that of humans. Scientifically, AI refers to the algorithms and computational models that allow computers to process data, spot patterns, and come to conclusions. It includes subfields like NLP, ML, and neural networks, which enable machines to gain knowledge, perform better, and adjust to new tasks. AI systems are designed to emulate human cognitive abilities, such as perception, reasoning, problem-solving, and language understanding. This allows for applications in a wide range of fields, including agriculture. By utilizing the learning capabilities of measuring and comprehending data, ML has begun to open up opportunities in the agricultural fields. ML refers to the process of building a mathematical model on sample data sets, also known as training data, to make predictions and decisions. It is a sub-branch of AI (Sevli and Kemaloglu, 2020) and is based on learning models as well as making predictions using data. The ML algorithms are used to train, classify, and test (sample) data for prediction. Misra et al. (2020) have categorized four major paradigms of ML algorithms in agriculture (Figure 1.1), which are explained in subsections.

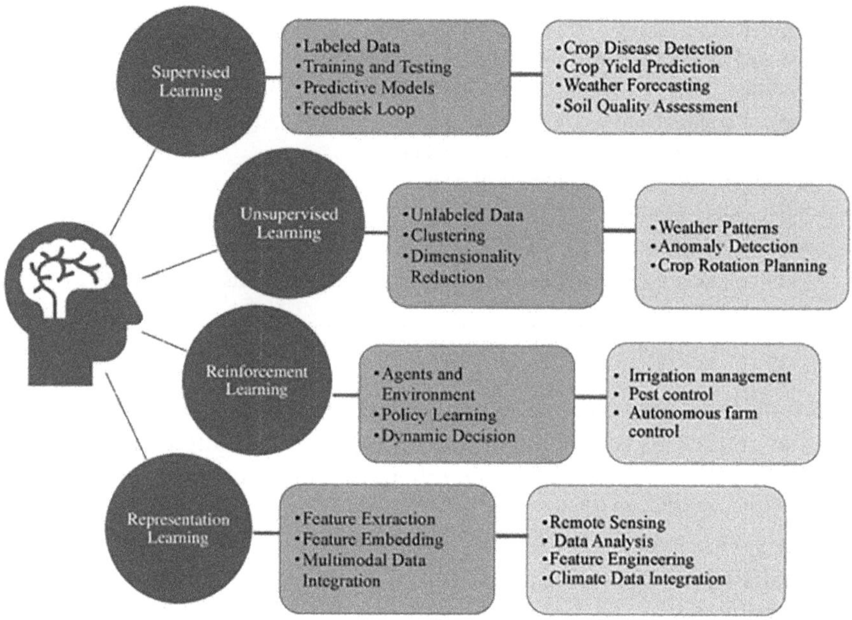

FIGURE 1.1 ML paradigms, along with their features and applications in agriculture.

1.4.1 SUPERVISED LEARNING

The input data and corresponding output labels are paired and trained using labeled data in this method. The algorithm learns to map the input data to the correct output by identifying patterns and relationships in the labeled training set. The "Plantix" app is an example of an existing agricultural model that makes use of supervised learning. This app analyzes images of affected plants to assist farmers in diagnosing plant diseases and nutrient deficiencies. A sizable dataset of labeled photos depicting a range of diseases and nutrient deficits is used to train the model on the physical traits and health of plants. After being trained, the model will accurately diagnose newly acquired plant images and help farmers take the necessary precautions to safeguard their crops. This application serves as a real-world illustration of how supervised learning can be used in agriculture to manage plant health.

1.4.2 UNSUPERVISED LEARNING

These algorithms are utilized when the input data is unlabeled and concentrate on uncovering any hidden relationships, structures, or patterns in the data. This algorithm lacks specific output labels to direct the learning process, in contrast to supervised learning. It includes clustering, where similar data points are grouped together, and dimensionality reduction, which aims to reduce the number of variables in the dataset while preserving its essential features. The International Center for Tropical Agriculture, Colombia, developed the "Climate Smart Agriculture Prioritization Tool" based on this algorithm. It assists in prioritizing agricultural interventions and strategies for climate-smart farming practices by identifying areas with comparable agro-climatic conditions.

1.4.3 REINFORCEMENT LEARNING

These algorithms continuously interact with the environment, sometimes affecting it, and receive rewards for each action; the objective here is to maximize the reward over a course of actions and iterations with the environment. The Microsoft "FarmBeats" project optimizes crop management and irrigation by utilizing reinforcement learning. The data gathered from farms via sensors are utilized to continuously interact with the environment and make decisions in real time. In order to maximize the rewards like crop yield and water conservation, the system learns as well as adjusts over time. Through constant learning and adjustment to the ever-changing farm environment, FarmBeats contributes to increased agricultural sustainability and productivity.

1.4.4 REPRESENTATION LEARNING

This learning, also referred to as feature learning, automatically identifies and extracts meaningful representations as well as features from raw data that are necessary for tasks like detection or classification. It adapts and learns useful features directly from the complex and high-dimensional datasets, in contrast to conventional methods where features are handcrafted by experts. The LeNet model to detect and classify diseases in the soybean plant makes use of this learning. This system uses deep learning (DL) methods, like convolutional neural networks (CNNs), to automatically identify and extract pertinent features from images in order to identify and categorize illnesses and other health concerns. Instead of manually creating features, the model gains the ability to identify distinct patterns and attributes within the images.

DL algorithms add depth and complexity to the model, enhancing the traditional ML model (Schmidhuber, 2015). The term deep learning was coined by Igor Aizenberg, who utilizes deep layers with multiple processing levels to model abstractions in DL networks, which consist of more than one hidden layer. By self-learning information that is passed through hidden layers, DL generates an output. Convolutional Neural Networks, Recurrent Neural Networks, Restricted Boltzmann Machines, and Deep Belief Networks are just a few of the algorithms it uses. The benefits of DL include processing unstructured data as much as possible, delivering high-quality results, and avoiding unnecessary costs. On the other hand, it has some drawbacks, like requiring a large amount of data and being expensive for both hardware and software. It has a variety of applications, such as face recognition, driverless vehicles, image processing, natural language processing, and personalized shopping planning.

1.5 RELATION BETWEEN AI, ML, AND DL

DL operates as a subset within the broader domain of ML, which in turn is a component of AI, as shown in Figure 1.2. The primary goal of AI is to build intelligent machines, whereas ML gives computers the ability to learn from data and gradually improve their performance. While DL focuses on multilayer neural networks that closely resemble the structure of the human brain, it facilitates the recognition of intricate patterns as well as features in the information they receive. DL techniques are highly beneficial in the AI field because of their exceptional performance in tasks like speech and image recognition, NLP, and decision-making. Algorithms of DL make use of hierarchical layers of representation to improve learning. This makes it possible for machines to comprehend and analyze data in ways that were previously difficult, thereby contributing to the advancement of AI technologies (Vázquez-Ramírez et al., 2023).

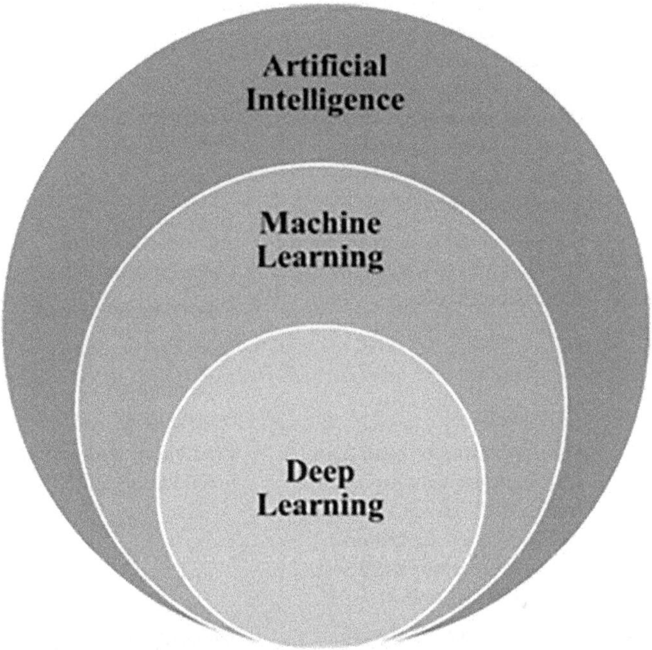

FIGURE 1.2 Overlapping between AI, ML, and DL.

1.6 INNOVATIONS OF AI IN CLIMATE RESILIENT AGRICULTURE

Climate-resilient agriculture innovations are being driven in large part by AI. Significant obstacles to agriculture are brought about by climate change, including severe weather events, altered growing seasons, and an increase in pests and diseases. AI tools can assist farmers in overcoming these obstacles and improving the sustainability of agriculture.

1.6.1 WEATHER FORECASTING AND CLIMATE MONITORING

AI is changing climate monitoring and weather forecasting, assisting agriculture, as well as reducing the effects of extreme weather. A brief overview of different AI methodologies used for predicting climate change parameters has been mentioned in Table 1.2.

TABLE 1.2 AI Methodologies Used for Estimating Different Climate Change Parameters

Climate Change Parameter	Methodologies Used	Application	References
CO_2	Principal component analysis (PCA), machine learning models: Linear regression (LR), random forest regression (RFR), decision tree regression (DTR), support vector regression (SVR), naïve bayes (NB), k-nearest neighbors (KNN)	• CO_2 level prediction • Emission source identification • Environmental policy planning	Kumar (2023)
Sea level rise	Adaptive neuro-fuzzy inference system (ANFIS), multiple linear regression (MLR)	• Accurate sea level change prediction • Coastal management • isaster prevention • Adaptability to coastal areas	Tur et al. (2021)
Global mean temperature changes	Data preparation techniques (differencing, power transforms, scaling), algorithms (KNN, Huber regression, Lasso regression, XGBoost, LightGBM, Prophet)	• Prediction of global mean temperature	Niyogi and Shrinivasan (2022)
Precipitation	Adaptive network-based fuzzy inference system optimized with particle swarm optimization (PSO-ANFIS), artificial neural networks (ANN), support vector machine (SVM)	• Daily rainfall prediction	Pham et al. (2020)
Cyclone	Artificial neural networks (ANN)	• Indian ocean cyclone	Chand et al. (2022)
Variability	Support vector machine (SVM), ANFIS, feedforward neural network (FFNN), gaussian models, long short-term memory (LSTM) neural network, decision tree regression (DTR), random forest regression (RFR), multilayer perceptron (MLP), ridge regressioN	• Concentration of greenhouse gases • Albedo effect • Increase in average surface temperature • Melting of ice bodies • Impact on hydro climatic systems	Ramírez et al. (2023)

Although there have been little steps forward in the field of numerical weather prediction (NWP), issues still arise with physical process representation, ensemble forecasting, and model initialization. Global action is required to prevent climate change, with regional initiatives like the European Green Deal and agreements like the Paris Agreement combating emissions. The accuracy of weather forecasting is improved by incorporating observations into NWP models, which are based on global satellite data. But AI can enhance current models using a residual learning technique, which is a supplement to NWP. Residual learning, which is frequently connected to deep learning architectures, employs shortcut connections or skip connections to bypass one or more layers in a neural network. In the context of NWP, deep neural networks intended for weather forecasting tasks can utilize residual learning. Rather than directly predicting the output, residual learning trains the network to predict the difference between the observed and expected outputs. The final prediction is then obtained by adding this residual, or the difference, to the block's input.

Benefits of new DL techniques as enhancements to traditional NWP methods:

- Customized observation parameters are the foundation of traditional NWP data assimilation, which is labor-intensive and frequently involves data thinning, which lowers the utilization of observations.
- DL methods simplify the process of learning from data, identifying hidden relationships, and making the best use of available observations.
- The ability of DL to fully replace conventional NWP methods is currently limited by its reliance on large amounts of training data. However, residual learning can enhance and innovate upon the physical models in NWP models.
- The latest 4D-Var NWP data assimilation process requires a lot of computation and iteration, whereas the training of the DL method optimizes the cost function only once in an offline mode, leading to operational computational efficiency. Because of its adaptability, DL can produce forecasts that are both probabilistic and deterministic, which streamlines computation and improves forecasting (Dewitte et al., 2021).

The impact of carbon emissions on global warming, the jet stream, and Arctic blocking patterns, often misrepresented in conventional models, is effectively addressed by ML-based sub-seasonal-to-seasonal (S2S) forecasting. CNNs outperform classical models in decadal climate prediction, overcoming the challenge of limited observational data through transfer learning. Earth system science benefits from the revolutionary power of DL since it reveals intricate linkages among massive datasets, providing valuable insights into the complex dynamics of climate change. AI is transforming our understanding of intricate environmental processes by improving satellite remote sensing, weather forecasting, and observation systems. Other applications of weather forecasting and climate monitoring with AI are described in the subsequent paragraphs:

1.6.1.1 AI AND CLIMATE MODELS

Climate models concentrate on long-term changes, especially in the oceans and land cover. They are similar to NWP models but do not incorporate real-time data. Anthropogenic carbon emissions are the primary cause of climate change, which leads to global warming and has an effect on the Arctic. Mechanisms including large-scale circulation modes in the climate system such as the El Niño-Southern Oscillation (ENSO), the Madden-Julian Oscillation (MJO), the Indian Ocean Dipole, and the North Atlantic Oscillation have a known influence on weather phenomena, including extreme events. These changes are difficult to accurately simulate by conventional climate models. S2S model forecasting systems have demonstrated enhanced abilities in identifying these mechanisms (White et al., 2017). CNNs have outperformed traditional models in decadal climate prediction, successfully predicting ENSO events well in advance. The problem of insufficient training data has been tackled by transfer learning, which applies AI models first on simulations and then refines them with available observational data.

1.6.1.2 AI INTEGRATION IN EARTH SYSTEM SCIENCE

AI applications are highly suitable for Earth system science, as it is enhanced by data from diverse sources. These techniques have gained recognition for their potential to supplement or replace traditional elements in these fields. AI combined with data from the Earth system promises to transform our understanding of complex environmental processes. For example, improving the observation system and atmospheric analysis to estimate precipitation. It can provide more precise and scalable solutions for a broad range of applications.

1.6.2 *CROP MODELING*

Crop models powered by AI can predict how different crop varieties will perform in a range of climatic circumstances. This aids farmers in modifying planting schedules and selecting the best crops for their area. ANNs are a good fit for modeling the complex factors influencing crop growth because of their well-known ability to capture complex nonlinear relationships. These models have key components like temperature, rainfall, and soil type, which underline their importance in comprehending crop development. CNNs are especially useful for spatial and temporal data analysis, which is essential for applications like disease detection and plant health evaluation. Deep neural networks (DNNs) and long short-term memory (LSTM) networks were also discovered to be extensively utilized in addition to CNNs. These models capture the dynamic nature of agricultural systems and forecast crop growth over time. Considering these characteristics, the aforementioned methods are revolutionizing the forecasting of crop yields, thereby offering a practical way to increase agricultural productivity and sustainability. The various DL techniques in crop growth modeling (Figure 1.3) are explained in subsections.

1.6.2.1 *DEEP NEURAL NETWORKS (DNN)*

DNNs are an expansion of traditional ANNs. Their ability to model intricate relationships within the data sets them apart, as they possess numerous hidden layers. Although DNNs and ANNs are similar, their depth allows them to identify complex patterns and representations in the data, but their differences become apparent when determining intricate and nonlinear relationships between variables that affect crop yield. For example, DNNs can show how intricate interactions between temperature, humidity, and irrigation levels affect crop growth, which may be difficult for ANNs with their shallower structures to capture. Because of this depth, DNNs are able

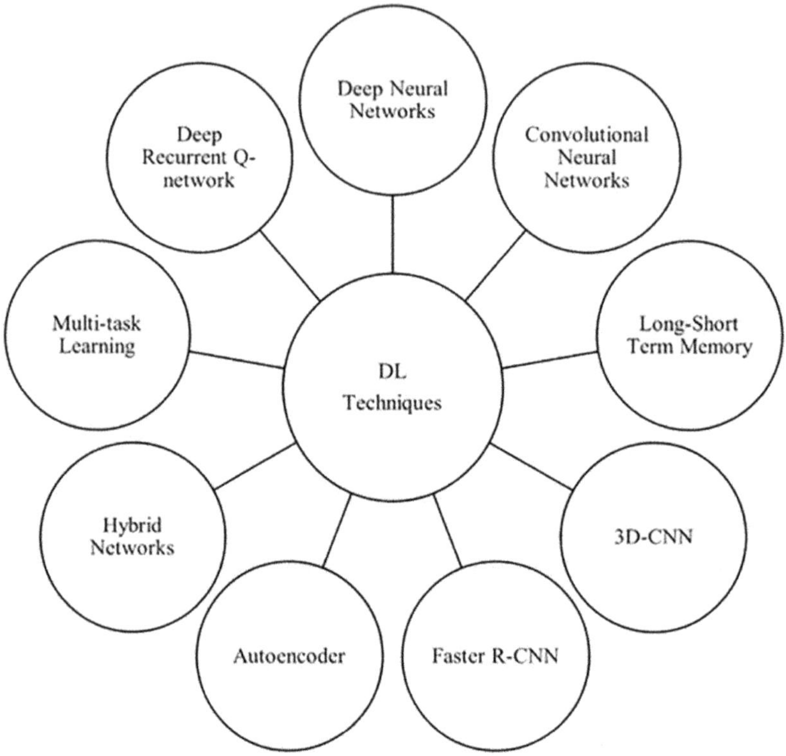

FIGURE 1.3 DL techniques in crop growth modeling.

to predict crop yields with greater accuracy and flexibility, which makes them an invaluable tool for farmers looking to optimize their practices in the face of shifting environmental factors and variables.

1.6.2.2 CONVOLUTIONAL NEURAL NETWORKS (CNN)

CNNs work well for analyzing spatial and image data. They are made up of three different kinds of layers: convolutional, pooling, and fully connected layers. Feature maps are the result of the filters that convolutional layers use to identify features in images. By pooling layers, overfitting is reduced and feature maps are down-sampled. Predictions are typically made using fully connected layers. In order to extract important growth-stage-related features, like leaf size and color changes, CNN's convolutional layers must first gather a diverse dataset of high-resolution crop images.

In order to reduce computational complexity, subsequent pooling layers down-sample these features while preserving important information and removing unnecessary details. Subsequently, fully connected layers integrate these attributes to categorize the stage of growth, providing accurate classification, even for minute modifications. Real-time updates on crop growth are ensured by this non-invasive, ongoing monitoring method, which helps with resource allocation. Because of their scalability, CNNs can effectively manage large-scale agricultural operations, which makes them a vital tool for precision agriculture in crop growth modeling.

1.6.2.3 LONG-SHORT TERM MEMORY (LSTM)

Long-short term memory (LSTM) is very useful in applications involving time-series data because they were created especially for sequence prediction problems. Vanilla LSTM, stacked LSTM, CNN-LSTM, encoder-decoder LSTM, bidirectional LSTM, and generative LSTM are examples of LSTM architectures. Because LSTMs have internal memory cells and gating mechanisms to control information flow, they can model sequential and variable-length data, which overcomes some of the limitations of traditional feedforward neural networks.

1.6.2.4 3D CONVOLUTIONAL NEURAL NETWORKS (3D CNNS)

This particular version of CNN functions in three dimensions, taking depth, length, and height into account. It is frequently used in applications like plant imaging and video analysis, where three-dimensional data is essential. 3D CNNs employ spatiotemporal feature capture through the use of 3D convolutions in their layers.

1.6.2.5 FASTER REGION-BASED CNN (FASTER R-CNN)

Faster R-CNN belongs to the R-CNN family that is optimized for object detection tasks. It introduces a Region Proposal Network, which is an improvement over earlier iterations such as R-CNN and Fast R-CNN. By interpreting features taken from the CNN, this network improves the speed and precision of plant, insect, and disease detection in pictures.

1.6.2.6 AUTOENCODER

The four primary parts of an autoencoder, which is an unsupervised learning model, are an encoder, a bottleneck, a decoder, and a reconstruction loss. By encoding data into a lower-dimensional space and then decoding it to reconstruct the original data, they seek to learn effective representations of the data. Applications for autoencoders include anomaly detection, feature learning, and climate change data compression in agriculture. Climate change data compression refers to the process of reducing the size or volume of climate-related datasets to make it more manageable for storage, transmission, and analysis without significant loss of important information while retaining essential information. Climate data often involves large amounts of information collected from various sources such as satellites, weather stations, and climate models. An autoencoder can learn to encode the relevant features of the data into a lower-dimensional space (the bottleneck) and then decode it to reconstruct the original data with minimal loss.

1.6.2.7 HYBRID NETWORKS

Hybrid networks are an innovative approach to utilize the unique strengths of various DL algorithms by combining them in creative ways. For example, in tasks like crop yield prediction, researchers can combine LSTMs and CNNs to handle both image and sequential data.

1.6.2.8 MULTI-TASK LEARNING (MTL)

It is an ML paradigm in which models are trained to execute several tasks at once in an effort to enhance the efficiency of each task. By enabling representations to be shared across tasks, MTL has the potential to enhance learning and boost model efficiency, which is useful for providing farmers with climate change and resilience-related advisories.

1.6.2.9 DEEP RECURRENT Q-NETWORK (DQN)

DQN is an algorithm for reinforcement learning that combines Q-learning with the capabilities of DNNs. It was designed to allow agents to observe their surroundings, act, and maximize rewards in order to tackle complex tasks.

These DL innovations provide a broad range of tools to address different data-driven problems in climate-resilient agriculture. They are indispensable parts of the contemporary ML packages due to their strength and adaptability in emerging climate change scenarios.

1.6.3 PEST AND DISEASE MANAGEMENT

AI techniques use image recognition and sensor data for the early detection of diseases and pests, as explained in Table 1.3. This makes it possible for farmers to proactively guard against crop damage.

TABLE 1.3 AI Techniques for Pest and Disease Management in Climate-Resilient Agricultural Practices

Sl. No.	AI Techniques	Characteristics of AI in Pest and Disease Management
1.	Random forest (RF) for pest detection	RF is a machine learning (ML) method used in agriculture to identify and detect pests using spectral images and fusion features. It has been effective in identifying diseases such as Fusarium head blight in wheat and evaluating crop damage (Chithambarathanu & Jeyakumar, 2023)
2.	Support vector machine (SVM) for pest identification	SVM applies image processing and automated pest sensing to identify plant species and detect pests and diseases with high accuracy. It can also differentiate between healthy and unhealthy plants based on characteristics such as thickness, color, structure, and smell. This method has been successfully used in pest management for strawberry plants (Ebrahimi et al., 2017).
3.	Pest identification with decision trees (DT)	DT employs classification techniques and advanced object classification algorithms, such as Convolutional Neural Networks (CNNs), to enhance pest detection. Researchers have used this method to identify *Alternaria solani* in potato crops (Van et al., 2020) and citrus Huanglongbing using hyperspectral imaging (Deng et al., 2019). This technique is also promising for applications such as pest and disease mapping, assessing plant chlorophyll sufficiency levels, and utilizing drones or tractors for precision agriculture.
4.	Naive bayes classifier (NBC) for pest identification	NBC is used for multi-class prediction in agricultural pest identification. It processes and classifies Landsat 8 OLI images to detect coffee berry necrosis. Additionally, it automates species identification and early disease detection in herbs, providing a portable and user-friendly solution for researchers and farmers. This method has also been applied to develop an ontology for classifying soil, insect, and weather categories based on agricultural data (Miranda et al., 2020)

TABLE 1.3 *(Continued)*

Sl. No.	AI Techniques	Characteristics of AI in Pest and Disease Management
5.	CNN for pest detection	CNN leverages image-based pest detection by analyzing photos of healthy and diseased plants. It accurately identifies and categorizes plant diseases and pests. Recent advancements have improved the fine-grained classification of diseased plants while reducing the number of parameters required for analysis (Wang et al., 2023; Zhao et al., 2022)
6.	LSTM for pest detection	LSTM networks are crucial for predicting and monitoring pest outbreaks. By analyzing environmental data from weather stations, they help farmers anticipate and manage potential pest threats. These models assess the relationship between pest activity and environmental conditions, providing early warnings that minimize crop damage and optimize economic outcomes (Domingues et al., 2022; Xiao et al., 2018)
7.	DBNs for identifying pests	DBNs analyze intricate features, textures, and patterns of pests in images. They provide highly accurate identification and are particularly useful for detecting subtle symptoms of diseases or pests that other methods may overlook. DBNs are commonly employed for detailed analysis and complex pattern recognition in pest and disease management (Cristin et al., 2020; Shoaib et al., 2023)

1.6.4 AUTOMATED IRRIGATION AND NUTRIENT MANAGEMENT

AI is able to control irrigation systems, thereby reducing water waste and ensuring ideal crop growth. Precision irrigation is a prime illustration of the potential of AI in the context of irrigation optimization. This method generates personalized irrigation schedules, guaranteeing that crops get the appropriate amount of water at the right time and place. By combining information from various sources, such as soil moisture sensors and weather forecasts, it is possible to avoid over-irrigation, reduce water loss from evaporation and runoff, and deliver water where it is most needed. Tace et al. (2022) emphasized irrigation optimization using K-Nearest Neighbors (KNN) to create models that had an extremely low root mean square error (RMSE) of 0.12 and a recognition rate of 98.3%. Zubaidi et al. (2020) assessed the effectiveness of an Internet of Things (IoT)-based integrated expert water management (IEWM) system, which showed an accuracy of 98.7%. On the other hand, conventional water management systems showed an accuracy of 87%, making IEWM an effective tool for agricultural water management.

AI also plays a key role in dynamic nutrient management, which is essential for sustainable agriculture. It evaluates variables like crop type, growth stage, soil nutrient levels, and environmental factors to determine the ideal time and location for fertilizer applications. Talaviya et al. (2020) examined applications of AI in agriculture, focusing on how AI improves weeding, spraying, and irrigation. Suchithra and Pai (2020) created a system that used an AI model, especially the extreme learning machine (ELM) algorithm, to predict and classify pH values and soil fertility indices according to a variety of soil properties. Their research produced accurate results, especially when it came to classifying pH values and predicting the potassium fertility index. This data-driven strategy minimizes environmental pollution from excessive fertilizer use, improves crop nutrient uptake, and lowers the risk of nutrient imbalances.

1.6.5 SUPPLY CHAIN OPTIMIZATION AND MARKET ANALYSIS

By expediting food delivery, cutting down on food waste, and optimizing transportation routes, AI is boosting the effectiveness of the agricultural supply chain. With its ability to analyze consumer demand and market trends, AI can assist farmers in making well-informed decisions about which crops to plant and where to sell their produce, which improves market analysis and supply chain optimization (Figure 1.4).

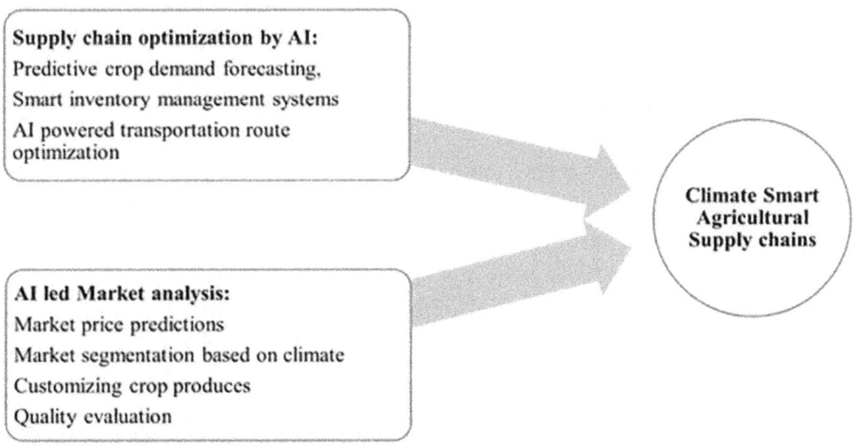

FIGURE 1.4 Smart agricultural supply chains for climate resilient agriculture.

1.6.6 ROBOTIC FARMING

AI-driven robots can carry out a range of farm tasks, such as planting, weeding, and harvesting, which lowers the need for human labor and boosts productivity. The life cycle of various agricultural operations where AI-enabled robotics are being used is explained in the following subsections.

1.6.6.1 WEED DETECTION

Agricultural weed detection is enhanced by advanced computer vision methods like YOLOv3, Mask R-CNN, Faster R-CNN, and semantic graphics. High-precision crop and weed differentiation is made possible by these techniques. The closeness of the vision sensors to the objects of interest, along with CNN for precise classification and image segmentation, is what makes them so effective (Tur et al., 2021).

1.6.6.2 CROP SCOUTING

Crop scouting is a systematic field monitoring approach in agriculture aimed at assessing the health, growth, and potential issues affecting crops. It involves the visual inspection of crops to identify pests, diseases, nutrient deficiencies, and other factors that may impact crop yield. AI plays a pivotal role in enhancing crop scouting through advanced image recognition and analysis. Scientific techniques like IR radiometers, OptRX sensors, color cameras, and spectral cameras are useful tools for measuring growth and health. Integrating neural networks for plant classification and CNN for object detection allows for the extraction of valuable information from sensor data.

AI-powered technologies, such as computer vision algorithms, can process large volumes of high-resolution satellite or drone imagery to automatically detect and classify crop features and anomalies. These algorithms can identify subtle visual cues indicative of plant stress, disease symptoms, or pest infestations with high accuracy. By automating the crop scouting process, AI enables rapid and consistent analysis of vast agricultural landscapes, providing farmers with timely and precise information. This allows for targeted interventions, optimizing resource use, minimizing crop losses, and ultimately improving overall agricultural productivity.

1.6.6.3 PHENOTYPING

To measure plant traits, phenotyping systems use vision systems that are primarily fitted with RGB cameras. In certain instances, extra sensors—such as stereo cameras—help to take particular measurements, like plant height and stem diameter. AI integration is required to extract trait information from the data collected by vision sensors. This necessitates the use of logistic regression and CNN to standardize measurements, where calibration is essential.

1.6.6.4 DISEASE, INSECT, DEFICIENCY DETECTION

Detection of diseases, insect infestations, and deficiencies in plants depends on a range of scientific methods. To improve detection capabilities, these include the use of RGB-D sensors, spectral cameras, color cameras, and thermal cameras. SVM, K-means, and neural networks are examples of advanced AI techniques that provide high-accuracy identification. Detection accuracy for many kinds of diseases or insect infestations can be higher than 90%. Additionally, in disease management, rule-based systems have been used to diagnose plant diseases and suggest control measures. Artificial neural network models have been utilized for disease control in various crops, including tomato, wheat, and jute. Hybrid systems, combining image processing and neural networks, have been used for disease classification and detection (Bannerjee et al., 2018).

1.6.6.5 ROBOTIC SPRAYING

RGB cameras are the main tool used in robotic spraying in agriculture; multispectral and Kinect sensors are also being investigated. Advanced guidance strategies are used to optimize spraying. In order to achieve effective spray coverage, these strategies incorporate experiments with data fusion algorithms, spectral indices, and genetic algorithms, as well as image processing techniques like segmentation and color transformation.

1.6.6.6 ROBOTIC HARVESTING

This approach addresses the problem of labor shortage caused by climate change, leading to the migration of labor away from agriculture and increasing labor costs (Saini et al., 2023). The vision component of these systems is crucial, and it frequently uses RGB cameras for fruit detection. It is common practice to process images using color indices. However, problems caused by fluctuating outdoor lighting conditions compel researchers to use non-linear operations like gamma correction and transformations to replace color spaces. Sensors that can provide three dimensions, like stereo cameras, make harvesting easier once the fruit has been successfully identified.

Modern farming practices are expected to undergo further revolution due to advancements in technology and AI algorithms. These scientific methods are revolutionizing agriculture by improving efficiency, reducing environmental impact, and addressing labor shortages (Fountas et al., 2022). AI applications are assisting farmers in promoting sustainable and climate-resilient farming practices, increasing agricultural productivity, decreasing resource waste, and adapting to changing climate conditions. AI is anticipated to become increasingly important in agriculture in the future as technology develops, helping with international efforts to combat climate change and ensure food security.

1.7 CHALLENGES IN ADOPTING AI

When considering the benefits of this technology for sustainable and climate-resilient farming, adopting AI may seem like a logical step for every farmer. But there are still some important limitations:

1. **Lengthy Technology Adoption Process:** Farmers need to be aware that AI is merely an advanced component of simpler technologies for processing, gathering, and monitoring field data.
2. **Infrastructure Development:** The proper technological infrastructure is required for AI to work. As a result, farms with some technology in place may still find it difficult to grow. Similar challenges are also confronted by software companies. Farmers should be gradually exposed to technology, beginning with simpler tools like an agro-trade platform.
3. **Inexperience with Cutting-Edge Technologies:** The agricultural industries in developing countries are different from those in Western Europe and the U.S. AI in agriculture is difficult to market where agricultural technology is uncommon. Its adoption by farmers will probably need assistance.

4. **Capacity Building of Stakeholders:** In the developing agricultural economies, tech giants need to adopt a proactive strategy for capacity building of farmers and supply chain stakeholders. For farmers and owners of agribusinesses who are eager to adopt cutting-edge solutions in addition to their products, they must provide training and continuous support.
5. **Privacy and Security Issues:** Due to the lack of established norms and regulations surrounding the use of AI in general, not just in agriculture, precision agriculture and smart farming raise a number of legal issues that frequently go unanswered. Threats to farmers' privacy and security, like cyberattacks and data leaks, may cause them serious problems.
6. **Cost:** Implementing AI and robotic systems can require significant investments in terms of hardware, software, and training. The initial setup costs and ongoing maintenance expenses may pose financial challenges for farmers, especially small-scale ones (Balaska et al., 2023).

1.8 ETHICAL CONCERNS RELATED TO AI IN CLIMATE SMART AGRICULTURE

One document that acknowledges the need to control the safety risks posed by advanced AI is the Bletchley Declaration. It was made public during the AI Safety Summit in November 2023 at Bletchley Park, United Kingdom, where 28 nations, including India, China, and the United States, signed the declaration. The Bletchley Declaration is significant because it is an important step toward minimizing the possible risks and negative effects of AI while optimizing its positive effects (Radanliev and Santos, 2023).

The use of AI in predicting climate change impacts raises ethical concerns such as privacy, data ownership, and the potential for unintended consequences (Rutenber et al., 2021). There is a risk of data being used nefariously by corrupt governments, competitors, or market traders (Ryan, 2019). Ryan (2022) identified several gaps in both the agricultural literature and AI ethics guidelines regarding the social and ethical impacts of AI in agriculture. Transparency, employment, and the limited discussion of solidarity and dignity were among the main gaps.

AI systems in agriculture often rely on large amounts of data, including personal and sensitive information. Ensuring that data is collected and used with consent and respecting the rights and privacy of farmers and individuals is essential. It is also vital to ensure that the benefits of AI in agriculture are accessible to all, including small-scale farmers and marginalized communities. It should neither exacerbate existing inequalities nor create new ones while also ensuring transparency in decision-making. The energy consumption and carbon footprint of AI technologies should be considered, especially if they are powered by fossil fuels, as striving for environmentally friendly AI solutions is important for climate-resilient agriculture. As these technologies become more prevalent in agriculture, there may be concerns about job displacement for farm workers. Therefore, governments, policymakers, and researchers must include retraining and providing alternative opportunities for those affected. Ensuring that AI solutions are sustainable in the long run, both environmentally and economically, is crucial for climate resilience. Farmers should be adequately informed about the capabilities and limitations of AI systems, ensuring they make informed decisions about their use. Respecting and incorporating the distinct cultural and ethical values related to agriculture in different communities and regions is crucial for AI applications.

1.9 CONCLUSION

The groundbreaking potential of AI in transforming agricultural practices in the era of climate change challenges has been highlighted in this chapter. As we delved further, it became clear that AI-based solutions like predictive analytics, smart irrigation, precision agriculture, and crop monitoring are essential for safeguarding the agriculture sector from the adverse effects of changing climate patterns. The dynamic nature of climate change necessitates constant advancements in AI technology. Funding research not only improves the effectiveness of currently available solutions but also creates novel techniques that can foresee and address new problems in agriculture. Furthermore, strategic actions must be taken to increase the efficacy of AI in reducing the negative impacts of climate change. The establishment of a robust infrastructural framework is paramount. AI innovations are based on a foundation of advanced computing capabilities, dependable data storage facilities, and high-speed internet connectivity. A smooth data flow guarantees prompt and accurate decision-making, giving farmers practical insights. The investment made in developing a skilled labor force is equally important. Quality labor with knowledge of AI technologies is indispensable. A skilled workforce equipped to leverage AI for sustainable agriculture should be developed through training programs, workshops, and educational initiatives. These competent professionals are essential for the creation, application, and improvement of AI solutions. Additionally, fostering a prompt and extensive knowledge base is also crucial. Collaborating with research institutes, meteorological agencies, and agricultural specialists makes it easier to gather a variety of pertinent data. In essence, a ray of hope in the face of climate change is the synergy between artificial intelligence and sustainable agriculture. We can harness the full potential of AI if we encourage a culture of innovation, make investments in manpower and infrastructure, and maintain an extensive knowledge base. Through these concerted efforts, we are bolstering the resilience of agriculture while also paving the path for a more sustainable and climate-resilient future.

KEYWORDS

- **agroforestry**
- **anthropogenic activities**
- **artificial intelligence**
- **carbon sequestration**
- **climate change**
- **crop rotation**
- **crop tolerance**
- **Internet of Things**
- **membrane lipid peroxidation**
- **oxidative stress**
- **remote sensing**
- **weather pattern analysis**

REFERENCES

Alvar-Beltrán, J., Elbaroudi, I., Gialletti, A., Heureux, A., Neretin, L., & Soldan, R. (2021). *Climate Resilient Practices: Typology and Guiding Material for Climate Risk Screening.* FAO.

Balaska, V., Adamidou, Z., Vryzas, Z., & Gasteratos, A. (2023). Sustainable crop protection via robotics and artificial intelligence solutions. *Machines, 11*(8), 774.

Bannerjee, G., Sarkar, U., Das, S., & Ghosh, I. (2018). Artificial intelligence in agriculture: A literature survey. *International Journal of Scientific Research in Computer Science Applications and Management Studies, 7*(3), 1–6.

Bibi, F., & Rahman, A. (2023). An overview of climate change impacts on agriculture and their mitigation strategies. *Agriculture, 13*(8), 1508. https://doi.org/10.3390/agriculture13081508.

Byers, E., Gidden, M., Leclère, D., Balkovic, J., Burek, P., Ebi, K., Greve, P., Grey, D., Havlik, P., Hillers, A., Johnson, N., Kahil, T., Krey, V., Langan, S., Nakicenovic, N., Novak, R., Obersteiner, M., Pachauri, S., Palazzo, A., & Riahi, K. (2018). Global exposure and vulnerability to multi-sector development and climate change hotspots. *Environmental Research Letters, 13*(5), 055012. https://doi.org/10.1088/1748-9326/aabf45.

Campbell, B. M., Thornton, P., Zougmoré, R., Van Asten, P., & Lipper, L. (2014). Sustainable intensification: What is its role in climate smart agriculture? *Current Opinion in Environmental Sustainability, 8*, 39–43.

Chand, C. P., Ali, M. M., Himasri, B., Bourassa, M. A., & Zheng, Y. (2022). Predicting Indian Ocean cyclone parameters using an artificial intelligence technique. *Atmosphere, 13*(7), 1157.

Chithambarathanu, M., & Jeyakumar, M. K. (2023). Survey on crop pest detection using deep learning and machine learning approaches. *Multimedia Tools and Applications, 82*(27), 42277–42310. https://doi.org/10.1007/s11042-023-15221-3.

Cristin, R., Kumar, B. S., Priya, C., & Karthick, K. (2020). Deep neural network-based Rider-Cuckoo search algorithm for plant disease detection. *Artificial Intelligence Review, 53*(7), 4993–5018. https://doi.org/10.1007/s10462-020-09813-w.

Deng, X., Huang, Z., Zheng, Z., Lan, Y., & Dai, F. (2019). Field detection and classification of citrus Huanglongbing based on hyperspectral reflectance. *Computers and Electronics in Agriculture, 167*, 105006.

Dewitte, S., Cornelis, J. P., Müller, R., & Munteanu, A. (2021). Artificial intelligence revolutionises weather forecast, climate monitoring and decadal prediction. *Remote Sensing, 13*(16), 3209.

Domingues, T., Brandão, T., & Ferreira, J. C. (2022). Machine learning for detection and prediction of crop diseases and pests: A comprehensive survey. *Agriculture, 12*(9), 1350. https://doi.org/10.3390/agriculture12091350.

Ebrahimi, M. A., Khoshtaghaza, M. H., Minaei, S., & Jamshidi, B. (2017). Vision-based pest detection based on SVM classification method. *Computers and Electronics in Agriculture, 137*, 52–58. https://doi.org/10.1016/j.compag.2017.03.016.

Eftekhari, M. S. (2022). Impacts of climate change on agriculture and horticulture. In S. A. Bandh (Ed.), *Climate Change: The Social and Scientific Construct* (pp. 117–131). Springer International Publishing. https://doi.org/10.1007/978-3-030-86290-9.

Fountas, S., Malounas, I., Athanasakos, L., Avgoustakis, I., & Espejo-Garcia, B. (2022). AI-assisted vision for agricultural robots. *AgriEngineering, 4*(3), 674–694.

Hatfield, J. L., Antle, J., Garrett, K. A., Izaurralde, R. C., Mader, T., Marshall, E., Nearing, M., Philip Robertson, G., & Ziska, L. (2020). Indicators of climate change in agricultural systems. *Climate Change, 163*, 1719–1732.

IPCC. (2014). *Climate Change 2014: Synthesis Report. Contribution of Working Groups I, II and III to the Fifth Assessment Report of the Intergovernmental Panel on Climate Change* (R. K. Pachauri & L. A. Meyer, Eds.). IPCC.

Islam, Z., Sabiha, N. E., & Salim, R. (2022). Integrated environment-smart agricultural practices: A strategy towards climate-resilient agriculture. *Economic Analysis and Policy, 76*, 59–72.

Kumar, S. (2023). A novel hybrid machine learning model for prediction of CO_2 using socio-economic and energy attributes for climate change monitoring and mitigation policies. *Ecological Informatics, 77*, 102253.

Liu, X., Fu, Z., Zhang, W., Xiao, S., Chen, H., & Wang, K. (2023). Soluble carbon loss through multiple runoff components in the shallow subsurface of a karst hillslope: Impact of critical zone structure and land use. *Catena, 222*, 106868.

Mahato, A. (2014). Climate change and its impact on agriculture. *International Journal of Scientific and Research Publications, 4*(4), 1–6.

Miranda, J. da R., Alves, M. de C., Pozza, E. A., & Santos Neto, H. (2020). Detection of coffee berry necrosis by digital image processing of Landsat 8 OLI satellite imagery. *International Journal of Applied Earth Observation and Geoinformation, 85*, 101983. https://doi.org/10.1016/j.jag.2019.101983.

Misra, N. N., Dixit, Y., Al-Mallahi, A., Bhullar, M. S., Upadhyay, R., & Martynenko, A. (2020). IoT, big data, and artificial intelligence in agriculture and food industry. *IEEE Internet of Things Journal, 9*(9), 6305–6324.

Nievola, C. C., Carvalho, C. P., Carvalho, V., & Rodrigues, E. (2017). Rapid responses of plants to temperature changes. *Temperature, 4*(4), 371–405.

Niyogi, D., & Srinivasan, J. (2022). The prediction of global mean temperature through machine learning. *Research Square.* https://doi.org/10.21203/rs.3.rs-1894261/v1.

Pham, B. T., Le, L. M., Le, T. T., Bui, K. T. T., Le, V. M., Ly, H. B., & Prakash, I. (2020). Development of advanced artificial intelligence models for daily rainfall prediction. *Atmospheric Research, 237,* 104845.

Qiu, J., Shen, Z., & Xie, H. (2023). Drought impacts on hydrology and water quality under climate change. *Science of The Total Environment, 858,* 159854.

Radanliev, P., & Santos, O. (2023). Ethics and responsible AI deployment. *Frontiers in Artificial Intelligence, 7.* https://doi.org/10.33774/apsa-2023-f1fkq.

Raghuvanshi, R., Raj, S., & Bhattacharjee, S. (2018). Climate smart agriculture and advisory services: Approaches and implications for future. *Research Report Brief 1.* Hyderabad, India: MANAGE–Center for Agricultural Extension Innovations, Reforms, and Agripreneurship, National Institute for Agricultural Extension Management (MANAGE).

Rajput, V. D., Minkina, T., Kumari, A., Harish, Singh, V. K., Verma, K. K., Mandzhieva, S., Sushkova, S., Srivastava, S., & Keswani, C. (2021). Coping with the challenges of abiotic stress in plants: New dimensions in the field application of nanoparticles. *Plants, 10*(6), 1221.

Rutenberg, I., Gwagwa, A., & Omino, M. (2021). Use and impact of artificial intelligence on climate change adaptation in Africa. In *African Handbook of Climate Change Adaptation* (pp. 1107–1126). Cham: Springer International Publishing.

Ryan, M. (2019). Ethics of using AI and big data in agriculture: The case of a large agriculture multinational. *Orbit Journal, 2*(2), 1–27.

Ryan, M. (2022). The social and ethical impacts of artificial intelligence in agriculture: Mapping the agricultural AI literature. *AI & Society, 38(6),* 1–13.

Sain, G., Loboguerrero, A. M., Corner-Dolloff, C., Lizarazo, M., Nowak, A., Martínez-Barón, D., & Andrieu, N. (2017). Costs and benefits of climate-smart agriculture: The case of the Dry Corridor in Guatemala. *Agricultural Systems, 151,* 163–173.

Saini, S., Burman, R. R., Padaria, R. N., Mahra, G. S., Bishnoi, S., Aditya, K., Nithyashree, M. L., Mallick, S., Mukherjee, S., & Padhan, S. R. (2023). Mapping the research trends of migration behavior in agricultural households: A bibliometric analysis. *Frontiers in Sustainable Food Systems, 7,* 1241716.

Schmidhuber, J. (2015). Deep learning in neural networks: An overview. *Neural Networks, 61,* 85–117.

Sevli, O., & Kemaloğlu, N. (2020). Turkish sign language digits classification with CNN using different optimizers. *International Advanced Researches and Engineering Journal, 4*(3), 200–207.

Shoaib, M., Shah, B., El-Sappagh, S., Ali, A., Ullah, A., Alenezi, F., Gechev, T., Hussain, T., & Ali, F. (2023). An advanced deep learning models-based plant disease detection: A review of recent research. *Frontiers in Plant Science, 14, 1158933.*

Suchithra, M. S., & Pai, M. L. (2020). Improving the prediction accuracy of soil nutrient classification by optimizing extreme learning machine parameters. *Information Processing in Agriculture, 7*(1), 72–82.

Tace, Y., Tabaa, M., Elfilali, S., Leghris, C., Bensag, H., & Renault, E. (2022). Smart irrigation system based on IoT and machine learning. *Energy Reports, 8,* 1025–1036.

Talaviya, T., Shah, D., Patel, N., Yagnik, H., & Shah, M. (2020). Implementation of artificial intelligence in agriculture for optimization of irrigation and application of pesticides and herbicides. *Artificial Intelligence in Agriculture, 4,* 58–73.

Taylor, A., Zotz, G., Weigelt, P., Cai, L., Karger, D. N., König, C., & Kreft, H. (2022). Vascular epiphytes contribute disproportionately to global centers of plant diversity. *Global Ecology and Biogeography, 31*(1), 62–74.

Tur, R., Tas, E., Haghighi, A. T., & Mehr, A. D. (2021). Sea level prediction using machine learning. *Water, 13*(24), 3566.

Van De Vijver, R., Mertens, K., Heungens, K., Somers, B., Nuyttens, D., Borra-Serrano, I., Lootens, P., Roldán-Ruiz, I., Vangeyte, J., & Saeys, W. (2020). In-field detection of *Alternaria solani* in potato crops using hyperspectral imaging. *Computers and Electronics in Agriculture, 168,* 105106.

Vázquez-Ramírez, S., Torres-Ruiz, M., Quintero, R., Chui, K. T., & Sanchéz-Mejorada, C. G. (2023). Analysis of climate change based on machine learning and endoreversible model. *Mathematics, 11(14), 3060.*

Wang, X., Zhang, S., Wang, X., & Xu, C. (2023). Crop pest detection by three-scale convolutional neural network with attention. *PLOS ONE, 18*(6), e0276456.

Wessolek, G., Bohne, K., & Trinks, S. (2023). Validation of soil thermal conductivity models. *International Journal of Thermophysics, 44*(2), 20.

White, C. J., Carlsen, H., Robertson, A. W., Klein, R. J., Lazo, J. K., Kumar, A., & Zebiak, S. E. (2017). Potential applications of sub seasonal-to-seasonal (S2S) predictions. *Meteorological Applications, 24*(3), 315–325.

Xiao, Q., Li, W., Chen, P., & Wang, B. (2018). Prediction of crop pests and diseases in cotton by long short-term memory network. In *Proceedings of the International Conference on Artificial Intelligence* (pp. 11–16).

Zhao, S., Liu, J., Bai, Z., Hu, C., & Jin, Y. (2022). Crop pest recognition in real agricultural environment using convolutional neural networks by a parallel attention mechanism. *Frontiers in Plant Science, 13, 839572.*

Zubaidi, Y. A., Hayder, K. L., Gaikwad, S. S., & Kamat, R. K. (2019). An integrated expert water management (IEWM) with IoT. *International Journal of Innovative Technology and Exploring Engineering, 9*(2), 3835–3839.

CHAPTER 2

Natural and Organic Farming: Sustainable Approaches Toward Environmental Sustainability

IMRAN RAMZAN,[1] YASIR HANIF MIR,[2,*] MALIK ASIF AZIZ,[1] GOUSIA GANI,[1] SEEMA NARGIS,[1] BASIT ABRAR,[1] AIMAN ASHRAF,[1] IMTIYAZ AHMAD MIR,[3] and SHAZIA SADIQ[2]

[1]*Division of Basic Sciences and Humanities, Faculty of Agriculture, Sher-e-Kashmir University of Agricultural Sciences and Technology of Kashmir, Jammu and Kashmir, India*

[2]*Division of Soil Science and Agricultural Chemistry, Faculty of Agriculture, Sher-e-Kashmir University of Agricultural Sciences and Technology of Kashmir, Jammu and Kashmir, India*

[3]*Division of Plant Pathology, Faculty of Agriculture, Sher-e-Kashmir University of Agricultural Sciences and Technology of Kashmir, Jammu and Kashmir, India*

*Corresponding author

ABSTRACT

Soil health and environmental integrity are paramount to ensuring sustainable agricultural production and preserving soil biodiversity. However, agricultural intensification has led to adverse environmental impacts such as soil degradation, eutrophication of water bodies, greenhouse gas (GHG) emissions, and biodiversity loss. Conventional farming relies on agrochemicals, which reduce soil fertility and contribute to environmental pollution. Residues of these synthetic chemicals persist

Climate-Resilient Agriculture for Sustainable Crop Production: New Adaptation and Mitigation Strategies. Sanjay Kumar, Shivani Ranjan, Sumit Sow, Smruti Ranjan Padhan, and Dhirendra Kumar Roy (Eds.)

DOI: 10.1201/9781998511150-2

in the soil and enter the food chain, ultimately affecting human health. Natural farming (NF) is a chemical-free method rooted in agroecology. It is a diverse farming system that integrates crops, trees, and livestock to support sustainable ecological functioning. Known as zero-budget farming due to its low input costs, NF reduces dependence on synthetic inputs by using locally available natural resources. Its goal is to minimize external inputs while maintaining soil fertility. Organic farming also plays a critical role in preserving soil biodiversity and promoting environmental sustainability. Based on agroecological principles, it emphasizes natural nutrient mobilization, biological pest control, and the use of farm-generated resources. Integrating natural and organic farming methods offers a pathway to sustainable agriculture, addressing global food security and long-term sustainability. This chapter explores both approaches, providing comprehensive insights into their principles and methodologies.

2.1 INTRODUCTION

Ever since the Green Revolution, food production has tremendously increased, thus increasing reliance on chemical inputs. The agricultural sector contributes over 15% to the national gross domestic product (GDP), making it a vital component of India's economy (Yadav et al., 2021). The adoption of Green Revolution technologies, such as intensive cropping systems, chemical fertilizers, high-yielding seed varieties, irrigation, and mechanization, has increased crop yields while also resulting in negative environmental, health, and economic consequences (Mariappan and Zhou, 2019; UN, 2015). The successful utilization of Green Revolution technologies has rendered the country food self-sufficient and a net exporter of various agricultural commodities (Bana et al., 2020). Nevertheless, this intensification has given rise to a host of adverse consequences. Such adverse impacts of the Green Revolution include reduced water quality, exposure to toxic chemicals, farmer indebtedness, soil compaction, land degradation, eutrophication of aquifers, GHG emissions, biodiversity loss, herbicide-resistant weeds, environmental pollution, declining productivity, nutrient deficiencies, rising production costs, diminishing farm profits, substandard food quality, and detrimental impacts on human and livestock health (Bana et al., 2016, 2020). Moreover, the accessibility and cost-effectiveness of synthetic inputs encounter challenges arising from the surge in coal and natural gas prices, the imposition of sanctions, limitations on exports, and the uncertainties surrounding Indian fertilizer subsidies (World Bank Group, 2022). Additionally, India's neoliberal economic policies have contributed to an ongoing agrarian crisis that has rendered small-scale farming financially unsustainable. The privatization of inputs and markets has not only made them prohibitively expensive for agrarian communities but has also restricted their accessibility, contributing to a cycle of debt for Indian farmers. Tragically, over the past two decades, more than 2,50,000 farmers in India have resorted to taking their own lives (Mariappan and Zhou, 2019). These factors underscore the urgent need for a transformation in food systems, with agroecology emerging as a solution that can ensure long-term food production and nutritional security. Therefore, recognizing these challenges, the United Nations (UN) emphasizes the importance of agricultural systems that work in harmony with nature, are adaptive to change, and minimize environmental impacts in the fight against hunger and malnutrition (UNEP, 2021).

Agroecology, which applies ecological principles to agricultural practices, has emerged as a comprehensive solution for transitioning towards sustainable food systems that foster benefits for people, the planet, and livelihoods. Farmers' movements worldwide advocate agroecology as the pivotal element to achieving nutritional autonomy. Agroecology is in harmony with various objectives of the UN Sustainable Development Goals for 2030, encompassing the eradication of poverty, hunger, and inequality, as well as the promotion of diverse nutritional alternatives, mitigating climate crises, and biodiversity preservation. The adoption of agroecological systems has the potential to significantly contribute to achieving sustainable development goal 2—zero hunger. Various sustainable alternatives to resource-intensive farming, such as organic farming (OF) and natural farming (NF), have been developed to enhance soil health, promote optimal crop growth, and address key agricultural challenges (Willer & Lernoud, 2017).

NF represents an ecological, chemical-free, and diversified farming approach that integrates crops, trees, and livestock to promote functional biodiversity. Based on the principles of working in harmony with nature and natural cycles, NF prioritizes the production of safe and healthy food while maintaining soil, human, and livestock health. OF, on the other hand, is another chemical-free agriculture approach that minimizes or excludes synthetic fertilizers, pesticides, growth regulators, genetically modified organisms, and livestock feed additives. NF and OF aim to produce chemical-free and environmentally friendly food, but they differ fundamentally in their approaches. Both systems are expected to contribute to achieving the dual objectives of global food security and environmental preservation (Ryss, 2020).

2.2 NATURAL FARMING AND ITS FOUR PILLARS

NF is an agricultural approach grounded in the principles of nature. It operates in harmony with the inherent biodiversity of each specific farming environment, fostering the intricate web of life that includes both flora and fauna, which collectively shape the ecosystem to support the growth of food crops. The concept of NF was championed by Masanobu Fukuoka (1913–2008), a Japanese farmer and philosopher. Fukuoka introduced these ideas in his influential work, "The One-Straw Revolution," and subsequently put them into practice on his family farm located on Shikoku Island. One of the fundamental tenets of NF is its minimal financial burden on farmers. This system eliminates the need for purchasing seeds, fertilizers, or chemical plant protection agents, while also discouraging excessive dependence on hired labor by minimizing intercultural operations. Under NF, farmers have the flexibility to produce their own seeds or share seeds with fellow farmers. The overarching philosophy of NF is to foster farmer self-reliance, liberating them from the dominance of moneylenders and the high costs associated with market-supplied inputs. In this system, all costs related to purchasing inputs are entirely eliminated, with any potential expenses offset by the profitability of the farming practices. In the present context characterized by a sharp increase in production costs, Zero Budget NF (ZBNF) is gaining prominence. ZBNF embodies an intensified version of Low External Input Sustainable Agriculture (LEISA), relying entirely on available inputs within the farm, with output from one farming system predominantly serving as inputs for another. ZBNF serves as a means to revitalize soil health, promoting sustainable production systems by adopting approaches such as crop diversification, nutrient recycling, microbial activities, and the facilitation of beneficial biological interactions (Palekar, 2006). This farming approach is supported by four foundational pillars, as outlined by Mr. Subash Palekar, including *Jivamrita/Jeevamrutha*, *Bijamrita/Beejamrutha*, *Acchadana*/Mulching, and *Whapasa*/Moisture.

2.2.1 JIVAMRITA/JEEVAMRUTHA

Jivamrita/Jeevamrutha represents a fermented microbial culture and biofertilizer employed in agriculture. Its role extends beyond nutrient provision for plants; it acts as a catalytic agent that invigorates soil microorganisms and enhances the activity of earthworms. Over the course of a 48-hour fermentation process, cow dung and urine foster both aerobic and anaerobic bacteria, actively breaking down organic materials. The resultant fermented culture not only enhances the population of these bacteria but also attracts and augments the functions of other advantageous soil microorganisms (Münster, 2017). A small amount of undisturbed soil is incorporated into the preparation to introduce native microorganisms. Upon application, these microbes engage in improving nutrient availability for crops and help in preventing fungal and bacterial plant diseases. Notably, *Jivamrita/Jeevamrutha* eliminates the need for external fertilizers by utilizing the inherent nutrient reserves of the soil, released through increased microbial activity. In situations where labor and water resources are scarce, a dry variant known as *Ghanajivamrita* can be prepared and stored for up to one year (Khadse and Rosset, 2019). It is suggested that *Jeevamrutha* is primarily required during the initial three years of transitioning, after which the system becomes self-sustaining.

2.2.2 BIJAMRITA/BEEJAMRUTHA

Bijamrita/Beejamrutha is a treatment specifically designed for seeds, seedlings, or any planting materials. It serves as an effective safeguard against fungal, soil-borne, and seed-borne diseases commonly affecting plants, especially post-monsoon. The composition of *Bijamrita* resembles that of *Jivamrita/Jeevamrutha*, incorporating local cow dung—a powerful natural fungicide, cow urine—a potent antibacterial agent, along with lime and soil. Similar to *Jivamrita/Jeevamrutha, Bijamrita* is enriched with advantageous bacteria that not only protect plants but also stimulate their growth (Sreenivasa et al., 2009). This treatment is applied to seeds by coating them, ensuring thorough mixing by hand, followed by proper drying before sowing. For leguminous seeds, a quick dip and subsequent drying is recommended.

2.2.3 ACCHADANA/MULCHING

Mulching, or *Acchadana*, offers several advantages when applied to soil. It serves to shield the topsoil from erosion, enhances soil aeration, and conserves soil moisture, thereby reducing water evaporation. Additionally, it partially inhibits weed emergence. There are three types of mulching which are discussed in subsections.

2.2.3.1 SOIL MULCHING

Soil mulching protects topsoil without disturbing it through tillage, promoting aeration and water retention.

2.2.3.2 STRAW MULCHING

Straw material typically consists of dried biomass waste from previous crops, such as paddy straw and groundnut husks (Ghosh, 2019; Keerthi et al., 2018). The approach emphasizes supplying dry organic material, encouraging decomposition to form humus through soil biota activity.

2.2.3.3 LIVE MULCHING

Diversifying crops is recommended, with a combination of monocotyledons and dicotyledons, ensuring the provision of vital nutrients to the soil and crops. For instance, nitrogen-fixing dicots like legumes, when intercropped with monocots such as rice and wheat, contribute to a balanced nutrient supply (Palekar, 2006).

2.2.4 WHAPASA/MOISTURE

Whapasa or moisture management seeks to enhance water utilization by minimizing both the quantity and regularity of irrigation. It focuses on providing the necessary vapor form of water for crop growth. *Whapasa* characterizes a condition in which air and water molecules coexist in the soil, promoting improved soil aeration and concurrently minimizing water consumption. This practice offers resilience against drought and challenges the conventional view of plant root water requirements. This approach promotes reduced irrigation, restricts watering to the midday period, and also encourages the adoption of alternate furrows. Farmers practicing NF report a significant reduction in irrigation needs. The preparation and implementation of these approaches are summarized in Table 2.1.

TABLE 2.1 Name, Formulation, and Implementation of Four Primary Techniques of ZBNF

SL No.	Name	Formulation	Implementation
1.	*Jivamrita/ Jeevamrutha*	To prepare *Jivamrita/Jeevamrutha*, begin by adding 200 liters of water into a barrel or container. Next, incorporate 10 kilograms of fresh cow dung sourced from locally bred Indian cows and 5–10 liters of mature cow urine. Following this, add 2 kilograms each of brown sugar (*jaggery*), pulse flour, and a handful of chemical-free soil. Mix these components thoroughly. The resulting mixture, known as *Jivamrita/Jeevamrutha*, should be kept in the shade for 48 hours to facilitate fermentation.	Application involves incorporating 200 liters of *Jivamrita/ Jeevamrutha* into irrigation water for one acre of land twice a month. Alternatively, it can be applied as a foliar spray in the form of a 10% solution.
2.	*Bijamrita/ Beejamrutha*	The preparation of *Bijamrita/Bejjamrutha* mirrors the process of *Jivamrita/ Jeevamrutha,* involving local Indian cow dung, urine, water, lime, and soil. Specifically, add 5 liters of cow urine and 5 kilograms of locally bred Indian cow dung to a container containing 20 liters of water. Incorporate 50 grams of lime and a handful of native soil, ensuring thorough mixing.	This solution is used for coating and hand-mixing seeds or for dipping leguminous seeds in *Bijamrita/ Beemamrutha* before drying them in the shade and subsequently sowing them.
3.	*Acchadana*/ mulching	Mulching practices encompass three distinct types: soil mulch, which involves covering topsoil with friable soil or dust; straw mulch, comprising dried residues of previous crops along with dead plant or animal materials; and live mulch, which involves the cultivation of mixed or intercrops, particularly combining monocot and dicot plants, such as a cereal-legume system.	These mulching methods are applied either before seed sowing or during cultivation to cover land areas, depending on the specific type of mulch used.
4.	*Whapasa*/ moisture	*Whapasa*, also known as moisture management, seeks to optimize water utilization by reducing both the volume and frequency of irrigation. This approach focuses on creating a soil environment that supplies water vapor to plant roots, thereby decreasing the dependence on extensive irrigation.	Ideal conditions for *Whapasa* are achieved by irrigating during the noon hours, adopting alternate furrows, and thereby encouraging the retention of air and water molecules within the soil.

Source: Modified from La Via Campesina (2016).

2.3 ORGANIC FARMING (OF) AND ITS PRINCIPLES

Over the past few decades, OF has gained significant attention as an environmentally sustainable method of agricultural production. It serves a pivotal role in mitigating the environmental issues linked to traditional farming practices. Beyond producing high-quality and nutritious food items, OF contributes to the improvement of soil quality (Isaacs, 2012). Rooted in ecological principles, OF minimizes the reliance on external inputs, reducing resource consumption and preventing nutrient overloading within the agricultural system. OF is an agricultural system that depends on environmentally sustainable practices such as the incorporation of green manures, utilization of biofertilizers, application of plant growth-promoting bacteria, the implementation of integrated pest management, and biological pest control strategies, as well as integrated nutrient management. This approach also adopts zero or minimum tillage, mulching, crop rotation techniques, biosolids, organic manures, and biopesticides. Notably, OF eliminates the use of synthetic chemicals in order to promote the preservation of land biodiversity, maintain ecosystem equilibrium, and reduce emissions and waste production (Hussain and Farooqi, 2021). This strategy reduces the likelihood of eutrophication from phosphorus and nitrogen and prevents excessive fertilization. As outlined by the Codex Alimentarius Commission, OF embodies a comprehensive approach, excluding the use of synthetic pesticides, fertilizers, and genetically modified organisms. Its aim is to curtail soil, water, and air pollution, concurrently enhancing the well-being and productivity of interconnected communities of plants, animals, and people.

According to the Food and Agriculture Organization (FAO), OF is an environmentally friendly approach that eliminates the usage of all synthetic inputs such as fertilizers, pesticides, hormones, and feed additives. The FAO officially incorporated organic agriculture into its agenda in March 1999, recognizing it as a strategy for sustainable development. The FAO is intrigued by organic agriculture due to its prospective contribution to rural development and the enhancement of global food security. During the International Conference on Organic Agriculture and Food Security in Rome in 2007, FAO member countries underscored the capability of organic agriculture to establish a more secure food supply, improve rural food accessibility, and preserve natural resources (Morgera et al., 2012).

The International Federation of Organic Agriculture Movements (IFOAM) outlines organic agriculture based on four fundamental principles (IFOAM, 1998). Figure 2.1 illustrates the four basic principles of organic farming.

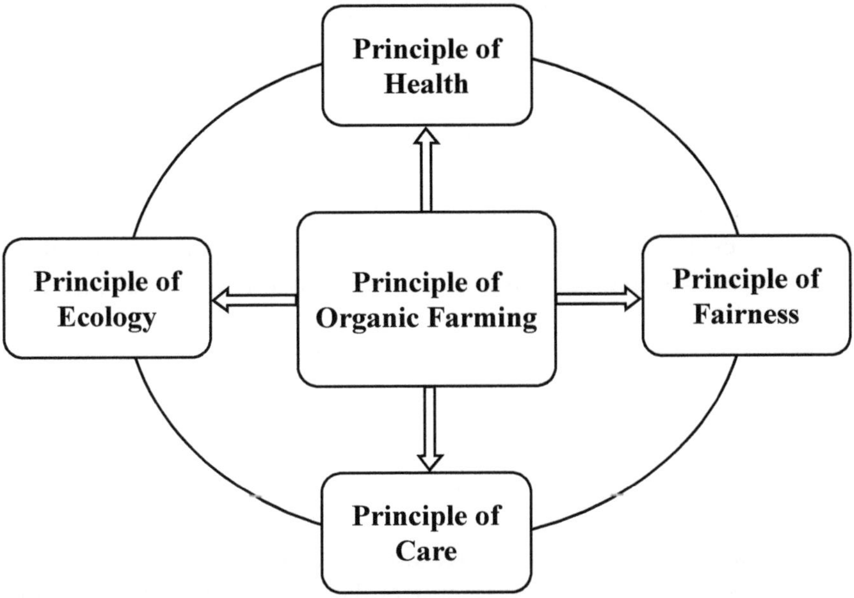

FIGURE 2.1 Four principles of organic farming.

2.3.1 THE PRINCIPLE OF HEALTH

In the context of organic agriculture, it is imperative to prioritize and elevate the well-being of soil, plants, animals, humans, and the overall planet. This approach underscores the interconnectedness of individual and community health with the well-being of ecosystems and the soils. Healthy soil, fostering the growth of robust crops, contributes directly to the health of both people and animals. Within organic agriculture, the essence of health is viewed as the integrity and completeness of living systems. Whether in the stages of farming, processing, distribution, or consumption, the pivotal role of organic agriculture remains dedicated to sustaining and improving the health of ecosystems, spanning from the minutest soil organisms to the well-being of humans.

2.3.2 THE PRINCIPLE OF ECOLOGY

The foundation of organic agriculture lies in aligning with ecological cycles and systems, aspiring to replicate and bolster their dynamics. Sustaining ecological equilibrium involves creating habitats, conserving agricultural and genetic diversity, and structuring farming systems that harmonize with the natural order.

2.3.3 THE PRINCIPLE OF FAIRNESS

This principle asserts that organic agriculture should be built on relationships that ensure equality in terms of shared life opportunities and the environment. The judicious use of environmental and natural resources in production and consumption must be governed in a manner that is both economically and socially fair, ensuring the preservation of these resources for the well-being of future generations.

2.3.4 THE PRINCIPLE OF CARE

Approaching organic agriculture requires a sense of responsibility and a precautionary mindset, aiming to protect the health of present and future generations and the overall prosperity of the environment. Within this principle, the significance of responsibility and caution extends to the planning, execution, and technological decisions in the domain of organic agriculture. A policy document released by the National Academy of Agricultural Sciences (NAAS) proposes the potential avoidance of synthetic pesticides, while acknowledging that the entire exclusion of fertilizers might not be globally applicable in all conditions (Deshmukh, 2010).

2.4 RESEMBLANCES AND DIFFERENCES IN THE TWO SYSTEMS

Both natural and organic farming share the fundamental principle of avoiding agricultural chemicals, which are regarded as harmful to agricultural practices. These approaches discourage the use of chemical fertilizers, fungicides, and pesticides, advocating instead for non-chemical methods and homemade pesticides to manage pests, diseases, and weeds below certain threshold levels. Both systems emphasize the use of local crop varieties and landraces for field crops, vegetables, and other crops. However, they differ significantly in their approaches to soil fertility and farming practices.

In organic farming, organic manures like compost, vermicompost, and farmyard manure are applied to manage soil fertility, replacing synthetic fertilizers. In contrast, natural farming refrains from using both chemical fertilizers and organic manures. Organic farming employs conventional farming practices such as tilling, plowing, and mixing manures. Conversely, natural farming avoids plowing, tillage, and weeding altogether. Furthermore, organic farming involves incorporating manures and compost into the soil, requiring effort and cost for their decomposition. NF, on the other hand, encourages organic matter decomposition by earthworms and other microfauna on the soil surface, steadily enriching the soil with essential plant nutrients. While organic farming may sometimes have a slight adverse environmental impact due to its intervention in natural systems, NF relies solely on local processes and biodiversity, having no impact on the surrounding environment. OF tends to be more resource-intensive, as it requires bulky organic manures, potentially impacting biodiversity. In contrast, NF operates with minimal human intervention and is a low-input agricultural approach, relying entirely on local biodiversity. Both systems are recognized as promising alternatives to chemical-based conventional farming practices, with the potential to mitigate the adverse effects of climate change.

2.5 OBJECTIVES OF NATURAL AND ORGANIC FARMING

- Enhance the economic viability of farming by increasing farmers' net incomes through cost reduction, reduced risks, comparable yields, and income from intercropping.
- Curtail production expenses substantially by inspiring farmers to create vital biological inputs through on-farm, natural, and locally accessible resources.
- Optimize biological cycles in the agricultural system, involving soil flora and fauna, microorganisms, plants, and animals.
- Engage in cooperative efforts with natural systems rather than exerting dominance over them.

- Conserve and enhance long-term soil fertility.
- Create an environment for livestock that allows them to exhibit their innate behaviors fully.
- Function predominantly within a self-contained system concerning nutrient elements and organic matter.
- Prevent any environmental pollution arising from agricultural practices.
- Preserve the genetic diversity of the farming system and its settings, encompassing the safeguarding of plants and wildlife habitats.
- Take into account the wider ecological repercussions of the agricultural system (Deshmukh, 2010).

2.6 COVERAGE OF ORGANIC FARMING IN INDIA

The area under natural and OF in India has been steadily growing in recent years, reflecting a shift toward more sustainable and environmentally friendly agricultural practices. India boasts a substantial portion of its agricultural land dedicated to organic and NF methods. This expansion can be attributed to increased awareness among farmers and consumers regarding the ecological and health benefits of these practices. Many small and marginal farmers across the country have adopted NF, including ZBNF. Additionally, OF has seen significant growth, with numerous farmers seeking organic certification for their produce. The adoption of these practices is not limited to specific regions but has spread throughout the country, encompassing a wide range of crops and agro-ecological zones. Government initiatives and subsidies, along with support from non-governmental organizations and private enterprises, have further encouraged the expansion of natural and OF in India. This transition signifies a commitment to more sustainable and eco-conscious agriculture that not only improves soil health and biodiversity but also offers consumers healthier and chemical-free food options. The adoption and coverage of OF, however, have not been uniform across all Indian states and union territories. While certain states have proactively embraced and expanded OF practices, others have been slower to do so. Table 2.2 illustrates the area coverage by OF among different states and union territories of India.

TABLE 2.2 Extent of Organic Farming Implementation Across States and Union Territories

SL. No	State/Union Territory#	Organic Area						
		Total Organic Area (000 ha) in 2019*	Percentage (%) of Organic Area Relative to the Net Sown Area**	Break-up of the Total Organic Area Based on Specific Schemes (%)^				
				NPOP		PKVY	MOVCDNER	State Schemes/ Non-Scheme @ (%)
				NPOP	In-Conversion			
1.	Madhya Pradesh	756	4.9	50.2	38.9	10.1	0.0	0.7
2.	Rajasthan	350	2.0	31.5	32.5	35.2	0.0	0.7
3.	Maharashtra	284	1.6	55.7	32.7	8.9	0.0	2.7
4.	Andhra Pradesh	144+	2.3	9.5	13.0	73.4	0.0	4.1
5.	Uttarakhand	128	18.2	15.7	13.0	70.2	0.0	1.1
6.	Odisha	118	2.6	62.0	19.2	17.6	0.0	1.2
7.	Karnataka	111	1.1	51.2	23.4	9.8	0.0	15.6
8.	Gujarat	103	1.0	58.2	32.5	1.9	0.0	7.3
9.	Uttar Pradesh	79	0.5	56.6	22.8	15.7	0.0	5.0
10.	Sikkim	155&	100.0	47.6	1.4	1.9	8.0	41.1
11.	Chhattisgarh	71	1.5	10.3	19.4	33.6	0.0	36.6
12.	Meghalaya	56	19.5	2.9	84.0	1.6	11.5	-
13.	Kerala	54	2.7	35.5	35.4	22.9	0.0	6.2
14.	Assam	43	1.5	35.7	30.4	10.3	16.3	7.3
15.	Jharkhand	31	2.2	9.	69.6	16.3	0.0	4.4
16.	Tamil Nadu	30	0.6	14.4	60.5	20.8	0.0	4.3
17.	Telangana	28	0.6	22.9	8.8	50.0	0.0	18.2

TABLE 2.2 *(Continued)*

SL. No	State/Union Territory#	Organic Area						
		Total Organic Area (000 ha) in 2019*	Percentage (%) of Organic Area Relative to the Net Sown Area**	Break-up of the Total Organic Area Based on Specific Schemes (%)^				
				NPOP		PKVY	MOVCDNER	State Schemes/ Non-Scheme @ (%)
				NPOP	In-Conversion			
18.	Jammu and Kashmir	26	3.4	68.3	29.0	2.2	0.0	0.5
19.	Goa	23	18.1	45.7	11.2	43.1	0.0	–
20.	Nagaland	23	6.0	12.0	24.1	2.1	56.9	4.9
21.	Arunachal Pradesh	22	9.8	2.8	38.9	1.7	38.4	18.1
22.	Manipur	19	5.0	1.3	27.3	3.1	65.4	2.9
23.	Himachal Pradesh	18	3.3	46.4	24.7	22.8	0.0	6.1
24.	Punjab	17	0.4	1.9	50.5	29.4	0.0	18.3
25.	Mizoram	14	10.0	0.0	48.8	4.7	46.1	0.4
26.	Bihar	12	0.2	0.0	28.8	69.9	0.0	1.2
27.	Delhi	10	45.5	0.0	0.0	99.8	0.0	0.1
28.	Dadar and Nagar Haveli	10	53	0.0	0.0	100.0	0.0	–
29.	Andaman and Nicobar	9	60	0.0	84.6	15.4	0.0	–
30.	West Bengal	9	0.2	56.4	14.8	27.2	0.0	1.6
31.	Tripura	9	3.4	2.4	27.2	11.7	58.8	–
32.	Haryana	7	0.2	33.0	53.1	5.8	0.0	8.1

TABLE 2.2 *(Continued)*

SL. No	State/Union Territory#	Organic Area						
		Total Organic Area (000 ha) in 2019*	Percentage (%) of Organic Area Relative to the Net Sown Area**	Break-up of the Total Organic Area Based on Specific Schemes (%)^				
				NPOP		PKVY	MOVCDNER	State Schemes/ Non-Scheme @ (%)
				NPOP	In-Conversion			
33.	Lakshadweep	3	~ 100	24.9	0.0	75.1	0.0	–
34.	Chandigarh	3	~ 100	0.0	0.0	39.4	0.0	60.6
35.	Daman and Diu	1	32	0.0	0.0	98.1	0.0	1.9
36.	Puducherry	1	08	0.5	0.0	99.5	0.0	–
	Total (India)	2777	–	39.5	30.3	21.5	2.6	6.1

Abbreviations: NPOP: National Program for Organic Production; PKVY: Paramparagat Kirshi Vikas Yojana; and MOVCDNER: Mission Organic Value Chain Development for North East Region.

#The table above arranges states/ Union Territories in descending order based on their coverage of organic area;

*The cumulative organic area derived from the data provided on state-wise basis;

^Information on schemes at the state level is obtained from the response to a parliamentary question in November 2019;

+According to the latest data, Andhra Pradesh constitutes 2.9% of the net sown area.

&Sikkim is a fully organic state, with 100% adherence to OF.

**The net sown area data for states pertains to the year 2014–2015, while for Union Territories, multiple sources have been consulted, providing net sown area figures from different years.

–Signifies the dearth of publicly available data on the subject.

Source: Modified from Khurana and Kumar (2020).

2.7 METHODS USED IN NATURAL AND ORGANIC FARMING

Natural, ecological, and non-chemical strategies for weed management encompass various practices such as planting weed-free seeds, adjusting planting periods, cultivating quick-growing cultivars with competitive traits, utilizing soil solarization, optimizing crop geometry and row orientation, implementing scientifically sound crop rotations, and employing the stale seed-bed method. Furthermore, within the context of NF, the focus on pest and disease management emphasizes ecologically safer preventive approaches due to the limited use of chemical interventions. Some of the methodologies used in natural and OF are discussed in subsections.

2.7.1 WEED MANAGEMENT

Weed management in natural and organic farming systems is of paramount importance and can be achieved through various strategies. Firstly, ensuring the use of high-quality, weed-free seed is crucial to minimize weed seed contamination and reduce the need for environmentally harmful herbicide applications (Kumar et al., 2021). Secondly, adjusting crop planting periods based on local conditions and target weed species can provide a competitive advantage to crops over weeds (Bana et al., 2020). Choosing cultivars with fast vegetative growth habits can inhibit early-season weed growth by effectively covering the ground surface (Colbach et al., 2019). Soil solarization, which involves raising soil temperatures with polythene sheets during the summer months, renders weed seeds nonviable and controls harmful insects and pathogens. Effective crop rotations and optimizing crop geometry and seed rates can disrupt the microclimate for weed growth, reducing weed populations and improving overall productivity (Choudhary et al., 2018). Additionally, varying row orientation and employing techniques like the stale seed-bed method can further reduce weed density and biomass (Borger et al., 2010; Hussain et al., 2017). These strategies offer ecological and non-chemical alternatives to herbicide use, contributing to sustainable and environmentally friendly agricultural practices, with the need for location-specific adaptations to address unique weed dynamics in different farming systems.

2.7.2 PEST MANAGEMENT

Pest and disease management strategies in natural farming rely on a range of ecological practices aimed at environmentally friendly pest and disease control. These approaches encompass intercropping systems, notably strip cropping, which leverages agroecosystem diversity to reduce pest pressures while promoting beneficial organisms (Maqsood et al., 2022; Rao et al., 2012). Companion crops are used in mixed cropping or push-pull systems to repel or attract pests away from the main crop. Trap crops strategically draw pests away from the primary crop, relying on a deep understanding of pest food preferences. Cover crops, like green manures and living mulches, offer shelter to beneficial insects, and insectary strips enhance their migration to the main crop (Meena et al., 2017). Flower strips of mixed flowering species provide continuous nutrients to natural enemies, enhancing aphid biocontrol (Tschumi et al., 2016) and reducing the need for insecticides. Windbreaks and shelterbelts foster natural enemy species and help control agricultural pests. Corridors facilitate the movement of predator species and enhance biodiversity. Furthermore, NF practices endorse the use of indigenous concoctions such as *neemastra, agniastra,* and *bramhastra*, among others, serving as effective bio-pesticides (Mishra, 2018). Table 2.3 presents the types of different formulations. These formulations are employed to control a variety of pests, including sucking pests, mealybugs, leaf rollers, and fruit, stem, and pod borers. These strategies underscore the importance of biodiversity, habitat manipulation, and strategic planting to create ecosystems that naturally regulate pests and diseases.

TABLE 2.3 Details of Ingredients and Their Respective Quantities Utilized in the Formulation of Alternative Bio-Pesticides

SL. No.	Insecticide/ Fungicide	Elements	Amount Used in Mixture
1.	Insecticide-I	Neem seed or leaf powder, water	20 kg, 200 liters
2.	Insecticide-II	Indian-bred cow urine, Indian-bred cow dung, neem leaves, water	5 kg, 10 liters, 10 kg, 200 liters
3.	Insecticide-III	Neem leaves, tobacco, garlic paste, and green chili paste soaked separately in cow urine for 10 days	10 kg, 3 kg, 3 kg, and 4 kg, respectively
4.	Fungicide-I	Cultured buttermilk fermented for five days, water	5 liters, 50 liters
5.	Fungicide-II	A scientifically devised mixture incorporates Indian-bred cow milk, black pepper powder, and water	5 liters, 200 g, and 200 liters, respectively

Source: Modified from Biswas (2020).

2.7.3 SOIL MANAGEMENT

Plants necessitate essential macronutrients like potassium, phosphorus, and nitrogen, alongside micronutrients for optimal growth. The incorporation of green manure and crop rotation is instrumental in providing nitrogen to the soil, primarily through leguminous plants, which fix atmospheric nitrogen through a symbiotic association with rhizospheric bacteria. While intercropping is intermittently utilized for insect and disease management, it provides extra nutrition to the soil. However, competition between the main crop and legumes can pose challenges, often requiring wider spacing between crop rows (Watson et al., 2002). Diversified farms that combine both crop cultivation and livestock rearing can operate as ley farms. In such systems, soil fertility is enhanced by incorporating grasses like alfalfa or white clover that fix atmospheric nitrogen. Once fertility is established, the cultivation shifts to cereal or cash crops. Farms without livestock may face greater challenges in maintaining soil fertility and often rely on external inputs, including grain legumes and green manure.

2.7.4 CONTROLLING OTHER ORGANISMS

In addition to weeds, organic farms encounter challenges from nematodes, arthropods (such as mites and insects), bacteria, and fungi, representing a diverse spectrum of organisms. Organic farmers employ diverse integrated pest management techniques to combat diseases and pests. It frequently leverages beneficial organisms to reduce pest populations, a practice known as biological pest control. Instances of advantageous insects encompass ladybugs, minute pirate bugs, and big-eyed bugs, although their mobility may pose challenges.

Natural insecticides approved for application on organic farms encompass *Bacillus thuringiensis, Pyrethrum,* neem, spinosad (a bacterial metabolite), and rotenone. The routine use of these pesticides is relatively limited, with fewer than 10% of organic farmers employing them regularly (Lotter, 2003). Often referred to as "green pesticides," these products are generally considered environmentally friendly and safe. However, it's important to note that *Pyrethrum* and rotenone can be controversial due to their mode of action, which affects the nervous system, similar to many conventional insecticides (Pottorff, 2010). Naturally derived fungicides approved for application on organic farms encompass *Trichoderma harzianum,* as well as the bacteria *Bacillus pumilus* and *Bacillus subtilis*. These options are particularly effective against diseases affecting plant roots.

2.8 NATURAL AND ORGANIC FARMING IN ECOSYSTEM CONSERVATION

2.8.1 ECOLOGICAL PRESERVATION

The excessive use of synthetic fertilizers exhausts soil nutrients and also leads to reduced yields and contamination of the ecosystem (Li et al., 2020). Various management practices, such as the application of biofertilizers, biopesticides, vermicompost, and farmyard manure, have been explored to mitigate the detrimental impacts of chemical inputs (Ray et al., 2020). Despite continuous scientific efforts, no single method has been proven to entirely substitute synthetic fertilizers and pesticides while ensuring sustainable outcomes. However, OF and NF appear to alleviate these adverse effects and contribute positively to climate change mitigation. They protect the environment from the harmful consequences associated with artificial inputs, particularly fertilizers, pesticides, and hormones, that discharge harmful chemicals into the soil and water, posing risks to both ecosystems and human health. Research has shown that OF and NF enhance biodiversity by minimizing soil disturbances and chemical applications (Nascimbene et al., 2012).

Natural farming has demonstrated partial enhancements in soil health, possibly attributed to the swift evolution of heterotrophic microbial communities, an increase in soil organic matter, and other associated factors (Saharan et al., 2023; Smith et al., 2020). OF exerts a notably positive impact on diverse elements of the environment, encompassing insects, birds, wildlife, and soil flora and fauna (Deshmukh, 2010). Non-crop species exhibit a preference for OF systems, demonstrating higher diversity and abundance, with an average of 30% more species inhabiting organic farms. This positive impact extends to various species, encompassing beetles, earthworms, spiders, butterflies, birds, mammals, soil microbes, and vegetation (Gabriel et al., 2006). Birds, in particular, tend to exhibit higher population densities and species diversity in organic farms (Deshmukh, 2010). OF effectively reduces the risk of poor yields, promoting overall biodiversity.

2.8.2 HEALTH BENEFITS

Global agriculture plays a vital role in providing food for a hungry world, fibers for textiles, animal feed, and even fuel for vehicles. However, biodiversity loss, nitrate leaching, groundwater pollution, and heavy metal accumulation in soils are common occurrences in regions with intensive agricultural practices (John and Babu, 2021). It contributes significantly to issues like water pollution and the unsustainable depletion of groundwater resources, which has led to the presence of agrochemical residues in drinking water and air that far exceed prescribed safety limits. Several studies have examined various potential impacts, including the presence of pesticide residues in food. To mitigate these environmental and health impacts, many farmers are turning to organic and NF methods, which aim to reduce the detrimental impacts on human health and the ecosystem. One notable outcome is that fruits and vegetables cultivated organically are prone to have fewer synthetic residues compared to conventionally grown counterparts. Since the health hazards of nitrate remain a topic of debate, studies have shown that organic products exhibit reduced levels of nitrate, a significant aspect as elevated nitrate intake has been linked to gastrointestinal cancer and methemoglobinemia in infants (Forman and Silverstein, 2012). Therefore, the adoption of organic and NF would pave avenues in overcoming the hazards of conventional farming.

2.8.3 SOCIAL BENEFITS

Organic agriculture contributes to the growth and sustainability of local communities. To remain competitive, farmers need to adjust to local conditions, effectively oversee labor, land, and resources to increase production. Collaboration among community farmers is essential to ensure the collective integrity of soil, water, and air resources. This not only strengthens community bonds but also promotes increased cooperation among organic farmers, facilitating valuable partnership establishments. By forming organized groups and cooperatives, farmers can exert influence in trade negotiations, consolidate their resources, and enhance access to markets. Evidence suggests that enhanced collaboration can lead to the emergence of new local businesses and increased engagement in local governance.

Additionally, consumer protection is a fundamental aspect of organic farming. Robust regulatory frameworks, with government oversight to validate organic certifications, are essential for instilling consumer confidence in the food they purchase (Morgera et al., 2012).

2.9 NATURAL AND ORGANIC FARMING IN CLIMATE CHANGE MITIGATION

The Intergovernmental Panel on Climate Change (IPCC, 2007) has identified global warming, resulting from anthropogenic GHG emissions such as carbon dioxide (CO_2), methane (CH_4), nitrous oxide (N_2O), hydrofluorocarbons, perfluorocarbons, and sulfur hexafluoride, as a leading cause of climate change. Agriculture, particularly in its contribution to CH_4 and N_2O emissions, and to a lesser extent CO_2 emissions, plays a substantial role in climate change. The interconnectedness of food security, agricultural productivity, and climate change arises from the direct influence of climate on a nation's ability to feed its population. On a global level, addressing the world's increasing food production needs while facing the serious long-term challenge of climate change is of paramount importance. To tackle climate change caused by agriculture, it is imperative to select agricultural forms and practices that maximize cost-effective production while minimizing detrimental impacts on the environment and climate. Organic and sustainable farming is currently gaining considerable worldwide support as a means of climate change mitigation. By employing organic and sustainable techniques to enhance soil health, farmers and ranchers can increase their resilience and mitigate the risks posed by climate change. These approaches are seen as a potential solution due to their capacity for soil carbon sequestration, reduced energy usage, and decreased GHG emissions (IFOAM, 2009). Soil serves as the primary carbon sink for atmospheric CO_2, and OF methods, involving the application of green manure, animal manure, composts, and crop rotations with intercropping and cover crops, contribute to soil carbon sequestration while simultaneously enhancing soil fertility and physical conditions for plant growth. Furthermore, organic agriculture bolsters resilience to climate change through its efficient water use, resistance to weather extremes, and reduced vulnerability to total crop failure. Shifting from conventional to organic crop production systems is expected to result in significantly reduced GHG emissions since organic production systems generally emit lower quantities of GHGs compared to conventional industrial farming systems. The worldwide implementation of organic agriculture holds the capacity to capture and store as much as 32% of the existing human-induced GHG emissions (Jordan et al., 2009). The FAO has also confirmed that organic systems emit fewer GHGs due to their reduced energy inputs and store related carbon in biomass compared to conventional farming. Moreover, organic agriculture significantly reduces both direct and indirect energy use in farming. Studies have indicated that organic farming systems require 20%–50% less energy than conventional farming systems (Muller et al., 2012). Aher et al. (2012) reported that organic yields are on par with conventional yields while utilizing 45% less energy, making OF more energy-efficient.

2.10 NATURAL AND ORGANIC FARMING IN FOOD AND ECONOMIC SECURITY

OF plays a pivotal role in enhancing food security, particularly in the context of the persistent global issue of hunger. While the global food supply is sufficient to satisfy the requirements of the entire populace, nearly 850 million people continue to suffer from malnutrition and food scarcity. Over the past decade, food prices have increased significantly, and conventional agricultural methods have resulted in a loss of genetic diversity in our food supply. This has left populations more vulnerable to the risks of food scarcities. OF holds promise in tackling these challenges. Advocates of conventional agriculture argue that OF leads to reduced crop yields. However, proponents of OF contend that yields remain comparable to those of traditional farming methods over the long term (Morgera et al., 2012). The diverse microbial community within the soil not only converts soil organic matter but also acts as both a source and reservoir of nutrients, contributing to plant growth, improving soil fertility, and enhancing crop yields (Li et al., 2020). Organic agriculture contributes to food security through several key mechanisms such as:

- Mitigation of acute food shortages during emergencies through improved ecosystem stability and diversification.
- Enhancing local nutrient consumption and the capacity to buy food via the promotion of sustainable expansion and commercialization of small-scale agriculture.

- Development of self-reliant food systems, particularly at the household scale.
- Improvement of micronutrient intake and dietary quality through crop diversification and the reintroduction of underutilized crop varieties.

Economic viability is another compelling aspect of OF. Numerous research findings indicate that organically grown crops require 97% fewer pesticides, exhibit a yield increase ranging from 95%–100%, and incur 50% lower costs in terms of energy and fertilizers. The augmented prices consumers are willing to invest in organic products, coupled with diminished spending on synthetic fertilizers and pesticides, contribute to heightened profitability. According to the FAO, "OF is at the forefront of implementing energy-saving practices based on organic principles." Utilizing principles that highlight farm-level sufficiency, the integration of externalities, and environmental stewardship, strategies can be developed to decrease the reliance on fossil fuel-based energy in OF. In summary, organic and NF significantly contribute to ecological preservation, food security, and economic viability, offering several noteworthy contributions, including:

- Fostering sustainable rural livelihoods by providing better returns on labor.
- Enhancing overall well-being through non-exploitative labor practices and fair wages that empower individuals.
- Contributing to rural development by revitalizing local economies.

Table 2.4 summarizes studies on farmers' feedback following the adoption of ZBNF and presents individual case studies.

2.11 CHALLENGES IMPEDING THE EXPANSION OF ORGANIC AND NATURAL FARMING

Despite the national OF policy in India having been in place for 15 years, the organic movement in the country remains a niche effort primarily driven by farmers and civil society groups, and widespread adoption of chemical-free farming practices within mainstream agriculture is still a challenge. Various barriers have hindered the growth of the organic and NF movement, with each major stakeholder group—government, farmers, and consumers—exhibiting their own concerns. These challenges have

TABLE 2.4 Feedback From Farmers Following the Transition to ZBNF Along with Individual Case Studies

Particulars	Findings	References
	Farmers' Responses	
Crop performance under changing climate	Crops cultivated using ZBNF techniques in Andhra Pradesh (AP), India demonstrated robust health and resilience to climate variations.	Bharucha et al. (2020)
Biodiversity	Practical observations indicated enhanced biodiversity through ZBNF practices. For example, in a study involving 480 samples from 13 AP districts, earthworm populations in ZBNF fields averaged 232 m^{-2}, in contrast to 32 m^{-2} in non-ZBNF plots. Additionally, populations of pollinators and natural enemies of pests exhibited improvement in the proximity of ZBNF fields.	Ryss (2018)
Economic shift and household transformation	Adoption of ZBNF led to a substantial increase in farm income, resulting in improvements in socio-economic status, lifestyle, and household transitions, such as the transition from mud houses to concrete ones.	Bharucha et al. (2020)
	Individual Case Study	
Krishna, a farmer from Anantapur, Andhra Pradesh, had been primarily cultivating papaya on most of his 2-hectare land for the past five years. In 2017, he became acquainted with ZBNF and was motivated by self-help groups offering loan facilities. Consequently, he adopted ZBNF practices for his papaya cultivation	Within a span of seven months, Krishna observed remarkable improvements in papaya yield and quality, resulting in higher market values compared to previous years. His success story inspired several other farmers to transition to ZBNF practices.	Bharucha et al. (2020)
Raheem, a farmer from Vandanmedu, Kerala, was practicing chemical-based farming methods for cultivating cardamom on his 14-acre land, resulting in high input costs due to synthetic pesticide use. Raheem decided to shift to ZBNF	Raheem effectively managed pests using botanical solutions and achieved sustainable cardamom production.	The Hindu (2010)

TABLE 2.4 *(Continued)*

Particulars	Findings	References
Irshad, a farmer from Prakasam, AP, adopted ZBNF in papaya cultivation on his 13-acre land	Irshad achieved a remarkable increase of 16 tons per acre in yield and a net return of rupees 1,28,000 more via ZBNF practices compared to non-ZBNF methods.	Tripathi et al. (2018)
Subham, a farmer from Andhra Pradesh, became disillusioned with the adverse impacts of hazardous synthetic fertilizers and pesticides on soil. He subsequently transitioned to ZBNF after practicing in a one-week training program	Subham noticed a significant improvement in soil health, accompanied by year-round high production and increased income.	Bharucha et al. (2020)
Sushil, a farmer from Muttukad, Kerala, commenced ZBNF practices on his seven-acre land for various crops, including pepper, nutmeg, cloves, and vegetables.	Sushil recorded substantial increase in crop yields as a result of ZBNF practices, along with pest-free crop cultivation.	The Hindu (2010)
Prem, a farmer from Guntur, AP, practiced conventional farming on his one-acre land for five years. However, the returns from his production were negated by input expenditures. In 2016, he adopted ZBNF and commenced cultivating bananas and yams, allocating half an acre to each crop.	Prem observed high crop yields and soil health improvement, resulting in a profit of rupees 1,55,000 via ZBNF practices.	Tripathi et al. (2018)

resulted in a lack of substantial political support, an outdated national organic policy lacking ambition and clear targets, minimal budget allocations for promoting OF compared to substantial subsidies for chemical fertilizers, and fragmentation of initiatives among different ministries. Furthermore, there is a deficiency of comprehensive perception and comprehension regarding the multifaceted advantages of natural and OF, particularly in terms of ecological, economic, and social aspects, as well as nutrition security and alignment with sustainable development goals. The scientific community predominantly concentrates on chemical-based agriculture and possesses inadequately equipped agricultural extension services to implement organic methods. Additionally, small-scale farmers face challenges with costly and complex organic certification systems, limited market linkages, and inadequate organic value chains. Issues such as the accessibility of high-quality organic seeds and bio-inputs, as well as outdated or absent organic policies in various states and Union Territories, further impede the progress of the OF movement in India. Table 2.5 summarizes the challenges impeding the expansion of organic and NF.

TABLE 2.5 Challenges Impeding the Expansion of Organic and Natural Farming

What Hinders the Central and State Governments from Actively Promoting Organic and NF?	What Causes Farmers to Hesitate in Embracing Organic or NF Approaches?	What Factors Contribute to the Majority of Consumers Refraining from Purchasing Organic or Natural Products?
• The inclination towards chemical farming. • Lack of alignment in the scientific community towards organic or NF. • Concerns about fewer output and food security. • Influence of the agrochemicals collectively contributes to the perception that organic and/or natural produce is not regarded as a comprehensive solution beyond providing pesticide-free food. • Insufficient documentation regarding holistic linkages. • Minimal focus on drawbacks of existing chemical-based models. • Inadequately trained and inexperienced extension machinery. • Limited demonstration of political will at the state level.	• Persistent inclination towards conventional farming. • Inadequate understanding of organic and/or natural approaches. • Uncertainty and apprehension about minimum returns. • Deficiency in guidance, support and risk coverage. • Unavailability of guaranteed markets providing lucrative prices. • Insufficient access to high-quality organic inputs. • Apprehensions regarding pest control. • Certification processes entails extensive paper work, providing burdensome and costly for small-scale farmers. • Reliance on livestock. • Waning interest of rural youth in agriculture, accompanied by a diminishing joint family support system.	• Organic goods frequently carry a premium— many consumers are either unwilling or financially unable to bear increased costs. • Organic products are not universaly accessible and available. • Doubts regarding the authencity of organic food in the market, with concerns about counterfeit orgnic produce. • Insufficient awareness or conviction regarding health connections (e.g., the conmplex relationship between cancer and pesticides). • Inadequate comprehension of the connections between organic or NF with environmetal integrity and sustainability. • Consumers may be unaware of food systems and producers heavily reliant on agrochemicals.

Source: Modified from Khurana and Kumar (2020).

2.12 FUTURE PROSPECTS

In order to achieve true self-reliance, a transformation in the agricultural sector is imperative, emphasizing increased farmer income, reduced chemical and pesticide use, and the production of healthy, resource-conserving food. To foster sustainability, a mass movement toward organic and natural farming is vital, requiring robust, coordinated efforts from both central and state governments, with unwavering political commitment and sufficient financial support. The essential measures for the adoption of these organic approaches encompass the establishment of a nationwide, well-funded program that integrates ministries, delineates fund allocation, and builds vibrant markets for farmers. Furthermore, there should be a shift from chemical to organic and biofertilizers, ensuring cost-effective production and distribution, and reallocation of subsidies. Robust scientific studies are essential to ascertain the diverse benefits of organic farming, encompassing biodiversity, water preservation, climate resilience, soil health, and increased yields, incomes, and health. Additionally, an advanced agricultural extension system should enable officials to lead the transition, integrating technology and community engagement. Organic certification processes must be made farmer-friendly, possibly through simpler alternatives, while states should actively support organic farming through initiatives like organic seeds, bio-inputs, farmer training, and market access, thereby facilitating the sale of organic produce and ensuring fair prices for farmers.

2.13 CONCLUSION

In light of the urgent need to increase food production globally by 70% by 2050 to accommodate the demands of a growing population, ensuring food security is a paramount challenge. The indiscriminate adoption of unproven large-scale farming practices and production technologies could jeopardize crop yields and food security. However, the promise of organic and natural farming emerges as a scientifically supported and environmentally responsible alternative, with techniques such as mulching, crop rotation, microbial cultures, intercropping, compost, vermicompost, and biofertilizers demonstrating their efficacy in enhancing productivity, especially in rainfed areas. Furthermore, the mitigation potential of organic agriculture, particularly in reducing greenhouse gas emissions and enhancing carbon sequestration, underscores its scientific rationale. Organic agriculture's focus on nutrient management, nitrous oxide emission reduction, and avoidance of mineral fertilizers contributes to reducing GHG emissions and protecting the environment. Notably, organic agriculture's adaptive approach makes it a resilient choice in the face of climate variability, promoting diversification, soil organic matter increase, and enhanced productivity, especially in marginal areas. Organic and natural farming offer a promising avenue to address the challenges of increasing food production sustainably and ensuring food security, all the while alleviating the environmental consequences associated with conventional farming.

KEYWORDS

- **agrarian crisis**
- **agricultural intensification**
- **agrochemicals**
- **agroecology**
- **chemical-free farming**
- **climate change mitigation**
- **eutrophication**
- **food security**
- **greenhouse gas emissions**
- **intensive cropping systems**
- **land degradation**
- **nutritional autonomy**
- **organic farming**
- **soil biodiversity**
- **sustainable agriculture**
- **sustainable food systems**

REFERENCES

Aher, S. B., Bhaveshananda, S., & Sengupta, B. (2012). Organic agriculture: Way towards sustainable development. *International Journal of Environmental Sciences, 3*(1), 209–216.

Bana, R. S., Pooniya, V., Choudhary, A. K., Rana, K. S., & Tyagi, V. K. (2016). Influence of organic nutrient sources and moisture management on productivity, bio-fortification and soil health in pearlmillet (*Pennisetum glaucum*) + clusterbean (*Cyamopsis tetragonaloba*) intercropping system of semi-arid India. *Indian Journal of Agricultural Sciences, 86*(11), 1418–1425.

Bana, R. S., Singh, D., Nain, M. S., Kumar, H., Kumar, V., & Sepat, S. (2020). Weed control and rice yield stability studies across diverse tillage and crop establishment systems under on-farm environments. *Soil and Tillage Research, 204*, 104729.

Bharucha, Z. P., Mitjans, S. B., & Pretty, J. (2020). Towards redesign at scale through zero budget NF in Andhra Pradesh, India. *International Journal of Agricultural Sustainability, 18*(1), 1–20.

Biswas, S. (2020). Zero budget natural farming in India: Aiming back to the basics. *International Journal of Environment and Climate Change, 10*(9), 38–52.

Borger, C. P., Hashem, A., & Pathan, S. (2010). Manipulating crop row orientation to suppress weeds and increase crop yield. *Weed Science, 58*(2), 174–178.

Choudhary, A. K., Bana, R. S., & Pooniya, V. (2018). *Integrated Crop Management Practices for Enhancing Productivity, Resource-Use Efficiency, Soil Health and Livelihood Security* (pp. 217). ICAR–Indian Agricultural Research Institute. ISBN 978-93-83168-32-3.

Colbach, N., Gardarin, A., & Moreau, D. (2019). The response of weed and crop species to shading: Which parameters explain weed impacts on crop production? *Field Crops Research, 238*, 45–55.

Deshmukh, S. M. (2010). Organic farming: Principles, prospects, and problems. Retrieved from:www.charoka.com/Organic-Farming-Principles-Prospects-and-Problems/b/60009.

Forman, J., & Silverstein, J. (2012). Organic foods: Health and environmental advantages and disadvantages. *Pediatrics, 130*(5), 25–79.

Gabriel, D., Roschewitz, I., Tscharntke, T., & Thies, C. (2006). Beta diversity at different spatial scales: Plant communities in organic and conventional agriculture. *Ecological Applications, 16*(5), 2011–2021.

Ghosh, M. (2019). Climate-smart agriculture, productivity, and food security in India. *Journal of Development Policy and Practice, 4*(2), 166–187.

Hussain, I., Khan, E. A., Hassan, G., Gul, J., Ozturk, M., Alharby, H., Hakeem, K. R., & Alamri, S. (2017). Integration of high seeding densities and crisscross row planting pattern suppresses weeds and increases grain yield of spring wheat. *Journal of Environmental Biology, 38*(5 Special Issue), 1139–1145.

Hussain, M. M., & Farooqi, Z. U. R. (2021). Allelopathic bacteria as an alternate weedicide: Progress and future standpoints. In *Microbiota and Biofertilizers: A Sustainable Continuum for Plant and Soil Health* (pp. 211–230).

IFOAM. (1998). *The IFOAM Basic Standards for Organic Production and Processing.* General Assembly, Argentina, IFOAM, Germany.

IFOAM. (2009). *High Sequestration Low Emission Food Secures Farming: Organic Agriculture a Guide to Climate Change and Food Security*. Retrieved from: http://orgprints.org/16769/1/IFOAM-CC-Guide-Web-20100210.pdf (accessed on 16 January 2025).

International Conference on Organic Agriculture and Food Security. (2007). *Organic Agriculture and Food Security, Rome, Italy* (Conference paper). Food and Agriculture Organization of the United Nations (FAO). Retrieved from: http://www.fao.org/tempref/docrep/fao/meeting/012/ah953e.pdf (accessed on 16 January 2025).

IPCC. (2007). From *Climate Change 2007: Chapter Agriculture*. Retrieved from: www.ipcc.ch/pdf/assessmentreport/ar4/wg3/ar4-wg3-chapter8.pdf.

Isaacs, J. R. (2012). Organic farming keeps carbon out of the atmosphere. Retrieved from: http://news.mongabay.com/2012/1128-isaacs-organic-carbon.

John, D. A., & Babu, G. R. (2021). Lessons from the aftermaths of the Green Revolution on food systems and health. *Frontiers in Sustainable Food Systems, 5*, 644559.

Jordan, R., Müller, A., & Oudes, A. (2009). *High Sequestration, Low Emission, Food Secure Farming: Organic Agriculture-A Guide to Climate Change & Food Security.*

Keerthi, P., Sharma, S. K., & Chaudhary, K. (2018). Zero budget NF: An introduction. In *Research Trends in Agricultural Sciences* (pp. 111–123). AkiNik Publications.

Khadse, A., & Rosset, P. M. (2019). Zero Budget NF in India–from inception to institutionalization. *Agroecology and Sustainable Food Systems, 43*(7, 8), 848–871.

Khurana, A., & Kumar, V. (2020). *State of Organic and Natural Farming: Challenges and Possibilities*. Center for Science and Environment, New Delhi.

Kumar, V., Bana, R. S., Singh, T., & Louhar, G. (2021). Ecological weed management approaches for wheat under rice–wheat cropping system. *Environmental Sustainability, 4*, 51–61.

La Via Campesina. (2016). *Zero Budget NF in India*. Family Farming Knowledge Platform of FAO. Retrieved from: http://www.fao.org/familyfarming/detail/en/c/429762/.

Li, R., Pang, Z., Zhou, Y., Fallah, N., Hu, C., Lin, W., & Yuan, Z. (2020). Metagenomic analysis exploring taxonomic and functional diversity of soil microbial communities in sugarcane fields applied with organic fertilizer. *BioMed Research International, 2020*, 9381506.

Lotter, D. W. (2003). Organic agriculture. *Journal of Sustainable Agriculture, 21*(4). Retrieved from: http://donlotter.net/lotter-organicag.pdf.

Maqsood, S., Buhroo, A. A., Sherwani, A., & Mukhtar, M. (2022). Optimizing control of flea beetles through ecological engineering of vegetable agroecosystem in Kashmir. *Plant Science Today, 9*(4), 785–793.

Mariappan, K., & Zhou, D. (2019). A threat of farmers' suicide and the opportunity in organic farming for sustainable agricultural development in India. *Sustainability, 11*(8), 2400.

Meena, A., Meena, M., Kumar, R., & Meena, B. M. (2017). Farm scaping: An ecological approach to insect pest management in agroecosystem. *Journal of Entomology and Zoology Studies, 5*(3), 598–603.

Mishra, S. (2018). *Zero Budget NF: Are This and Similar Practices the Answers*. Nabakrushna Choudhury Center for Development Studies (NCDS), Odisha, India.

Morgera, E., Caro, C. B., & Duran, G. M. (2012). *Organic Agriculture and The Law*. FAO, Rome.

Muller, A., Olesen, J. E., Davis, J., Dytrtova, K., Gattinger, A., Lampkin, N., & Niggli, U. (2012). Reducing global warming and adapting to climate change: The potential of organic agriculture. *Working Papers in Economics, 526*, University of Gothenburg, Department of Economics.

Münster, D. (2017). Zero budget NF and bovine entanglements in South India. *RCC Perspectives, 1*, 25–32.

Nascimbene, J., Marini, L., & Paoletti, M. G. (2012). Organic farming benefits local plant diversity in vineyard farms located in intensive agricultural landscapes. *Environmental Management, 49*, 1054–1060.

Palekar, S. (2006). *The Principles of Spiritual Farming*. Zero Budget NF Research, Development & Extension Movement, Amravati, Maharashtra, India. Retrieved from: http://www.vedicbooks.net/principles-spiritual-farming-volume-p-14779.html (accessed on 16 January 2025).

Pottorff, L. P. (2010). Some pesticides permitted in organic gardening. *Colorado State University Cooperative Extension.*

Rao, M. S., Rao, C. R., Srinivas, K., Pratibha, G., Sekhar, S. M., & Vani, G. S. (2012). Intercropping for management of insect pests of castor, *Ricinus communis*, in the semi-arid tropics of India. *Journal of Insect Science, 12*, 1536–2442.

Ray, P., Lakshmanan, V., Labbé, J. L., & Craven, K. D. (2020). Microbe to microbiome: A paradigm shift in the application of microorganisms for sustainable agriculture. *Frontiers in Microbiology, 11*, 622926.

RySS. (2018). Abstract data on earthworm populations in ZBNF vs. non-ZBNF fields per square meter. RySS unpublished internal data.

RySS. (2020). *Zero Budget NF: Official Website of ZBNF Programme of Rythu Sadhikara Samstha, Government of Andhra Pradesh.* Retrieved from: http://apzbnf.in/ (accessed on 16 January 2025).

Saharan, B. S., Tyagi, S., Kumar, R., Vijay, O. H., Mandal, B. S., & Duhan, J. S. (2023). Application of Jeevamrit improves soil properties in zero budget NF fields. *Agriculture, 13*(1), 196.

Smith, J., Yeluripati, J., Smith, P., & Nayak, D. R. (2020). Potential yield challenges to scale-up of zero budget NF. *Nature Sustainability, 3*(3), 247–252.

Sreenivasa, M. N., Naik, N., & Bhat, S. N. (2010). Beejamrutha: A source for beneficial bacteria. *Karnataka Journal of Agricultural Sciences, 22*(5), 1038–1040.

The Hindu. (2010). Zero budget farming a success.

Tripathi, S., Shahidi, T., Nagbhushan, S., & Gupta, N. (2018). Zero budget NF for the sustainable development goals, Andhra Pradesh, India. *Council on Energy, Environment and Water* (CEEW), Issue brief, 2nd edition.

Tschumi, M., Albrecht, M., Collatz, J., Dubsky, V., Entling, M. H., Najar-Rodriguez, A. J., & Jacot, K. (2016). Tailored flower strips promote natural enemy biodiversity and pest control in potato crops. *Journal of Applied Ecology, 53*(4), 1169–1176. https://doi.org/10.1111/1365-2664.12630.

United Nations (UN). (2015). *Transforming Our World: The 2030 Agenda for Sustainable Development*. United Nations Department of Economic and Social Affairs: Sustainable Development. https://sdgs.un.org/2030agenda (accessed on 16 January 2025).

United Nations Environment Programme (UNEP). (2021). *Making Peace with Nature: A Scientific Blueprint to Tackle the Climate, Biodiversity, and Pollution Emergencies*. United Nations Environment Programme.

Watson, C. A., Atkinson, D., Gosling, P., Jackson, L. R., & Rayns, F. W. (2002). Managing soil fertility in organic farming systems. *Soil Use and Management, 18*, 239–247. https://doi.org/10.1111/j.1475-2743.2002.tb00160.x.

Willer, H., & Lernoud, J. (2017). *The World of Organic Agriculture: Statistics and Emerging Trends 2017* (pp. 1–336). Research Institute of Organic Agriculture FiBL and IFOAM-Organics International.

World Bank Group. (2022). *Commodity Markets Outlook: The Impact of the War in Ukraine on Commodity Markets*. World Bank.

Yadav, P., Jaiswal, D. K., & Sinha, R. K. (2021). Climate change: Impact on agricultural production and sustainable mitigation. In *Global Climate Change* (pp. 151–174). Elsevier. https://doi.org/10.1016/B978-0-12-823269-1.00012-9.

CHAPTER 3

Response of Soil Microbial Communities to Climate Change

LUCIANA RAPHAEL CHAPPA,[1,*] HARUN IRERI GITARI,[2] EMMANUELY ZEPHALINE NUNGULA,[3] VEDASTO VICTOR NGAIZA,[4] YAMIKANI HARRY MAKWINJA,[5] SHIVANI RANJAN,[6] SUMIT SOW,[6] ASHWAQ M. ALNEMARI,[7] and MAHMOUD F. SELEIMAN[8]

[1]*Mwalimu Nyerere University of Agriculture and Technology (MNUAT); Department of Agricultural sciences. Butiama-Mara Tanzania*

[2]*Department of Agricultural Sciences, Mwalimu Julius Nyerere University of Agriculture and Technology, Butiama-Mara, Tanzania*

[3]*Department for Environment and Sustainable Development, Mzumbe University, Morogoro, Tanzania*

[4]*School of Life Sciences and Bioengineering, Nelson Mandela Institution of Sciences and Technology, Arusha, Tanzania*

[5]*Department of Forestry, Ministry of Natural Resources and Climate Change, Blantyre, Malawi*

[6]*Department of Agronomy, Dr. Rajendra Prasad Central Agricultural University, Pusa, Samastipur, Bihar, India*

[7]*Biology Department, College of Science and Humanities, Prince Sattam bin Abdulaziz University, Al-Kharj, Saudi Arabia*

[8]*Plant Production Department, College of Food and Agriculture Sciences, King Saud University, Riyadh, Saudi Arabia*

*Corresponding author

Climate-Resilient Agriculture for Sustainable Crop Production: New Adaptation and Mitigation Strategies. Sanjay Kumar, Shivani Ranjan, Sumit Sow, Smruti Ranjan Padhan, and Dhirendra Kumar Roy (Eds.)

DOI: 10.1201/9781998511150-3

ABSTRACT

Climate change, a complex phenomenon characterized by long-term alterations in Earth's climate, arises from human-induced factors such as population growth, pollution, and land degradation. This escalation contributes to increased greenhouse gases (GHGs)—including nitrous oxide (N_2O), methane (CH_4), and carbon dioxide (CO_2)—which intensify the greenhouse effect and trigger unprecedented environmental events such as droughts, fires, heatwaves, floods, and biodiversity loss. Soil microbial communities, comprising bacteria, fungi, archaea, and protozoa, form dynamic ecosystems within the soil matrix. These microorganisms play a vital role in maintaining soil health, nutrient cycling, and ecosystem processes. Their interactions with plants are essential for nutrient acquisition, utilization, and plant growth. Beneficial microbes enhance soil quality, plant health, and agricultural sustainability. The diversity of these communities greatly influences ecosystem resilience and stability, yet climate change elicits varied responses from them. This chapter explores how soil microbial communities react to extreme weather events, including drought, flooding, salinization, permafrost thaw, temperature fluctuations, fire, and elevated CO_2. It also examines their role in climate change mitigation and the management of climate-related impacts. The chapter identifies a significant research gap and calls for expanded studies, particularly those considering plant–microbe interactions in the context of climate adaptation.

3.1 INTRODUCTION

Climate change refers to the intricate phenomenon expressed by extended changes in Earth's climate, including shifts in precipitation patterns, temperature, and other climatic variables (Pielke, 2004). As reported by Sage (2020), continuous population growth, a steady increase in the number of pollutants and toxic substances created by humans, degradation of the land used for farming, carbon dioxide accumulation in the atmosphere brought on by burning fossil fuels, and greenhouse gases are collectively putting Earth's ecosystems under siege (Hassan et al., 2021; Nungula et al., 2024; Sairaam et al., 2023). Greenhouse gases (GHGs) play a crucial role in driving climate change by trapping heat in the atmosphere.

Carbon dioxide (CO_2), methane (CH_4), and nitrous oxide (N_2O) contribute to the greenhouse effect, leading to global warming (Kannadhasan and Nagarajan, 2023). Human activities, notably industrial processes, deforestation, and the burning of fossil fuels, greatly contribute to the rise in levels of greenhouse gases, intensifying the greenhouse effect and accelerating global warming (Pachauri et al., 2014; Pielke, 2004). The increase in greenhouse gases (GHGs), as reported by Zandalinas et al. (2021), Masson-Delmotte et al. (2021), and Wagner et al. (2021), can lead to severe droughts, wildfires, heat waves, floods, and biodiversity loss.

The impact of climate change is evident across various sectors, as indicated by several research findings. Ecosystems face significant challenges, with climate-induced stress affecting plant and animal species as well as soil microorganisms. Sage (2020) highlights the vulnerability of ecosystems to climate change, resulting in shifts in species distributions and altered ecological dynamics. The decline in insect populations, as demonstrated by Wagner et al. (2021), highlights the extensive impact of climate change on biodiversity. Climate change, as reported by Sage (2020) and Savary and Willocquet (2020), could further jeopardize global food security and production, ultimately causing chaos, hunger, and possibly a catastrophic extinction. Soil microbial communities, as defined by Adomako et al. (2022), Britannica (2016), Foster (1988), Tisdall (1994), and Welbaum et al. (2004) are dynamic ecosystems within the soil matrix, consisting of a diverse assemblage of microbes, including bacteria, protozoa, fungi, and archaea. These communities form intricate networks of interactions and functions that play a pivotal role in shaping soil health and driving essential ecosystem processes. Chandra and Enespa (2019) and Van Der Heijden et al. (2008) reported that soil microbial communities encompass a varied assembly of microorganisms, and their actions possess the capacity to impact plant growth and productivity either positively or negatively.

The function of soil microbes in plant nutrition is fundamental, with interactions between microorganisms and plants playing a significant role in nutrient uptake and utilization (Jacoby et al., 2017). Beneficial microorganisms further contribute to soil quality and plant health, highlighting their positive impact on enhancing overall sustainability in agricultural systems (Mirriam et al., 2022; Ortiz and Sansinenea, 2022). The pivotal function of soil microorganisms' communities in nutrient cycling and soil fertility, demonstrated by Welbaum et al. (2004), encompasses their active involvement in decomposition, mineralization, and nitrogen fixation.

This involvement promotes the availability of necessary nutrients for the growth of plants and contributes fundamentally to the soil's productivity. Moreover, Nyawade et al. (2019) and Tisdall (1994) highlighted the substantial contribution of soil microbes to the stabilization and formation of soil aggregates, thereby influencing soil structure and physical properties like water retention and root penetration. Abawi and Widmer (2000) and Singh et al. (2009) demonstrated that these microbial communities play a critical role in disease suppression, acting as natural barriers against soilborne pathogens and contributing to plant health. The diversity and composition of these communities, as emphasized by Zak et al. (2003), significantly impact ecosystem stability and resilience, showing correlations between microbial diversity and enhanced ecosystem functions in various environmental conditions.

3.2 RESPONSE OF SOIL MICROBIAL COMMUNITIES TOWARDS CLIMATE CHANGE

The response of soil microbial communities to climate change involves changes in their composition, metabolic activities, and ecological functions due to shifts in factors such as temperature, precipitation patterns, and greenhouse gas emissions (Abirami et al., 2021; Castro et al., 2010; de Nijs et al., 2019; Jansson and Hofmocke, 2020; Lladó et al., 2017). Similarly, an experiment was conducted by Castro et al. (2010) to investigate the actions of various climate change drivers in shaping soil microbial communities. They utilized a comprehensive climate manipulation experiment that altered atmospheric CO_2 levels (+300 ppm, baseline), precipitation (dry and wet conditions), and temperature (+3°C, baseline) within an established old-field ecosystem that had been running for three and a half years at the time of data collection. Previous studies associated with this project have already showcased the individual and combined impacts of these treatments on plant community biomass and composition (Engel et al., 2009; Kardol et al., 2010), nitrogen fixation (Garten et al., 2008), soil microbial activity (Kardol et al., 2010), respiration (Wan et al., 2007), and soil carbon content (Garten et al., 2009). This interaction influences nutrient cycling, carbon sequestration, and overall soil health (Figure 3.1), highlighting the crucial role of microbial communities in the context of environmental changes (Castro et al., 2010; Jansson and Hofmocke, 2020; Lladó et al., 2017).

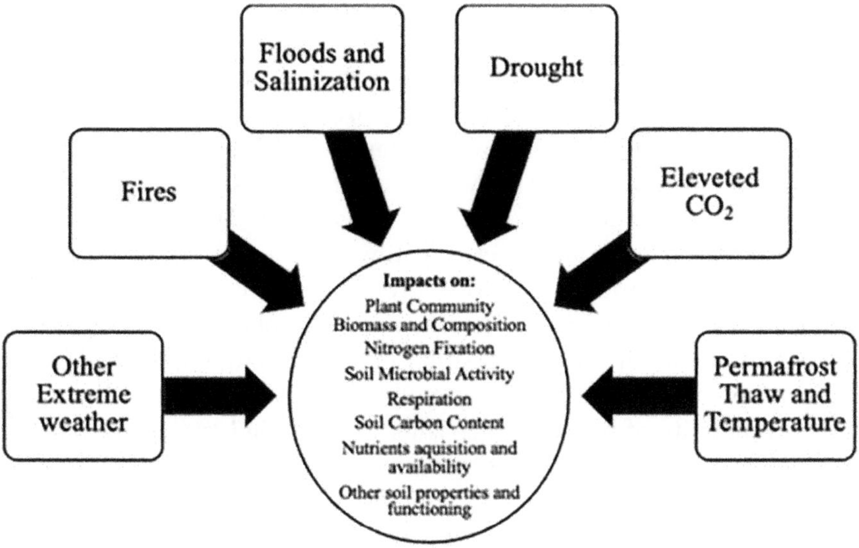

FIGURE 3.1 Impact of various extreme weather events on soil.

3.2.1 RESPONSE OF SOIL MICROBIAL COMMUNITIES TOWARDS DROUGHT

Drought poses a significant challenge globally, with well-documented effects on plants in both agricultural and natural environments (Seleiman et al., 2021). Schimel et al. (2007) confirmed that the impacts of hydrological stress on soils and their interrelated organisms, particularly soil microorganisms, are less clearly understood, despite being a prevalent environmental stressor for them. Numerous studies highlight the repercussions of water stress on microorganism communities and soil properties. Prolonged and severe water stress, in extreme cases, can result in desertification, but its consequences on soil health may manifest earlier. Approximately 20% of the Earth's soil is presently experiencing acute degradation and reduced productivity, often attributed to drought-related factors (Alexander and Dudley, 2017; Nungula et al., 2023). Drought is a prevalent stress that affects microorganism processes and community, growth of crops, and crop productivity as well (Abdul Rahman et al., 2021; Siebielec et al., 2020). Drought's detrimental impacts on soil microbes may result in the disruption of the cycling of nutrients (N, P, C, and others), reduced enzyme activity, and decreased fertility of the soil, which can affect crop yields, particularly for crops that are prone to drought, and have a fiscal impact (Abdul Rahman et al., 2021; Nasar et al., 2023; Nguyen et al., 2018).

Under water scarcity stress, the state of the soil is closely connected to plant development and growth (Heydarzadeh et al., 2021). The morphology, biochemistry, and physiology of the plants are all affected by drought directly (Hanaka et al., 2021). Moreover, it decreases the sprouting of seeds and the development of seedlings (Khan et al., 2019). Observations of plant reactions to drought were equally noted at the metabolomic and transcriptomic tiers (You et al., 2019). Prolonged and intense drought stress disrupts the plant roots' ability to reach soil microbes, thereby causing a substantial impact on the composition of their microbiome, resulting in the secretion of root exudates, alterations to root structure, and disruption in the availability of beneficial nutrients. Drought induces changes in the plant root's microbiome, promoting various gram-positive species, chiefly actinobacteria, replacing the predominance of Gram-negative taxonomy typically found in the root zone (Breitkreuz et al., 2021).

Typically, soil microbial communities can boost a plant's capability to endure drought through various mechanisms. These mechanisms include the generation of polysaccharides that enhance water retention and soil structure, the synthesis of proline (pro) deaminase, and indoleacetic acid (IAA), to enhance plants' resilience to drought and aid in better water distribution via fungal mycelia (Milošević et al., 2012). Remarkably, the adverse impact of drought on microbes can be alleviated through organic matter incorporation into the soil (Pascual et al., 1998). Furthermore, drought-induced phytohormones like jasmonic acid (JA), 1-aminocyclopropane-1-carboxylic acid (ACC), and abscisic acid (ABA) can alter microbial activity, specifically respiration rates, and the structure of microbiological colonies in the soil. However, extended exposure to drought conditions can modify how soil microbial communities respond to these hormones (Sayer et al., 2021). Notwithstanding the unfavorable impacts of drought-induced changes, the challenging environmental circumstances can stimulate noteworthy adaptations in both plants and microbes, enabling them to reproduce and survive. Such adjustments might cause new functional groupings to arise in the ecosystem or turn into useful instruments for improving plant breeding and farming techniques (da Silva et al., 2013; Sayer et al., 2021). It may be possible to investigate the traits of drought-tolerant microorganisms to create efficient biopreparations that promote growth in plants (Hanaka et al., 2021).

3.2.2 RESPONSE OF SOIL MICROBIAL COMMUNITIES TOWARDS FLOODS AND SALINIZATION

Flooding leads to anoxic soils (oxygen-deprived soils) and saturated soil; thereafter, once the floodwaters recede, salt residues persist, creating a lasting impact of salinization on the soil that hinders long-term crop development (Glenn et al., 1998; Warrence et al., 2002). Gould et al. (2020) and Hanley et al. (2020) reported that the soil salinization effect on crops and plants is widely recognized, but the concurrent impact of both flooding and salinity remains less understood and documented. Understanding the ecology of the underground microbial elements in these systems is crucial for potentially lessening these impacts. Disruption of these components can have severe ramifications for the amount of biomass above ground (Meena et al., 2017). In their research, Berga et al. (2017) and Momrikotb et al. (2022) confirmed that salinity resulting from the influx of seawater, coupled with a decrease in oxygen potential, acts as a stressor on soil microorganism communities. During stressful periods, changes in microbial biodiversity are commonly seen. Nevertheless, there has been a less comprehensive examination of the overall functionality of the community. Furthermore, Hanley et al. (2020) proved that the surge of water and an elevation in salinity have the potential to disrupt crucial functions of soil microbes.

Long-term flooding causes oxygen-deprived soil, which affects the microbiological cycling of nutrients and soil enzyme activity (Chaudhary et al., 2016; Chambers et al., 2016; Fitch et al., 2022; Wright et al., 2015). Therefore, understanding the capacity of flood-prone communities to withstand such pressures is crucial. In the event of an impact, assessing their resilience and ability to recover becomes essential, as published by Shade et al. (2012). The capacity of soil microbe communities to endure or recuperate from changes in environments may, nonetheless, hinge on their previous encounters with stress (Allison and Martiny, 2018; Shade et al., 2012). Coastal habitats characterized by fringing, like salt marshes, have evolved to tolerate frequent exposure to saltwater flooding. Nonetheless, salt marshes face diverse tidal flooding along the shore, leading to the unique adaptation of species to specific conditions (Lee et al., 2016).

The variation in soil microbial composition and environmental factors is induced by the shift of flooding duration and hydrologic gradients, as reported by Hu et al. (2014) and Yu and Ehrenfeld (2010). Similarly, Ponnamperuma (1984) and Stams and Plugge (2010) published that oxygen-depleted conditions caused by flooding will have an impact on ecosystems and soil properties. According to Zhou et al. (2002), inundated soils have a lower diversity of bacterial species. Most biogeochemical cycles are dominated by microorganisms, and changes in community function and structure can lead to disruptions in these processes. As published by previous studies by Bardgett et al. (1999) and Unger et al. (2009), distinctive responses to changes in soil moisture are exhibited by each group of phospholipid fatty acids (PLFA). Under aerobic and anaerobic conditions, frequent occurrences are noted for monounsaturated fatty acids and branched fatty acids, respectively, as confirmed by Bossio et al. (2006). Certain research investigations used focused methodologies to investigate the effects of flooding on microbial communities. Using PLFA and denaturing gel gradient electrophoresis (DGGE), Bodelier et al. (2012) discovered that the number of methanotrophs rose with the increase in flooding at four different points along a river. Similarly, Kemnitz et al. (2004) detected a rise in methanogenic microorganism diversity in samples from rivers by employing terminal-restriction fragment length polymorphism (T-RFLP). Meanwhile, Unger et al. (2009) observed a reduction in the ratio of bacteria to fungi with flooding through the use of PLFA. These investigations offer valuable insights into the influence of flooding on microbial diversity and community structure. Nevertheless, a more profound comprehension of the consequences of environmental stressors on the entire community is necessary.

Rhythmic patterns of alternating draining and flooding have the potential to disrupt the dynamics of microbial communities. The induction of anoxic conditions during flooding might result in the suppression or mortality of certain populations, creating opportunities for the development of others (Denef et al., 2001). The fluctuation between these states prevents the community from establishing stability with either predominantly anaerobic or aerobic populations, and those organisms that survive display adaptability to both conditions. The duration of flooding plays a pivotal role, influencing the community as redox potentials take time to decrease during anoxia (Mohanty et al., 2013; Wang et al., 1993). The sequence of events involves denitrification, followed by iron and sulfur reduction, and ultimately methanogenesis (Patrick and Jugsujinda, 1992; Reddy and Patrick, 1975). The reversal of these processes through drainage elevates redox potential, hindering subsequent reduction processes.

Research by Baldwin and Mitchell (2000) suggests that nitrification and denitrification decline after desiccation periods but rebound following rewetting. Morillas et al. (2015) noted a decrease in nitrification with an increased frequency of dry/wet cycles. Conrad (2007) reported that anaerobic soils may house methanogens, which are Archaea that produce methane under rigidly anaerobic conditions. Flooding could stimulate their proliferation. Methanotrophs, found in both anaerobic and aerobic conditions, metabolize CH_4. Given methane's 100-year global warming potential compared to carbon dioxide, Myhre et al. (2013) reported that understanding factors influencing CH_4 flux is crucial for evaluating climate change risks. Investigations in rice paddies (Ratering and Conrad, 1998; Sigren et al., 1997; Yagi et al., 1996) reveal a sharp decline in CH_4 emissions with short-term drainage, attributed to methanogens' intolerance to even low oxygen levels (Conrad, 2007). However, when flooding resumes, methane emissions remain repressed, likely due to the oxidation of reduced sulfate and ferric iron during drainage (Patrick and Jugsujinda, 1992), providing substrates for sulfate/iron-reducing bacteria. The response of microbial communities to frequent flooding and drainage in pasture soil remains an unexplored area. While flooding induces an oxygen-deprived state in the soil, the presence of oxygen during and after drainage can potentially bring the community back to its original state (Ponnamperuma, 1984).

Most changes in the soil's chemical, biological, and physical processes resulting from flooding can be reversed through drying and draining. However, the speed of this reversal depends on various factors, including the rate of flood subsidence, species proliferation rates, available metabolic substrates, and redox potentials. Bacteria that thrive in aerobic conditions, whether obligate aerobic or facultative anaerobic, can endure periods of anoxia or hypoxia, as observed in *Methylosinus trichosporium* (Roslev and King, 1994) and *Mycobacterium smegmatis* (Berney et al., 2014). Frequent flooding followed by drainage tends to inhibit the growth of many bacterial species rather than eliminate them. Additionally, many aerobic bacterial species prefer a moist environment, providing a conducive environment for their growth during intermittent flooding (Fredrickson et al., 2008; Potts, 1994; Roberson et al., 1993).

3.2.3 MICROBIAL COMMUNITY RESPONSES TO PERMAFROST THAW AND TEMPERATURE

Permafrost thaw exerts a profound impact on ecosystems, introducing shifts in soil hydrology and carbon turnover (Meena et al., 2023). Notably, the fungal community undergoes substantial transformations, marked by the decline of mycorrhizal fungi and an increase in potential fungal pathogens. This, in turn, shapes plant responses to permafrost thaws (Schütte et al., 2019). The disruption caused by permafrost thaw extends to microbial communities, influencing levels of dissolved organic carbon and nitrogen (Coolen et al., 2011). While permafrost historically maintained subfreezing temperatures and limited microbial activity, ancient permafrost, some millions of years old, contains viable microorganisms that could reinvigorate upon thawing, accelerating physiological activities (Coolen et al., 2011). Recent findings suggest that ancient permafrost not only hosts metabolically active microbes but also stores considerable carbon and nutrients usable by indigenous microbial communities (Coolen et al., 2011).

A consequential rise in total dissolved nitrogen is observed (Inglese et al., 2017). Freeze-thaw cycles play a pivotal role in impacting microbial communities by reducing the fungi-to-bacteria ratio and enzyme activities (Liu et al., 2020). These cycles induce shifts in bacterial communities, with potential ramifications for soil organic matter decomposition and the permafrost carbon feedback (Monteux et al., 2018). Simultaneously, changes in archaeal communities contribute to increased methane emissions and alterations in soil characteristics (Wei et al., 2018). The responses of microbial communities to permafrost thaw are intricate, wielding influence over carbon release and community compositions (Chen et al., 2020). Temperature emerges as a key factor, wielding a strong influence on soil microbes. It affects enzymatic activity, metabolism, and respiration (Li et al., 2021). Higher temperatures create a favorable environment for fast-growing microbial taxa adapted to heat stress (Donhauser et al., 2020). Consequently, microbial activities, particularly respiration, accelerate, raising concerns about potential carbon loss from soil (Walker et al., 2018). Understanding the intricate relationships among permafrost thaw, temperature variations, and microbial communities is indispensable. Such insights are critical for predicting ecosystem responses and formulating effective soil management strategies.

3.2.4 MICROBIAL COMMUNITY RESPONSES TO INCREASED FIRE FREQUENCY OR INTENSITY

Fires are on the rise globally, driven by prolonged and drier fire seasons coupled with unsustainable land management practices. The impacts of fires on soil microorganisms involve various compounding disturbances, as highlighted by Knelman et al. (2019). Additionally, fires contribute directly to climate feedback by releasing significant amounts of carbon dioxide into the atmosphere, as indicated by Sun et al. (2015). The duration and intensity of fires play a crucial role in shaping the characteristics of soil organic carbon. Longer fire durations and higher fire intensities lead to increased heat transfer to the subsoil, particularly when the organic layer of soil ignites, producing more heat. In soils with elevated moisture levels, the heating process may be deferred, but it can still result in the pasteurization-induced death of more microorganisms, as noted by Hart et al. (2005). Intensive high fires can reduce soil aeration and damage soil aggregate structure. The aftermath of fires includes land degradation and erosion, causing soil to wash into waterways, thereby exacerbating the impact on soil ecosystems. Furthermore, Hinojosa et al. (2016) and Nave et al. (2011) highlighted that fires lead to a substantial reduction in nitrogen stocks and soil carbon. Hart et al. (2005) observed that, following a fire event, there is a documented decrease in microbial biomass due to a depletion of resources necessary for supporting microbial growth.

Multiple factors play a crucial role in determining the impact of fires on soil microorganisms and soil organic carbon. These factors include plant inputs as well as landscape topography. Research conducted in the Nome Creek area of Alaska after an intense fire event revealed a substantial loss of soil carbon and soil water from the mountain slope. The consequences extended to the thawing of the underlying permafrost soil layer, as documented by Taş et al. (2014). This impact was accompanied by a reduction in the abundance of most soil microbiome members, although certain bacteria, such as those belonging to the AD3 candidate phylum, were enriched post-fire. In contrast, a study focusing on temperate pine-dominated forests demonstrated that fires in such ecosystems led to new inputs into the system. These inputs included plant litter and charcoal, as outlined by Hart et al. (2005). Hart et al. (2005) reported that the straightforward impact of fire on soil microbes involves processes like cell lysis and combustion as well as protein denaturation, leading to a decline in microbes' biomass. Despite this, there is a short-term surge in available nutrients immediately post-fire, which can result in increased microbial activity.

Goberna et al. (2012) conducted a study indicating elevated microbial activity and biomass after fire treatment, with a higher abundance of archaea.

It was suggested that this increase could be attributed to the higher heat resistance of archaeal cell walls. In another study, following a fire event, Hinojosa et al. (2016) observed a decrease in the proportion of fungi and an increase in Actinobacteria. Nevertheless, due to variations in heat sensitivity among different bacterial, fungal, and archaeal groups, making generalizations across diverse ecosystems is challenging, and the long-term functional consequences of these shifts remain uncertain. The soil microbiome can experience indirect consequences as a result of fires. Changes in plant cover triggered by fires play a substantial role in impacting the soil microbiome, mainly due to their connection with plant roots. Other non-direct impacts encompass heightened solar penetration, deposition of charcoal and alkaline ash, chemical modifications in the soil, and alterations in albedo due to soil darkening (Hart et al., 2005). Fire-produced charcoal, known for its chemical and biological stability, was demonstrated in one study to sustain the activity of nitrifying bacteria in boreal forests post-fire (DeLuca and Sala, 2006). Elevating soil pH, which has a major impact on soil microbial variety and richness, is another compounding effect of fires (Fierer and Jackson, 2006). After a fire, fungal diversity in boreal woods was shown to rise, possibly as a result of the elevated pH, but then progressively decreased (Sun et al., 2015). When fires occur alongside other extremities in the climate, for instance, drought, the microbial community of soil may suffer severe consequences and display diminished tolerance to possible subsequent disruptions (Hinojosa et al., 2016). Bowker et al. (2004) confirmed that, yet, drawing inferences about the consequences of fires on soil microbes can be challenging since certain ecosystems, like biological soil crusts in grasslands that have adapted to fire, naturally show resistance to low-intensity fires.

3.2.5 MICROBIAL COMMUNITY RESPONSES TO ELEVATED CO_2

Microbial community responses to elevated CO_2 have been extensively investigated in various field experiments, providing valuable insights into the projected shifts in climate. Dunbar et al. (2012) reported that free-air CO_2 enrichment experiments conducted across different ecosystem types revealed shifts in microorganisms. For instance, a 10-year cross-biome study demonstrated ecosystem-specific responses, including increases in Acidobacteria with elevated CO_2. In the context of Australian grasslands,

Hayde et al. (2012) reported changes in fungi and archaea, as well as specific groups of bacteria in response to elevated CO_2. Several studies have contributed to understanding how phylogenetic changes reflect the characteristics of microbial communities within ecosystems. Allison (2012) underscored the significance of adopting a strategy centered on traits, offering a structure to incorporate microbial physiology into the broader field of ecosystem ecology. The exploration of potential functions of the soil microbiome under elevated CO_2 involves screening gene abundances in metagenomes. The BioCON grassland experiment, as published by Tu et al. (2017), revealed elevated CO_2-stimulated increases in gene families associated with decomposition, dissimilatory nitrate reduction, and nitrogen fixation. Similarly, in arid grasslands exposed to elevated CO_2, Yu et al. (2018) showed increased microbial genes involved in decomposition, CH_4 metabolism, carbon fixation, nitrogen fixation, denitrification, and nitrogen mineralization. Plant–microbe interactions, as reported by Bengtson et al. (2012), Bréche et al. (2018), and Carol et al. (2011), play a crucial role in carbon exchange between the soil and atmosphere under elevated CO_2 conditions.

Distinct plant varieties exhibit varied reactions to increased CO_2 levels, impacting both the volume and nature of carbon contributions to the rhizosphere. For example, Hayde et al. (2012) illustrated that C4 plants, which demonstrate higher photosynthetic efficiency compared to C3 plants, might distribute a greater amount of carbon to rhizosphere-linked microorganisms beneath the soil, leading to alterations in community composition. The impact of increased CO_2 levels on the cycling of soil organic matter (SOM) entails modifications in gene functions. As indicated by Qio et al. (2024), the augmentation of rhizodeposition in response to heightened CO_2 can initiate microbial decomposition of the existing soil organic carbon (SOC) reservoirs, a phenomenon termed "priming." Nevertheless, accurately forecasting the equilibrium between the accrual of carbon through mineral association and soil aggregation and the expedited decomposition triggered by priming remains an intricate task, as deliberated by Drake et al. (2011) and Van Groenigen et al. (2014). Untangling the combined impacts of elevated CO_2 and warming proves challenging, as evidenced in investigations like the one conducted by Hayde et al. (2012). The enduring consequences on soil carbon stocks hinge on the availability of water and nutrients, influencing processes such as photosynthesis, microbial decomposition, and the overall accumulation of soil carbon. It is imperative to comprehend how alterations in CO_2 interact with various environmental factors, including temperature, precipitation, and nutrient levels, as underscored by multiple researchers (Bréche et al., 2018; Hayde et al., 2012; Scharlemann et al., 2014). This underscores the need for ongoing and prospective extended field studies that encompass the array of significant climate-related changes anticipated in a given region.

3.3 CONTRIBUTION OF SOIL MICROBES TO CLIMATE CHANGE MITIGATION

Soil microbes play a crucial role in the global carbon cycle and have the potential to influence climate change dynamics significantly (Table 3.1). Understanding the mechanisms by which soil microbes interact with the environment and contribute to climate change mitigation is essential for developing effective strategies for sustainable land management. One of the primary mechanisms by which soil microbes mitigate climate change is through carbon sequestration (Dutta and Dutta, 2016; Mukhtar et al., 2023). Soil microbes, such as bacteria and fungi, decompose organic matter and release CO_2 into the atmosphere during respiration (Hussain et al., 2022; Mukhtar et al., 2023). However, they also incorporate carbon into their biomass and facilitate the formation of stable organic carbon compounds in the soil (Dutta and Dutta, 2016; Hussain et al., 2022; Mukhtar et al., 2023). This process, known as carbon sequestration, helps to remove CO_2 from the atmosphere and store it in the soil, thereby mitigating the greenhouse effect (Dutta and Dutta, 2016; Maheshwari et al., 2022; Mukhtar et al., 2023).

Soil microbes such as *Rhizobium* spp. and *Azotobacter* spp. are pivotal in nitrogen fixation (Table 3.1), the process of converting atmospheric nitrogen into forms usable by plants (Aasfar et al., 2021; Jena et al., 2022; Mirriam et al., 2023; Shridhar, 2012). These microbes form symbiotic relationships with plants or exist as free-living organisms in soil (Dutta and Dutta, 2016; Shridhar, 2012). By facilitating the conversion of atmospheric nitrogen into ammonium, they enhance soil fertility, promote plant growth, and contribute to carbon sequestration through increased plant biomass (Grover et al., 2023; Islam et al., 2008; Mohanty and Swain, 2018; Rashid et al., 2015). Soil bacteria like *Methylococcus capsulatus* and *Methylosinus trichosporium* are adept at oxidizing methane, a potent greenhouse gas, in aerobic conditions (Kalyuzhnaya et al., 2019;

Mancinelli, 1995; Nazaries et al., 2013; Roslev et al., 1997). By utilizing methane as a carbon and energy source, these microbes effectively reduce its concentration in the atmosphere, thus mitigating its impact on global warming and climate change (Dutta and Dutta, 2016; Kalyuzhnaya et al., 2019; Mancinelli, 1995; Nazaries et al., 2013; Roslev et al., 1997).

Soil microbes are key players in decomposing organic matter (Table 3.1), releasing carbon dioxide and essential nutrients into the soil (Hussain et al., 2022; Mukhtar et al., 2023). While this process contributes to CO_2 emissions, it also influences soil carbon stocks and nutrient availability, thereby affecting plant growth and ecosystem resilience to climate change (Dutta and Dutta, 2016; Hussain et al., 2022; Mukhtar et al., 2023). Bacteria like *Thiobacillus* spp. are known for their ability to oxidize sulfur compounds in soil environments (Friedrich et al., 2001; Usharani et al., 2019; Zhi-Hui et al., 2010). By mediating the oxidation of sulfur, these microbes influence sulfur cycling within ecosystems, impacting the emissions of sulfur-containing greenhouse gases such as hydrogen sulfide and sulfur dioxide (Usharani et al., 2019; Zhi-Hui et al., 2010).

TABLE 3.1 Examples of Soil Microbial Processes and Their Impacts on Climate Change

Microbial Process	Impact on Climate Change
Carbon sequestration	Removes CO_2 from the atmosphere, mitigating the greenhouse effect.
Nitrogen fixation	Influences nutrient cycling, affecting plant growth and carbon sequestration.
Methane oxidation	Reduces methane emissions, a potent greenhouse gas.
Organic matter decomposition	Releases CO_2 and nutrients, influencing soil carbon stocks.
Sulfur oxidation	Affects sulfur cycling, influencing greenhouse gas emissions.
Iron reduction	Influences carbon and nutrient cycling in anaerobic environments.
Ammonia oxidation	Converts ammonia to nitrate, impacting nitrogen cycling and greenhouse gas emissions.
Hydrogen production	Aids in redox reactions, influencing carbon and nutrient cycling.
Phosphorus solubilization	Enhances phosphorus availability, impacting plant growth and carbon sequestration.

Iron-reducing bacteria such as *Geobacter sulfurreducens* and *Shewanella oneidensis* play crucial roles in anaerobic carbon and nutrient cycling (Bücking et al., 2012; Cutting et al., 2009). These microbes facilitate the reduction of iron compounds in soil environments, influencing the availability of carbon and nutrients for microbial and plant growth, and ultimately impacting ecosystem resilience to climate change. Ammonia-oxidizing bacteria such as *Nitrosomonas europaea* and *Nitrosospira multiformis* oxidize ammonia to nitrite in the first step of nitrification (Norton et al., 2008). While this process enhances nitrogen availability for plant uptake and microbial processes, it also contributes to the emission of nitrous oxide, a potent greenhouse gas, highlighting the complex interplay between microbial activities and climate change. This process also impacts nitrogen cycling, which promotes plant growth and plays a crucial role in climate change mitigation.

Hydrogen-producing bacteria such as *Clostridium* spp. and *Ruminococcus albus* generate hydrogen gas as a metabolic byproduct during fermentation processes (Dey et al., 2022; Nandi and Sengupta, 1998). This hydrogen gas can participate in redox reactions, influencing carbon and nutrient cycling within soil environments, and thereby impacting ecosystem dynamics and resilience to climate change. Phosphorus-solubilizing bacteria such as *Bacillus megaterium* and *Pseudomonas putida* produce organic acids and enzymes to solubilize insoluble forms of phosphorus in soil (Setiawati et al., 2020). By enhancing phosphorus availability, these microbes promote plant growth and carbon sequestration, ultimately contributing to climate change mitigation efforts.

3.4 MANAGEMENT OF THE IMPACT OF CLIMATE CHANGE ON THE SOIL MICROBIAL COMMUNITIES

Soil microbial communities can be managed through various strategies, which fall into different categories (Figure 3.2). One approach involves altering soil conditions to regulate the diversity, biomass, and activity of specific microbial groups. This encompasses adjustments in pH levels, aeration, soil moisture, and the diversity and availability of nutrients (Chernov and Semenov, 2021). Agricultural practices such as the application of organic and mineral fertilizers, cover cropping, no-till agriculture, mechanical tillage, agroforestry, integrated water and soil management, changes in land use systems, and the implementation of diverse crop rotations are employed to achieve this regulation (Chernov and Semenov, 2021; Maitra et al., 2023; Nyawade et al., 2019). Additionally, ecological conservation practices modify soil conditions to support soil microbial communities while addressing the impacts of climate change. These practices include techniques such as soil carbon sequestration, biodiversity conservation, soil bioengineering, soil bioremediation, urban green infrastructure, permaculture design, and ecological restoration (Sahoo et al., 2024). By preserving soil structure, moisture levels, and nutrient cycling, these methods enhance the resilience of soil microbial communities to climate change-induced stresses.

Another approach entails the deliberate introduction of microorganisms into the soil to perform or stimulate specific processes actively. These microorganisms can play roles in facilitating plant nutrition and growth, dissolving phosphates, fixing nitrogen, or breaking down xenobiotics (Chernov and Semenov, 2021). This strategy involves the development of various biofertilizers or microbial inoculants, which consist of one or more strains of microorganisms chosen for their ability to carry out desired functions. This section explains some practices and outlines several practices used to manage the impact of climate change on soil microbial communities (Figure 3.2).

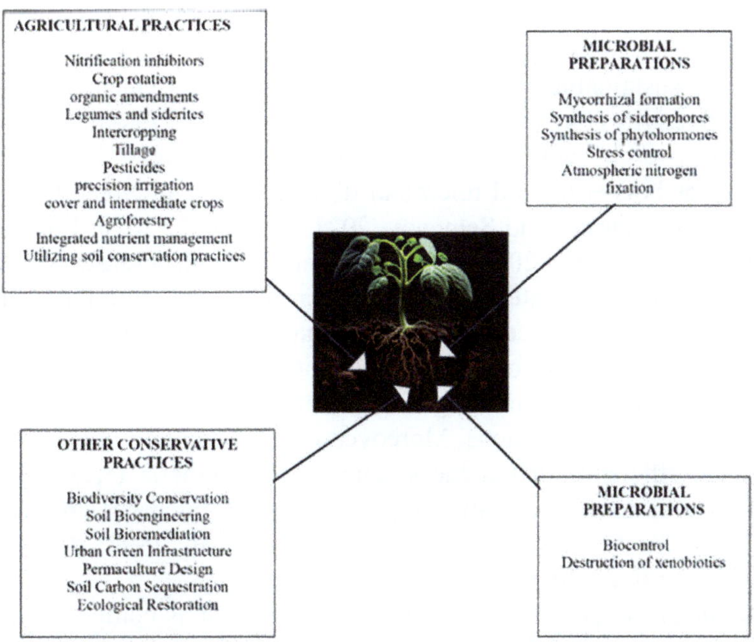

FIGURE 3.2 Strategies for the management of the impact of climate.

In agricultural landscapes, implementing conservation tillage practices, such as no-till or reduced tillage, stands out as a fundamental strategy (Alhammad et al., 2023; Kisaka et al., 2023). These practices minimize soil disturbance and erosion, thus preserving crucial microbial habitats and organic matter (Balser et al., 2010; Chen et al., 2020; Chernov and Semenov, 2021; Lori et al., 2020; Mathew et al., 2012; Piton et al., 2021; Shao et al., 2023). By maintaining optimal soil structure and moisture levels, conservation tillage fosters microbial diversity and activity while curbing the loss of vital soil carbon and nutrients (Lori et al., 2020; Mathew et al., 2012). As a result, these practices contribute significantly to bolstering soil health and resilience against the adverse impacts of climate change (Balser et al., 2010; Chen et al., 2020; Lori et al., 2020; Piton et al., 2021).

Another effective tactic is the introduction of cover crops during fallow periods or intercropping phases. Cover crops play a pivotal role in enriching the soil with living roots and organic matter, thereby boosting microbial biomass and activity (Balser et al., 2010; Chernov and Semenov, 2021; Finney et al., 2017; Lori et al., 2020). This enrichment facilitates soil carbon sequestration, nutrient cycling, and erosion control, ultimately nurturing robust and adaptable microbial communities. Consequently, the integration of cover crops into agricultural systems serves as a linchpin for promoting sustainable soil management practices and enhancing overall land productivity (Finney et al., 2017; Lori et al., 2020; Piton et al., 2021).

Rotating crops within fields emerges as yet another indispensable strategy for bolstering soil microbial diversity and functionality (Balser et al., 2010; Chernov and Semenov, 2021; Lori et al., 2020; Lupwayi et al., 1998; Piton et al., 2021). By disrupting pest and disease cycles and diversifying root exudates and residues, crop rotation stimulates microbial activity and enhances soil health (Lupwayi et al., 1998; Venter et al., 2016). The varying crop species in rotation foster unique microbial communities, thereby enriching soil fertility and fortifying its resilience to environmental fluctuations. Moreover, crop rotation optimizes nutrient utilization efficiency and reduces reliance on synthetic fertilizers, thus fostering a harmonious relationship between soil microbes and agricultural productivity (Lori et al., 2020; Lupwayi et al., 1998; Piton et al., 2021; Venter et al., 2016).

The incorporation of organic amendments, such as compost, manure, or green manures, stands as a cornerstone for enhancing soil fertility and microbial activity (Gitari et al., 2019). These organic inputs enrich the soil with essential nutrients, improve its structure, and bolster water retention capacity, thereby creating an optimal environment for diverse microbial communities to thrive (Balser et al., 2010; Chernov and Semenov, 2021; Lori et al., 2020; Piton et al., 2021). Consequently, organic amendments play a pivotal role in fostering sustainable soil management practices and promoting long-term agricultural productivity.

Adopting precision irrigation techniques (Figure 3.2), such as drip or micro-sprinkler irrigation, represents a crucial step in optimizing water use efficiency while minimizing soil waterlogging and nutrient leaching (Chernov and Semenov, 2021; Lori et al., 2020; Neupane and Guo, 2019; Piton et al., 2021). By delivering water precisely to the root zone, precision irrigation maintains optimal soil moisture levels conducive to microbial growth and activity. This, in turn, fosters a healthy soil microbiome even amidst changing precipitation patterns, thus contributing to enhanced soil health and agricultural sustainability (Balser et al., 2010; Lori et al., 2020; Neupane and Guo, 2019; Piton et al., 2021).

Implementing agroforestry practices, which involve integrating trees or shrubs into agricultural landscapes, offers multifaceted benefits for soil microbial communities (Jose, 2012; Maitra et al., 2024; Udawatta et al., 2021). Through the provision of additional organic inputs, shade, and habitat diversity, agroforestry systems create favorable conditions for soil microbes to thrive (Chernov and Semenov, 2021; Jose, 2012; Udawatta et al., 2021). Moreover, tree roots foster the formation of stable soil aggregates, while carbon sequestration in aboveground biomass and soil organic matter supports microbial-mediated processes, thereby promoting long-term soil health and sustainability (Jose, 2012; Udawatta et al., 2021).

Integrated nutrient management, characterized by a balanced approach to nutrient inputs through combining organic and inorganic fertilizers with biological amendments, plays a crucial role in enhancing soil fertility and productivity (Chappa et al., 2023; Faridvand et al., 2021; Lori et al., 2020; Piton et al., 2021; Selim, 2020). By fostering synergistic interactions between microbes and plants, integrated nutrient management optimizes nutrient utilization efficiency, reduces nutrient losses, and minimizes environmental impacts (Adesemoye et al., 2018; Lori et al., 2020; Piton et al., 2021). Consequently, this approach contributes significantly to sustaining soil health and agricultural productivity in the face of climate change challenges.

Utilizing soil conservation practices, such as contour farming, terracing, and grassed waterways, represents a cornerstone in preventing soil erosion and degradation (Balser et al., 2010; Chernov and Semenov, 2021; Habig and Swanepoel, 2015; Lori et al., 2020; Piton et al., 2021). These practices effectively safeguard microbial habitats and ecosystem services critical for sustainable agriculture and environmental resilience. By maintaining soil structure and organic matter, soil conservation practices nurture a conducive environment for soil microbes to thrive, thus promoting long-term soil health and productivity (Balser et al., 2010; Lori et al., 2020; Piton et al., 2021).

Regular soil health monitoring, involving the assessment of key indicators such as microbial biomass, enzyme activities, and microbial community composition, provides invaluable insights into soil microbial dynamics and ecosystem functioning (Lori et al., 2020; Piton et al., 2021; Schloter et al., 2018). By evaluating soil health status, farmers can tailor management practices accordingly, thus fostering microbial resilience and ecosystem sustainability in response to climate change impacts (Balser et al., 2010; Chernov and Semenov, 2021; Piton et al., 2021). Consequently, regular soil health monitoring serves as a cornerstone for informed decision-making and adaptive management strategies in agricultural systems.

3.5 THE IDENTIFIED RESEARCH GAPS

The research gap lies in the insufficient understanding of how drought impacts soil microbial communities. While plant responses to drought are well-documented, the specific consequences on soil properties and microorganisms remain unclear. Further research is needed to elucidate the intricate relationships within soil ecosystems under water stress, enhancing our comprehension of soil health dynamics. The existing research on soil microbial communities' response to floods and salinization highlights critical gaps requiring further exploration. While acknowledging the individual impacts of flooding and salinity, there is a limited understanding of their concurrent influence on microbial communities. Current studies primarily focus on changes in biodiversity during stress, leaving a gap in understanding broader community functions. Long-term flooding's effects on nutrient cycling and soil enzyme activity are recognized, but the resilience and recovery mechanisms of flood-prone microbial communities need deeper investigation. The duration of flooding's impact on redox potentials and microbial processes is poorly understood, presenting a research gap. Additionally, the response of microbial communities to frequent flooding and drainage in pasture soil remains underexplored. While studies note the reversible nature of flooding-induced changes, factors influencing the speed of community reversal are not well-documented. The adaptability of aerobic bacterial species to intermittent flooding and the inhibition rather than elimination during frequent flooding and drainage demand further investigation. Furthermore, the existing literature on permafrost thaw and its impact on microbial communities lacks comprehensive studies on the specific mechanisms governing microbial responses to temperature variations. There is a notable gap in understanding how different temperature regimes influence the enzymatic activity, metabolism, and respiration of soil microbes. Additionally, the interactions between permafrost thaw, temperature changes, and the intricate relationships within microbial communities remain insufficiently explored. Furthermore, the research gap lies in the need for comprehensive, long-term field studies that encompass the multifaceted interactions between elevated CO_2 levels and various environmental factors like temperature, precipitation, and nutrient availability. While existing research, including studies by Dunbar et al. (2012) and Hayde et al. (2012), provides valuable insights into microbial community responses to elevated CO_2, there remains a lack of integrated, long-term studies examining the combined effects of CO_2 elevation and other climatic variables on soil microbial communities and carbon cycling. The intricacies of how these factors interact and influence the equilibrium between carbon accrual and decomposition, particularly in the context of changing environmental conditions, remain a significant research frontier necessitating further investigation and long-term monitoring.

3.6 CONCLUSION

The response of soil microbial communities to climate change underscores a pivotal interconnection between environmental shifts and agricultural sustainability. Climate change, driven by escalating greenhouse gas levels due to human activities, poses multifaceted challenges, including extreme weather events and biodiversity loss. Soil microbes, encompassing diverse microorganisms like bacteria, fungi, archaea, and protozoa, play an indispensable role in soil health, nutrient cycling, and plant productivity.

Despite the potential positive contributions of soil microbes to soil fertility and agricultural sustainability, the adverse impacts of climate change cannot be overlooked. The phenomenon disrupts the delicate balance within these microbial communities, altering their composition, diversity, and functions. Salinization, exacerbated by changing climate patterns, poses a threat to soil microbial communities' equilibrium, affecting their ability to sustain nutrient cycling and soil health. Additionally, certain microbial activities, like methane production, contribute to greenhouse gas emissions, exacerbating climate change. Safeguarding soil microbial communities is pivotal for mitigating the adverse effects of climate change on agricultural ecosystems. Prioritizing sustainable agricultural practices that foster microbial diversity and resilience becomes imperative. Moreover, efforts to mitigate climate change by reducing greenhouse gas emissions and promoting environmental conservation are vital to preserving these critical ecosystems. Research endeavors should focus on developing crop varieties that not only withstand changing climatic conditions but also support and enhance soil microbial communities. Understanding the intricate interactions between plants and microbes in varying climatic scenarios can facilitate the development of agricultural strategies that harness the potential of these communities for sustainable food production. Furthermore, advocating for policies that support sustainable farming practices, such as crop rotation, reduced tillage, and organic amendments, can bolster soil microbial diversity and functionality. These measures not only enhance soil fertility but also contribute to carbon sequestration, mitigating climate change impacts. Emphasizing education and awareness among farmers, policymakers, and the general public about the significance of soil microbial communities in sustainable agriculture is crucial. Encouraging the adoption of practices that preserve and enhance these communities fosters a holistic approach to addressing climate change challenges in agriculture. The intricate relationship between soil microbial communities, climate change, and agricultural sustainability demands comprehensive strategies and collaborative efforts. Preserving and bolstering soil microbial diversity and functionality is fundamental for resilient and sustainable agricultural systems in the face of a changing climate. Prioritizing research, policy, and educational initiatives that recognize the pivotal role of soil microbes can pave the way toward more resilient, productive, and environmentally sustainable agricultural practices.

KEYWORDS

- **agricultural sustainability**
- **climate change mitigation**
- **climate-induced stress**
- **drought**
- **ecological dynamics**
- **ecosystem resilience**
- **fossil fuels**
- **greenhouse gases**
- **microbial diversity**
- **nitrogen fixation**
- **pollutants**
- **salinization**
- **soil fertility**
- **water retention**

REFERENCES

Aasfar, A., Bargaz, A., Yaakoubi, K., Hilali, A., & Meftah Kadmiri, I. (2021). Nitrogen-fixing *Azotobacter* species as potential soil biological enhancers for crop nutrition and yield stability. *Frontiers in Microbiology, 12*, 628379.

Abawi, G. S., & Widmer, T. L. (2000). Impact of soil health management practices on soilborne pathogens, nematodes, and root diseases of vegetable crops. *Applied Soil Ecology, 15*(1), 37–47.

Abdul Rahman, N. S. N., Abdul Hamid, N. W., & Nadarajah, K. (2021). Effects of abiotic stress on soil microbiome. *International Journal of Molecular Sciences, 22*(16), 9036.

Abirami, B., Radhakrishnan, M., Kumaran, S., & Wilson, A. (2021). Impacts of global warming on marine microbial communities. *Science of the Total Environment, 791*, 147905.

Adair, C. E., Reich, P. B., Trost, J. J., & Hobbie, S. E. (2011). Elevated CO_2 stimulates grassland soil respiration by increasing carbon inputs rather than by enhancing soil moisture. *Global Change Biology, 17*(12), 3546–3563.

Adesemoye, A. O., Torbert, H. A., & Kloepper, J. W. (2008). Enhanced plant nutrient use efficiency with PGPR and AMF in an integrated nutrient management system. *Canadian Journal of Microbiology, 54*(10), 876–886.

Adomako, M. O., Roiloa, S., & Yu, F. H. (2022). Potential roles of soil microorganisms in regulating the effect of soil nutrient heterogeneity on plant performance. *Microorganisms, 10*(12), 1–17. https://doi.org/10.3390/microorganisms10122399.

Alexander, S., & Dudley, N. (2017). *The Global Land Outlook.* United Nations Convention to Combat Desertification.

Alhammad, B. A., Roy, D. K., Ranjan, S., Padhan, S. R., Sow, S., Nath, D., Seleiman, M. F., & Gitari, H. (2023). Conservation tillage and weed management influencing weed dynamics, crop performance, soil properties, and profitability in rice-wheat-green gram system in Eastern Indo-Gangetic Plains. *Agronomy, 13*(7), 1953. https://doi.org/10.3390/agronomy13071953.

Allison, S. D. (2012). A trait-based approach for modelling microbial litter decomposition. *Ecology Letters, 15*(9), 1058–1070.

Allison, S. D., & Martiny, J. B. (2008). Resistance, resilience, and redundancy in microbial communities. *Proceedings of the National Academy of Sciences, 105*, 11512–11519.

Baldwin, D. S., & Mitchell, A. M. (2000). The effects of drying and re-flooding on the sediment and soil nutrient dynamics of lowland river–floodplain systems: A synthesis. *Regulated Rivers: Research & Management, 16*, 457–467.

Balser, T. C., Gutknecht, J. L., & Liang, C. (2010). How will climate change impact soil microbial communities? In *Soil Microbiology and Sustainable Crop Production* (pp. 373–397). Springer.

Bardgett, R. D., Kandeler, E., Tscherko, D., Hobbs, P. J., Bezemer, T. M., Jones, T. H., & Thompson, L. J. (1999). Below-ground microbial community development in a high-temperature world. *Oikos, 85*(2), 193–203.

Bengtson, P., Barker, J., & Grayston, S. J. (2012). Evidence of a strong coupling between root exudation, C and N availability, and stimulated SOM decomposition caused by rhizosphere priming effects. *Ecology and Evolution, 2*(8), 1843–1852.

Berga, M., Zha, Y., Székely, A. J., & Langenheder, S. (2017). Functional and compositional stability of bacterial metacommunities in response to salinity changes. *Frontiers in Microbiology, 8*, 948.

Berney, M., Greening, C., Conrad, R., Jacobs, W. R., & Cook, G. M. (2014). An obligately aerobic soil bacterium activates fermentative hydrogen production to survive reductive stress during hypoxia. *Proceedings of the National Academy of Sciences of the United States of America, 111*, 11479–11484.

Bodelier, P. L., Bar-Gilissen, M.-J., Meima-Franke, M., & Hordijk, K. (2012). Structural and functional response of methane-consuming microbial communities to different flooding regimes in riparian soils. *Ecology and Evolution, 2*, 106–127.

Bossio, D. A., & Scow, K. M. (1998). Impacts of carbon and flooding on soil microbial communities: Phospholipid fatty acid profiles and substrate utilization patterns. *Microbial Ecology, 35*, 265–278.

Bowker, M. A., Belnap, J., Rosentreter, R., & Graham, B. (2004). Wildfire-resistant biological soil crusts and fire-induced loss of soil stability in Palouse prairies, USA. *Applied Soil Ecology, 26*(1), 41–52.

Bréchet, L. M., Lopez-Sangil, L., George, C., Birkett, A. J., Baxendale, C., Castro Trujillo, B., & Sayer, E. J. (2018). Distinct responses of soil respiration to experimental litter manipulation in temperate woodland and tropical forest. *Ecology and Evolution, 8*(7), 3787–3796.

Breitkreuz, C., Herzig, L., Buscot, F., Reitz, T., & Tarkka, M. (2021). Interactions between soil properties, agricultural management, and cultivar type drive structural and functional adaptations of the wheat rhizosphere microbiome to drought. *Environmental Microbiology, 23*(10), 5866–5882.

Britannica, T. Editors of Encyclopaedia. (2016). Soil organism. *Encyclopedia Britannica.* https://www.britannica.com/science/soil-organism (accessed on 16 January 2025).

Bücking, C., Schicklberger, M., & Gescher, J. (2012). The biochemistry of dissimilatory ferric iron and manganese reduction in *Shewanella oneidensis.* In *Microbial Metal Respiration: From Geochemistry to Potential Applications* (pp. 49–82). Springer.

Castro, H. F., Classen, A. T., Austin, E. E., Norby, R. J., & Schadt, C. W. (2010). Soil microbial community responses to multiple experimental climate change drivers. *Applied and Environmental Microbiology, 76*(4), 999–1007.

Chambers, L. G., Guevara, R., Boyer, J. N., Troxler, T. G., & Davis, S. E. (2016). Effects of salinity and inundation on microbial community structure and function in a mangrove peat soil. *Wetlands, 36*, 361–371.

Chandra, P., & Enespa. (2019). Soil–microbes–plants: Interactions and ecological diversity. In *Plant Microbe Interface* (pp. 145–176). Springer.

Chappa, L. R., Mugwe, J., Gitari, H., & Maitra, S. (2023). Upholding sunflower (*Helianthus annuus*) yield and profitability while maintaining soil fertility under intercropping with sunn hemp and mineral fertilizer application. *International Journal of Bioresource Science, 10*(1), 31–49. https://doi.org/10.30954/10.30954/2347-9655.01.2023.4.

Chaudhary, D. R., Rathore, A. P., & Jha, B. (2016). Effects of seawater irrigation on soil microbial community structure and physiological function. *International Journal of Environmental Science and Technology, 13*, 2199–2208.

Chen, H., Dai, Z., Veach, A. M., Zheng, J., Xu, J., & Schadt, C. W. (2020). Global meta-analyses show that conservation tillage practices promote soil fungal and bacterial biomass. *Agriculture, Ecosystems & Environment, 293*, 106841.

Chen, Y., Liu, F., Kang, L., Zhang, D., Kou, D., Mao, C., & Yang, Y. (2020). Large-scale evidence for microbial response and associated carbon release after permafrost thaw. *Global Change Biology, 27*(14), 3218–3229. https://doi.org/10.1111/gcb.15487.

Chernov, T. I., & Semenov, M. V. (2021). Management of soil microbial communities: Opportunities and prospects (a review). *Eurasian Soil Science, 54*, 1888–1902.

Conrad, R. (2007). Microbial ecology of methanogens and methanotrophs. In D. L. Sparks (Ed.), *Advances in Agronomy* (pp. 1–63). Academic Press.

Coolen, M. J., van de Giessen, J., Zhu, E. Y., & Wuchter, C. (2011). Bioavailability of soil organic matter and microbial community dynamics upon permafrost thaw. *Environmental Microbiology, 13*(8), 2299–2314.

Cutting, R. S., Coker, V. S., Fellowes, J. W., Lloyd, J. R., & Vaughan, D. J. (2009). Mineralogical and morphological constraints on the reduction of Fe (III) minerals by *Geobacter sulfurreducens. Geochimica et Cosmochimica Acta, 73*(14), 4004–4022.

da Silva, E. C., de Albuquerque, M. B., de Azevedo Neto, A. D., & da Silva Junior, C. D. (2013). Drought and its consequences to plants—From individual to ecosystem. In *Responses of Organisms to Water Stress* (pp. 18–47). Springer.

de Nijs, E. A., Hicks, L. C., Leizeaga, A., Tietema, A., & Rousk, J. (2019). Soil microbial moisture dependences and responses to drying–rewetting: The legacy of 18 years drought. *Global Change Biology, 25*(3), 1005–1015.

DeLuca, T. H., & Sala, A. (2006). Frequent fire alters nitrogen transformations in ponderosa pine stands of the inland northwest. *Ecology, 87*(10), 2511–2522.

Denef, K., Six, J., Bossuyt, H., Frey, S. D., Elliott, E. T., Merckx, R., & Paustian, K. (2001). Influence of dry–wet cycles on the interrelationship between aggregate, particulate organic matter, and microbial community dynamics. *Soil Biology and Biochemistry, 33*, 1599–1611.

Dey, S., Kasai, T., & Katayama, A. (2022). Promotion of biological H2 (Bio-H2) production by the nitrogen-fixing anaerobic microbial consortia using humin, a solid-phase humic substance. *Journal of Bioscience and Bioengineering, 134*(2), 144–152.

Donhauser, J., Niklaus, P. A., Rousk, J., Larose, C., & Frey, B. (2020). Temperatures beyond the community optimum promote the dominance of heat-adapted, fast growing, and stress-resistant bacteria in alpine soils. *Soil Biology and Biochemistry, 148*, 107873.

Drake, J. E., Gallet-Budynek, A., Hofmockel, K. S., Bernhardt, E. S., Billings, S. A., Jackson, R. B., & Finzi, A. C. (2011). Increases in the flux of carbon belowground stimulate nitrogen uptake and sustain the long-term enhancement of forest productivity under elevated CO_2. *Ecology Letters, 14*(4), 349–357.

Dunbar, J., Eichorst, S. A., Gallegos-Graves, L. V., Silva, S., Xie, G., Hengartner, N. W., & Kuske, C. R. (2012). Common bacterial responses in six ecosystems exposed to 10 years of elevated atmospheric carbon dioxide. *Environmental Microbiology, 14*(5), 1145–1158.

Dutta, H., & Dutta, A. (2016). The microbial aspect of climate change. *Energy Ecology & Environment, 1*, 209–232.

Engel, E. C., Weltzin, J. F., Norby, R. J., & Classen, A. T. (2009). Responses of an old-field plant community to interacting factors of elevated [CO_2], warming, and soil moisture. *Journal of Plant Ecology, 2*(1), 1–11.

Faridvand, S., Rezaei-Chiyaneh, E., Battaglia, M., Ireri, G. H., Raza, M. A., & Siddique, K. H. M. (2021). Application of bio and chemical fertilizers improves yield, and essential oil quantity and quality of Moldavian balm (*Dracocephalum moldavica* L.) intercropped with mung bean (*Vigna radiata* L.). *Food and Energy Security, 11*, e319. https://doi.org/10.1002/fes3.319.

Fierer, N., & Jackson, R. B. (2006). The diversity and biogeography of soil bacterial communities. *Proceedings of the National Academy of Sciences, 103*(3), 626–631.

Finney, D. M., Buyer, J. S., & Kaye, J. P. (2017). Living cover crops have immediate impacts on soil microbial community structure and function. *Journal of Soil and Water Conservation, 72*(4), 361–373.

Fitch, A. A., Blount, K., Reynolds, L., & Bridgham, S. D. (2022). Partial recovery of microbial function in restored coastal marshes of Oregon, USA. *Soil Science Society of America Journal, 86*(3), 831–846.

Foster, R. C. (1988). Microenvironments of soil microorganisms. *Biology and Fertility of Soils, 6*(3), 189–203. https://doi.org/10.1007/BF00260816.

Fredrickson, J. K., Li, S. W., Gaidamakova, E. K., Matrosova, V. Y., Zhai, M., Sulloway, H. M., Scholten, J. C., Brown, M. G., Balkwill, D. L., & Daly, M. J. (2008). Protein oxidation: Key to bacterial desiccation resistance? *ISME Journal, 2*, 393–403.

Friedrich, C. G., Rother, D., Bardischewsky, F., Quentmeier, A., & Fischer, J. (2001). Oxidation of reduced inorganic sulfur compounds by bacteria: Emergence of a common mechanism? *Applied and Environmental Microbiology, 67*(7), 2873–2882.

Garten, C. T., Classen, A. T., & Norby, R. J. (2009). Soil moisture surpasses elevated CO_2 and temperature as a control on soil carbon dynamics in a multi-factor climate change experiment. *Plant and Soil, 319*, 85–94.

Garten, C. T., Classen, A. T., Norby, R. J., Brice, D. J., Weltzin, J. F., & Souza, L. (2008). Role of N2-fixation in constructed old-field communities under different regimes of [CO_2], temperature, and water availability. *Ecosystems, 11*, 125–137.

Gitari, H. I., Gachene, C. K. K., Karanja, N. N., Kamau, S., Nyawade, S., & Schulte-Geldermann, E. (2019). Potato-legume intercropping on a sloping terrain and its effects on soil physico-chemical properties. *Plant and Soil, 438*, 447–460. https://doi.org/10.1007/s11104-019-04036-7.

Glenn, E. P., Brown, J. J., & O'Leary, J. W. (1998). Irrigating crops with seawater. *Scientific American, 279*(2), 76–81.

Goberna, M., García, C., Insam, H., Hernández, M. T., & Verdú, M. (2012). Burning fire-prone Mediterranean shrublands: Immediate changes in soil microbial community structure and ecosystem functions. *Microbial Ecology, 64*, 242–255.

Gould, I. J., Wright, I., Collison, M., Ruto, E., Bosworth, G., & Pearson, S. (2020). The impact of coastal flooding on agriculture: A case-study of Lincolnshire, United Kingdom. *Land Degradation & Development, 31*(12), 1545–1559.

Grover, M., Yaadesh, S., & Jayasurya, A. (2023). Associative nitrogen fixers—Options for mitigating climate change. In *Bioinoculants: Biological Options for Mitigating Global Climate Change* (pp. 217–237). Springer Nature Singapore.

Habig, J., & Swanepoel, C. (2015). Effects of conservation agriculture and fertilization on soil microbial diversity and activity. *Environmental Science, 2*(3), 358–384.

Hanaka, A., Ozimek, E., Reszczy, E., Jaroszuk-Sciseł, J., & Stolarz, M. (2021). Plant tolerance to drought stress in the presence of supporting bacteria and fungi: An efficient strategy in horticulture. *Horticulturae, 7*, 390.

Hanley, M. E., Hartley, F. C., Hayes, L., & Franco, M. (2020). Simulated seawater flooding reduces oilseed rape growth, yield, and progeny performance. *Annals of Botany, 125*(2), 247–254.

Hart, S. C., DeLuca, T. H., Newman, G. S., MacKenzie, M. D., & Boyle, S. I. (2005). Post-fire vegetative dynamics as drivers of microbial community structure and function in forest soils. *Forest Ecology and Management, 220*(1–3), 166–184.

Hassan, M. J., Raza, M. A., Rehman, S. U., Ansar, M., Gitari, H., Khan, I., Wajid, M., Ahmed, M., Shah, G. A., Peng, Y., & Li, Z. (2020). Effect of cadmium toxicity on growth, oxidative damage, antioxidant defense system, and cadmium accumulation in two sorghum cultivars. *Plants, 9*(11), 1575. https://doi.org/10.3390/plants9111575.

Heydarzadeh, S., Hosseinpour, A., Gitari, H., Maitra, S., & Rahimi, A. (2021). Improving quantitative and qualitative properties of cotton through the use of organic and chemical fertilizers under water deficit stress conditions. *International Journal of Bioresource Science, 8*(2), 69–80. https://doi.org/10.30954/2347-9655.02.2021.2.

Hinojosa, M. B., Parra, A., Laudicina, V. A., & Moreno, J. M. (2016). Post-fire soil functionality and microbial community structure in a Mediterranean shrubland subjected to experimental drought. *Science of the Total Environment, 573*, 1178–1189.

Hu, Y., Wang, L., Tang, Y., Li, Y., Chen, J., Xi, X., & Sun, Y. (2014). Variability in soil microbial community and activity between coastal and riparian wetlands in the Yangtze

River estuary—Potential impacts on carbon sequestration. *Soil Biology and Biochemistry, 70*, 221–228.

Hussain, M. M., Farooqi, Z. U. R., Rasheed, F., & Din, W. M. U. (2022). Role of microorganisms as climate engineers: Mitigation and adaptations to climate change. In *Climate Change and Microbes* (pp. 1–24). Apple Academic Press.

Inglese, C. N., Christiansen, C. T., Lamhonwah, D., Moniz, K., Montross, S. N., Lamoureux, S., & Walker, V. K. (2017). Examination of soil microbial communities after permafrost thaw subsequent to an active layer detachment in the high Arctic. *Arctic, Antarctic, and Alpine Research, 49*(3), 455–472. https://doi.org/10.1657/AAAR0016-066.

Islam, M. Z., Sharif, D. I., & Hossain, M. A. (2008). A comparative study of *Azotobacter* spp. from different soil samples. *Journal of Soil and Nature, 2*(3), 16–19.

Jacoby, R., Peukert, M., Succurro, A., Koprivova, A., & Kopriva, S. (2017). The role of soil microorganisms in plant mineral nutrition—Current knowledge and future directions. *Frontiers in Plant Science, 8*, 1–19. https://doi.org/10.3389/fpls.2017.01617.

Jansson, J. K., & Hofmockel, K. S. (2020). Soil microbiomes and climate change. *Nature Reviews Microbiology, 18*(1), 35–46.

Jena, J., Maitra, S., Hossain, A., Pramanick, B., Ireri, G. H., Praharaj, S., Shankar, T., Palai, J. B., Rathore, A., Mandal, T. K., & Jatav, H. S. (2022). Role of legumes in cropping system for soil ecosystem improvement. In H. S. Jatav & V. D. Rajput (Eds.), *Ecosystem Services: Types, Management, and Benefits* (pp. 415–433). Nova Science Publishers. https://doi.org/10.52305/PFZA6988.

Jose, S. (2012). Agroforestry for conserving and enhancing biodiversity. *Agroforestry Systems, 85*, 1–8.

Kalyuzhnaya, M. G., Gomez, O. A., & Murrell, J. C. (2019). The methane-oxidizing bacteria (methanotrophs). In *Hydrocarbon-Degrading Microbes* (pp. 245–278). Taxon Genomics Ecophysiology.

Kannadhasan, S., & Nagarajan, R. (2023). Greenhouse gases: Challenges, effects, and climate change. In *Biomass and Bioenergy Solutions for Climate Change Mitigation and Sustainability* (pp. 65–74). IGI Global.

Kardol, P., Cregger, M. A., Campany, C. E., & Classen, A. T. (2010). Soil ecosystem functioning under climate change: Plant species and community effects. *Ecology, 91*(3), 767–781.

Kemnitz, D., Chin, K.-J., Bodelier, P., & Conrad, R. (2004). Community analysis of methanogenic archaea within a riparian flooding gradient. *Environmental Microbiology, 6*, 449–461.

Khan, M. N., Zhang, J., Luo, T., Liu, J., Ni, F., Rizwan, M., Fahad, S., & Hu, L. (2019). Morpho-physiological and biochemical responses of tolerant and sensitive rapeseed cultivars to drought stress during early seedling growth stage. *Acta Physiologiae Plantarum, 41*, 1–13.

Kisaka, M. O., Shisanya, C., Cournac, L., Manlay, J. R., Gitari, H., & Muriuki, J. (2023). Integrating no-tillage with agroforestry augments soil quality indicators in Kenya's dry-land agroecosystems. *Soil and Tillage Research, 227*, 105586. https://doi.org/10.1016/j.still.2022.105586.

Knelman, J. E., Schmidt, S. K., Garayburu-Caruso, V., Kumar, S., & Graham, E. B. (2019). Multiple, compounding disturbances in a forest ecosystem: Fire increases susceptibility of soil edaphic properties, bacterial community structure, and function to change with extreme precipitation event. *Soil Systems, 3*(2), 40.

Lee, J. S., Kim, J. W., Lee, S. H., Myeong, H. H., Lee, J. Y., & Cho, J. S. (2016). Zonation and soil factors of salt marsh halophyte communities. *Journal of Ecology and Environment, 40*(1), 1–4.

Li, J., Bååth, E., Pei, J., Fang, C., & Nie, M. (2021). Temperature adaptation of soil microbial respiration in alpine, boreal, and tropical soils: An application of the square root (Ratkowsky) model. *Global Change Biology, 27*(6), 1281–1292.

Liu, M., Feng, F., Cai, T., & Tang, S. (2020). Soil microbial community response differently to the frequency and strength of freeze–thaw events in a *Larix gmelinii* forest in the Daxing'an mountains, China. *Frontiers in Microbiology, 11*, 1164. https://doi.org/10.3389/fmicb.2020.01164.

Lladó, S., López-Mondéjar, R., & Baldrian, P. (2017). Forest soil bacteria: Diversity, involvement in ecosystem processes, and response to global change. *Microbiology and Molecular Biology Reviews, 81*(2), e00063-16.

Lori, M., Piton, G., Symanczik, S., Legay, N., Brussaard, L., Jaenicke, S., & Foulquier, A. (2020). Compared to conventional, ecological intensive management promotes beneficial proteolytic soil microbial communities for agro-ecosystem functioning under climate change-induced rain regimes. *Scientific Reports, 10*(1), 7296.

Lupwayi, N. Z., Rice, W. A., & Clayton, G. W. (1998). Soil microbial diversity and community structure under wheat as influenced by tillage and crop rotation. *Soil Biology and Biochemistry, 30*(13), 1733–1741.

Maheshwari, N., Thakur, I. S., & Srivastava, S. (2022). Role of carbon-dioxide sequestering bacteria for clean air environment and prospective production of biomaterials: A sustainable approach. *Environmental Science and Pollution Research, 29*(26), 38950–38971.

Maitra, S., Bhattacharya, U., Pramanick, B., Sagar, L., Gaikward, D. J., Pramanick, B., Sairam, M., Sahoo, U., Jatav, H. S., Ireri, G. H., Shankar, T., Kandileri, A. M., Rezaei-Chiyaneh, E., & Hossain, A. (2024). Agroforestry: A resource conserving technology for efficient utilization of agricultural inputs, leading to food and environmental security. In H. S. Jatav, V. D. Rajput, T. Minkina, E. D. Van Hullebusch, & A. Dutta (Eds.), *Agroforestry to Combat Global Challenges* (Vol. 36, pp. 1–16). Springer. https://doi.org/10.1007/978-981-99-7282-1_2.

Mancinelli, R. L. (1995). The regulation of methane oxidation in soil. *Annual Review of Microbiology, 49*(1), 581–605.

Masson-Delmotte, V. P., Zhai, P., Pirani, S. L., Connors, C., Péan, S., Berger, N., & Scheel Monteiro, P. M. (2021). IPCC, 2021: Summary for policymakers. In *Climate Change 2021: The Physical Science Basis* (pp. 1–41). Contribution of Working Group I to the Sixth Assessment Report of the Intergovernmental Panel on Climate Change.

Mathew, R. P., Feng, Y., Githinji, L., Ankumah, R., & Balkcom, K. S. (2012). Impact of no-tillage and conventional tillage systems on soil microbial communities. *Applied and Environmental Soil Science, 2012*, 1–7.

Meena, K. K., Sorty, A. M., Bitla, U. M., Choudhary, K., Gupta, P., Pareek, A., & Minhas, P. S. (2017). Abiotic stress responses and microbe-mediated mitigation in plants: The omics strategies. *Frontiers in Plant Science, 8*, 172.

Meena, M., Yadav, G., Sonigra, P., Nagda, A., Mehta, T., Swapnil, P., & Kumar, S. (2023). Multifarious responses of forest soil microbial community toward climate change. *Microbial Ecology, 86*(1), 49–74.

Milošević, N. A., Marinković, J. B., & Tintor, B. B. (2012). Mitigating abiotic stress in crop plants by microorganisms. *Zbornik Matice Srpske za Prirodne Nauke,* (123), 17–26.

Mirriam, A., Mugwe, J., Nasar, J., Kisaka, O., Ranjan, S., & Ireri, H. (2023). Role of phosphorus and inoculation with *Bradyrhizobium* in enhancing soybean production. *Advances in Agriculture, 2023*, 3231623. https://doi.org/10.1155/2023/3231623.

Mirriam, A., Mugwe, J., Raza, M. F., Seleiman, M. A., Maitra, S., & Gitari, H. I. (2022). Aggrandizing soybean yield, phosphorus uses efficiency and economic returns under phosphatic fertilizer application and inoculation with *Bradyrhizobium*. *Journal of Soil Science and Plant Nutrition, 22*, 5086–5098. https://doi.org/10.1007/s42729-022-00985-8.

Mohanty, S. R., Kollah, B., Sharma, V. K., Singh, A. B., Singh, M., & Rao, A. S. (2013). Methane oxidation and methane-driven redox process during sequential reduction of a flooded soil ecosystem. *Annals of Microbiology, 64*, 65–74.

Mohanty, S., & Swain, C. K. (2018). Role of microbes in climate-smart agriculture. In *Microorganisms for Green Revolution: Microbes Sustain Agro-ecosystems* (pp. 129–140).

Mombrikotb, S. B., Van Agtmaal, M., Johnstone, E., Crawley, M. J., Gweon, H. S., Griffiths, R. I., & Bell, T. (2022). The interactions and hierarchical effects of long-term agricultural stressors on soil bacterial communities. *Environmental Microbiology Reports, 14*(5), 711–718.

Monteux, S., Weedon, J. T., Blume-Werry, G., Gavazov, K., Jassey, V. E., Johansson, M., & Dorrepaal, E. (2018). Long-term *in situ* permafrost thaw effects on bacterial communities and potential aerobic respiration. *The ISME Journal, 12*(9), 2129–2141.

Morillas, L., Durán, J., Rodríguez, A., Roales, J., Gallardo, A., Lovett, G. M., & Groffman, P. M. (2015). Nitrogen supply modulates the effect of changes in drying–rewetting frequency on soil C and N cycling and greenhouse gas exchange. *Global Change Biology, 21*, 3854–3863.

Mukhtar, H., Wunderlich, R. F., Muzaffar, A., Ansari, A., Shipin, O. V., Cao, T. N. D., & Lin, Y. P. (2023). Soil microbiome feedback to climate change and options for mitigation. *Science of the Total Environment,* 163412.

Myhre, G., Shindell, D., Bréon, F. M., Collins, W. D., Fuglestvedt, J., Huang, J., Koch, D., Lamarque, J. F., Lee, D., & Mendoza, B. (2013). Anthropogenic and natural radiative forcing. In T. Stocker, D. Qin, G. K. Plattner, M. Tignor, S. K. Allen, J. Boschung, A. Nauels, Y. Xia, V. Bex, & P. M. Midgley (Eds.), *Climate Change 2013: The Physical Science Basis. Contribution of Working Group I to the Fifth Assessment Report of the Intergovernmental Panel on Climate Change* (pp. 659–740). Cambridge University Press.

Nandi, R., & Sengupta, S. (1998). Microbial production of hydrogen: An overview. *Critical Reviews in Microbiology, 24*(1), 61–84.

Nasar, J., Zhao, C. J., Khan, R., Gul, H., Gitari, H., Shao, Z., Abbas, G., Haider, I., Iqbal, Z., Ahmed, W., Rehman, R., Liang, Q. P., Zhou, X. B., & Yang, J. (2023). Maize-soybean intercropping at optimal N fertilization increases the N uptake, N yield and N use efficiency of maize crop by regulating the N assimilatory enzymes. *Frontiers in Plant Science, 13*, 1077948. https://doi.org/10.3389/fpls.2022.1077948.

Nave, L. E., Vance, E. D., Swanston, C. W., & Curtis, P. S. (2011). Fire effects on temperate forest soil C and N storage. *Ecological Applications, 21*(4), 1189–1201.

Nazaries, L., Murrell, J. C., Millard, P., Baggs, L., & Singh, B. K. (2013). Methane, microbes and models: Fundamental understanding of the soil methane cycle for future predictions. *Environmental Microbiology, 15*(9), 2395–2417.

Neupane, J., & Guo, W. (2019). Agronomic basis and strategies for precision water management: A review. *Agronomy, 9*(2), 87.

Nguyen, L. T., Osanai, Y., Anderson, I. C., Bange, M. P., Tissue, D. T., & Singh, B. K. (2018). Flooding and prolonged drought have differential legacy impacts on soil nitrogen cycling, microbial communities, and plant productivity. *Plant and Soil, 431*, 371–387.

Norton, J. M., Klotz, M. G., Stein, L. Y., Arp, D. J., Bottomley, P. J., Chain, P. S., & Starkenburg, S. R. (2008). Complete genome sequence of *Nitrosospira multiformis*, an ammonia-oxidizing bacterium from the soil environment. *Applied and Environmental Microbiology, 74*(11), 3559–3572.

Nungula, E. Z., Mugwe, J., Nasar, J., Massawe, H. J., Karuma, A. N., Maitra, S., Seleiman, M. F., Dindaroglu, T., Khan, N., & Gitari, H. I. (2023). Land degradation unmasked as the key constraint in sunflower (*Helianthus annuus*) production: Role of GIS in revitalizing this vital sector. *Cogent Food & Agriculture, 9*(2), 2267863. https://doi.org/10.1080/23311932.2023.2267863.

Nungula, E. Z., Raza, M. A., Nasar, J., Maitra, S., Seleiman, M. F., Ranjan, S., Padhan, S. R., Sow, S., Gaikwad, D. J., & Gitari, H. I. (2024). Cadmium in soil and plants: A review. In A. K. Jha & N. Kumar (Eds.), *Cadmium Toxicity in Water: Challenges and Solutions* (pp. 15–34). Springer. https://doi.org/10.1007/978-3-031-54005-9_2.

Nyawade, S. O., Karanja, N. N., Gachene, C. K. K., Gitari, H. I., Schulte-Geldermann, E., & Parker, M. L. (2019). Short-term dynamics of soil organic matter fractions and microbial activity in smallholder legume intercropping systems. *Applied Soil Ecology, 142*, 123–135. https://doi.org/10.1016/j.apsoil.2019.04.015.

Ortiz, A., & Sansinenea, E. (2022). The role of beneficial microorganisms in soil quality and plant health. *Sustainability, 14*(9), 5358. https://doi.org/10.3390/su14095358.

Pachauri, R. K., Allen, M. R., Barros, V. R., Broome, J., Cramer, W., Christ, R., & van Ypserle, J. P. (2014). *Climate Change 2014: Synthesis Report*. Contribution of working groups I, II, and III to the fifth assessment report of the Intergovernmental Panel on Climate Change. IPCC.

Pascual, J. A., Hernandez, T., Garcia, C., & Ayuso, M. (1998). Enzymatic activities in an arid soil amended with urban organic wastes: Laboratory experiment. *Bioresource Technology, 64*(2), 131–138.

Patrick, W. H., & Jugsujinda, A. (1992). Sequential reduction and oxidation of inorganic nitrogen, manganese, and iron in flooded soil. *Soil Science Society of America Journal, 56*(3), 1071–1076.

Pielke Jr, R. A. (2004). What is climate change? *Energy & Environment, 15*(3), 515–520.

Piton, G., Foulquier, A., Martinez-García, L. B., Legay, N., Arnoldi, C., Brussaard, L., & De Deyn, G. B. (2021). Resistance–recovery trade-off of soil microbial communities under altered rain regimes: An experimental test across European agroecosystems. *Journal of Applied Ecology, 58*(2), 406–418.

Ponnamperuma, F. N. (1984). Effects of flooding on soils. In G. Meurant (Ed.), *Flooding and Plant Growth* (pp. 47–62). Academic Press.

Potts, M. (1994). Desiccation tolerance. *Microbiology Reviews, 58*(4), 755–805.

Rashid, M. H. O., Krehenbrink, M., & Akhtar, M. S. (2015). Nitrogen-fixing plant-microbe symbioses. *Sustainable Agriculture Reviews, 15*, 193–234.

Ratering, S., & Conrad, R. (1998). Effects of short-term drainage and aeration on the production of methane in submerged rice soil. *Global Change Biology, 4*(4), 397–407.

Reddy, K. R., & Patrick, W. H. (1975). Effect of alternate aerobic and anaerobic conditions on redox potential, organic matter decomposition, and nitrogen loss in a flooded soil. *Soil Biology and Biochemistry, 7*(1), 87–94.

Roberson, E. B., Chenu, C., & Firestone, M. K. (1993). Microstructural changes in bacterial exopolysaccharides during desiccation. *Soil Biology and Biochemistry, 25*(10), 1299–1301.

Roslev, P., & King, G. M. (1994). Survival and recovery of methanotrophic bacteria starved under oxic and anoxic conditions. *Applied and Environmental Microbiology, 60*(7), 2602–2608.

Roslev, P., Iversen, N., & Henriksen, K. A. J. (1997). Oxidation and assimilation of atmospheric methane by soil methane oxidizers. *Applied and Environmental Microbiology, 63*(3), 874–880.

Sage, R. F. (2020). Global change biology: A primer. *Global Change Biology, 26*(1), 3–30.

Sahoo, U., Maitra, S., Hossain, A., Santosh, D. T., Nath, S., Sairam, M., Sagar, L., Jena, J., Pattanayak, S., Gitari, H. I., & Rezaei-Chiyaneh, E. (2024). Biochar for soil health improvement in the present context of climate change: A reality or fantasy. In S. Fahad, M. Adnan, R. Zhou, T. Nawaz, & S. Saud (Eds.), *Biochar-Assisted Remediation of Contaminated Soils Under Changing Climate* (pp. 1–16). Elsevier. https://doi.org/10.1016/B978-0-443-21562-9.00001-3.

Sairaam, M., Maitra, S., Prahara, S., Nath, S., Shankar, T., Sahoo, U., Santosh, D. T., Sagar, L., Panda, M., Priya, G. S., Ashwini, T. R., Gaikwad, D. J., Hossain, A., Pramanick, B., Jatav, H. S., Gitari, H. I., & Aftab, T. (2023). An insight into the consequences of emerging contaminants in soil and water and plant responses. In T. Aftab (Ed.), *Emerging Contaminants and Plants* (pp. 1–21). Springer. https://doi.org/10.1007/978-3-031-22269-6_1.

Savary, S., & Willocquet, L. (2020). Modeling the impact of crop diseases on global food security. *Annual Review of Phytopathology, 58*, 313–341.

Sayer, E. J., Crawford, J. A., Edgerley, J., Askew, A. P., Hahn, C. Z., Whitlock, R., & Dodd, I. C. (2021). Adaptation to chronic drought modifies soil microbial community responses to phytohormones. *Communications Biology, 4*(1), 516.

Scharlemann, J. P., Tanner, E. V., Hiederer, R., & Kapos, V. (2014). Global soil carbon: Understanding and managing the largest terrestrial carbon pool. *Carbon Management, 5*(1), 81–91.

Schimel, J., Balser, T. C., & Wallenstein, M. (2007). Microbial stress-response physiology and its implications for ecosystem function. *Ecology, 88*(6), 1386–1394.

Schloter, M., Nannipieri, P., Sørensen, S. J., & van Elsas, J. D. (2018). Microbial indicators for soil quality. *Biology and Fertility of Soils, 54*, 1–10.

Seleiman, M. F., Aslam, M. T., Alhammad, B. A., Hassan, M. U., Maqbool, R., Chattha, M. U., Khan, I., Gitari, H. I., Uslu, O. S., Roy, R., & Battaglia, M. L. (2021). Salinity stress in wheat: Effects, mechanisms and management strategies. *Phyton-International Journal of Experimental Botany, 91*(4), 667–694. https://doi.org/10.32604/phyton.2022.017365.

Selim, M. M. (2020). Introduction to the integrated nutrient management strategies and their contribution to yield and soil properties. *International Journal of Agronomy, 2020*, 1–6.

Setiawati, T. C., Pandutama, M. H., Mandala, M., & Arta, C. (2020, October). The capacity of soil bacteria, *Bacillus sprif* and *Pseudomonas sprif*, in solubilizing soil phosphate and potassium. In *IOP Conference Series: Earth and Environmental Science* (Vol. 583, No. 1, p. 012040). IOP Publishing.

Shade, A., Peter, H., Allison, S. D., Baho, D. L., Berga, M., Bürgmann, H., & Handelsman, J. (2012). Fundamentals of microbial community resistance and resilience. *Frontiers in Microbiology, 3*, 417.

Shao, Z., Mwakidoshi, E. R., Muindi, E. M., Soratto, R. P., Ranjan, S., Padhan, S. R., Wamukota, A. W., Sow, S., Wasonga, D. O., Nasar, J., Seleiman, M. F., & Ireri, G. H. (2023). Synthetic fertilizer application coupled with bioslurry optimizes potato (*Solanum tuberosum*) growth and yield. *Agronomy, 13*, 2162. https://doi.org/10.3390/agronomy13082162.

Shridhar, B. S. (2012). Nitrogen fixing microorganisms. *International Journal of Microbiology Research, 3*(1), 46–52.

Siebielec, S., Siebielec, G., Klimkowicz-Pawlas, A., Gałązka, A., Grządziel, J., & Stuczyński, T. (2020). Impact of water stress on microbial community and activity in sandy and loamy soils. *Agronomy, 10*(9), 1429.

Sigren, L. K., Lewis, S. T., Fisher, F. M., & Sass, R. L. (1997). Effects of field drainage on soil parameters related to methane production and emission from rice paddies. *Global Biogeochemical Cycles, 11*(2), 151–162.

Singh, B. K., Dawson, L. A., Macdonald, C. A., & Buckland, S. M. (2009). Impact of biotic and abiotic interaction on soil microbial communities and functions: A field study. *Applied Soil Ecology, 41*(3), 239–248. https://doi.org/10.1016/j.apsoil.2008.10.003.

Stams, A. J. M., & Plugge, P. (2010). The microbiology of methanogenesis. In *Methane and Climate Change* (pp. 14–26). Earthscan.

Sun, H., Santalahti, M., Pumpanen, J., Köster, K., Berninger, F., Raffaello, T., & Heinonsalo, J. (2015). Fungal community shifts in structure and function across a boreal forest fire chronosequence. *Applied and Environmental Microbiology, 81*(22), 7869–7880.

Taş, N., Prestat, E., McFarland, J. W., Wickland, K. P., Knight, R., Berhe, A. A., & Jansson, J. K. (2014). Impact of fire on active layer and permafrost microbial communities and metagenomes in an upland Alaskan boreal forest. *The ISME Journal, 8*(9), 1904–1919.

Tisdall, J. M. (1994). Possible role of soil microorganisms in aggregation in soils. *Plant and Soil, 159*, 115–121.

Tu, Q., He, Z., Wu, L., Xue, K., Xie, G., Chain, P., & Zhou, J. (2017). Metagenomic reconstruction of nitrogen cycling pathways in a CO_2-enriched grassland ecosystem. *Soil Biology and Biochemistry, 106*, 99–108.

Udawatta, R. P., Rankoth, L. M., & Jose, S. (2021). Agroforestry for biodiversity conservation. In *Agroforestry Ecosystem Services* (pp. 245–274). Springer.

Unger, I. M., Kennedy, A. C., & Muzika, R. M. (2009). Flooding effects on soil microbial communities. *Applied Soil Ecology, 42*, 1–8.

Usharani, K., Naik, D. Y., & Manjunatha, R. (2019). Sulphur oxidizing bacteria: Oxidation, mechanism and uses–a review. *Trends in Biosciences, 12*(3), 10965–10972.

Van Der Heijden, M. G., Bardgett, R. D., & Van Straalen, N. M. (2008). The unseen majority: Soil microbes as drivers of plant diversity and productivity in terrestrial ecosystems. *Ecology Letters, 11*(3), 296–310.

Van Groenigen, K. J., Qi, X., Osenberg, C. W., Luo, Y., & Hungate, B. A. (2014). Faster decomposition under increased atmospheric CO_2 limits soil carbon storage. *Science, 344*(6183), 508–509.

Venter, Z. S., Jacobs, K., & Hawkins, H. J. (2016). The impact of crop rotation on soil microbial diversity: A meta-analysis. *Pedobiologia, 59*(4), 215–223.

Wagner, D. L., Grames, E. M., Forister, M. L., Berenbaum, M. R., & Stopak, D. (2021). Insect decline in the Anthropocene: Death by a thousand cuts. *Proceedings of the National Academy of Sciences, 118*(2), e2023989118.

Walker, T. W., Kaiser, C., Strasser, F., Herbold, C. W., Leblans, N. I., Woebken, D., & Richter, A. (2018). Microbial temperature sensitivity and biomass change explain soil carbon loss with warming. *Nature Climate Change, 8*(10), 885–889. https://doi.org/10.1038/s41558-018-0259-x.

Wan, S., Norby, R. J., Ledford, J., & Weltzin, J. F. (2007). Responses of soil respiration to elevated CO2, air warming, and changing soil water availability in a model old-field grassland. *Global Change Biology, 13*(11), 2411–2424.

Wang, Z. P., DeLaune, R. D., Patrick, W. H., & Masscheleyn, P. H. (1993). Soil redox and pH effects on methane production in a flooded rice soil. *Soil Science Society of America Journal, 57*, 382.

Warrence, N. J., Bauder, J. W., & Pearson, K. E. (2002). Basics of salinity and sodicity effects on soil physical properties. *Department of Land Resources and Environmental Sciences, Montana State University-Bozeman, MT, 129*, 1–29.

Wei, S., Cui, H., Zhu, Y., Lu, Z., Pang, S., Zhang, S., & Su, X. (2018). Shifts of methanogenic communities in response to permafrost thaw results in rising methane emissions and soil property changes. *Extremophiles, 22*, 447–459. https://doi.org/10.1007/s00792-018-1007-x.

Welbaum, G. E., Sturz, A. V., Dong, Z., & Nowak, J. (2004). Managing soil microorganisms to improve productivity of agro-ecosystems. *Critical Reviews in Plant Sciences, 23*(2), 175–193. https://doi.org/10.1080/07352680490433295.

Wright, A. J., Ebeling, A., De Kroon, H., Roscher, C., Weigelt, A., Buchmann, N., & Eisenhauer, N. (2015). Flooding disturbances increase resource availability and productivity but reduce stability in diverse plant communities. *Nature Communications, 6*(1), 6092.

Yagi, K., Tsuruta, H., Kanda, K., & Minami, K. (1996). Effect of water management on methane emission from a Japanese rice paddy field: Automated methane monitoring. *Global Biogeochemical Cycles, 10*, 255–267.

You, J., Zhang, Y., Liu, A., Li, D., Wang, X., Dossa, K., & Zhang, X. (2019). Transcriptomic and metabolomic profiling of drought-tolerant and susceptible sesame genotypes in response to drought stress. *BMC Plant Biology, 19*(1), 1–16.

Yu, H., Deng, Y., He, Z., Van Nostrand, J. D., Wang, S., Jin, D., & Zhou, J. (2018). Elevated CO2 and warming altered grassland microbial communities in soil top-layers. *Frontiers in Microbiology, 9*, 1790.

Yu, S., & Ehrenfeld, J. G. (2010). Relationships among plants, soils and microbial communities along a hydrological gradient in the New Jersey Pinelands, USA. *Annals of Botany, 105*(1), 185–196.

Zak, D. R., Holmes, W. E., White, D. C., Peacock, A. D., & Tilman, D. (2003). Plant diversity, soil microbial communities, and ecosystem function: Are there any links? *Ecology, 84*(8), 2042–2050. https://doi.org/10.1890/02-0433.

Zandalinas, S. I., Fritschi, F. B., & Mittler, R. (2021). Global warming, climate change, and environmental pollution: Recipe for a multifactorial stress combination disaster. *Trends in Plant Science, 26*(6), 588–599.

Zhi-Hui, Y. A. N. G., Stöven, K., Haneklaus, S., Singh, B. R., & Schnug, E. (2010). Elemental sulfur oxidation by *Thiobacillus* spp. and aerobic heterotrophic sulfur-oxidizing bacteria. *Pedosphere, 20*(1), 71–79.

Zhou, J., Xia, B., Treves, D. S., Wu, L.-Y., Marsh, T. L., O'Neill, R. V., Palumbo, A. V., & Tiedje, J. M. (2002). Spatial and resource factors influencing high microbial diversity in soil. *Applied and Environmental Microbiology, 68*, 326–334.

CHAPTER 4

Economic and Environmental Implications of SSNM Adoption

HARSHIT MISHRA,[1,*] DIVYANSHI MISHRA,[1] DHARMINDER,[2] MAINAK GHOSH,[3] SUMIT SOW,[2] and SHIVANI RANJAN[2]

[1]*Department of Agricultural Economics, College of Agriculture, Acharya Narendra Deva University of Agriculture and Technology, Kumarganj, Ayodhya, Uttar Pradesh, India*

[2]*Department of Agronomy, Dr. Rajendra Prasad Central Agricultural University, Pusa, Samastipur, Bihar, India*

[3]*Department of Agronomy, Bihar Agricultural University, Sabour, Bhagalpur, Bihar, India*

*Corresponding author

ABSTRACT

The adoption of site-specific nutrient management (SSNM) techniques in agriculture carries significant economic and environmental implications. SSNM, a precision farming approach, optimizes nutrient application by accounting for soil conditions, crop types, and growth stages across different field zones. This chapter offers a comprehensive overview of SSNM, examining both its economic and environmental dimensions. A primer on sustainable soil nutrient management lays the foundation by emphasizing its role in enhancing productivity while reducing environmental degradation. The economic benefits of SSNM are explored, highlighting increased yields and profitability for farmers. The discussion extends to its influence on commodity markets and international trade. The chapter also stresses the importance of policy and institutional support, noting that government

Climate-Resilient Agriculture for Sustainable Crop Production: New Adaptation and Mitigation Strategies. Sanjay Kumar, Shivani Ranjan, Sumit Sow, Smruti Ranjan Padhan, and Dhirendra Kumar Roy (Eds.)

DOI: 10.1201/9781998511150-4

initiatives and organizational frameworks are crucial for widespread adoption. Environmentally, SSNM contributes to biodiversity conservation and supports ecosystem health, enhancing resilience in the face of climate change. Challenges to SSNM adoption—technological, educational, and economic—are critically assessed. Concluding with a forward-looking perspective, the chapter envisions SSNM as a transformative tool for global agriculture, capable of balancing economic prosperity with environmental sustainability. Overall, it provides valuable insights into the dynamic relationship between productivity, ecological stewardship, and the future of nutrient management.

4.1 INTRODUCTION

In the quest for global food security and sustainable agricultural systems, the adoption of innovative and responsible farming practices has become imperative. Among these practices, SSNM has emerged as a pivotal strategy that not only addresses the challenges of feeding a growing population but also ensures the long-term health of our planet (Pampolino et al., 2007; Reidsma et al., 2012). The foundation of sustainable agriculture lies in maintaining soil fertility and optimizing nutrient use efficiency. SSNM represents a holistic approach that seeks to strike a balance between crop productivity, environmental preservation, and economic viability (Yamano et al., 2016). SSNM embodies the principles of precision agriculture, meticulously tailoring nutrient application to the unique demands of crops, considering soil properties, crop varieties, and developmental stages (Aryal et al., 2018; Laborte et al., 2009; Ranjan et al., 2022). The economic implications of SSNM adoption are substantial and far-reaching. As agricultural systems transition to more sustainable practices, a series of direct and indirect economic effects emerge (Hengsdijk et al., 2005; Rodriguez et al., 2012).

By optimizing nutrient use, farmers witness increased crop yields, enhanced product quality, and reduced input costs. These gains not only bolster individual farm incomes but also contribute to regional and national economic growth (Chivenge et al., 2021a; Mariana et al., 2012; Sida et al., 2023).

The adoption of SSNM extends beyond the agricultural sector, influencing market dynamics and global trade patterns. The increased output and improved quality of crops resulting from SSNM practices can lead to shifts in supply and demand, affecting commodity prices and trade flows (Aryal et al., 2018; Mishra and Singh, 2023; Sarkar et al., 2017). Moreover, consumer preferences for sustainably produced food products further shape market trends. The success of SSNM depends on supportive policies and strong institutional backing. Governments and international organizations play a pivotal role in incentivizing SSNM adoption through policy frameworks that promote sustainable agricultural practices (Singh and Mishra, 2023; Yi et al., 2018). Additionally, extension services, research institutions, and agribusinesses contribute by disseminating knowledge and providing technical assistance to farmers (Parihar et al., 2017). The environmental consequences of SSNM adoption are as profound as its economic consequences. By minimizing nutrient runoff and soil degradation, SSNM contributes to the preservation of biodiversity and ecosystem health (Mishra, 2022; Senyolo et al., 2018). Furthermore, SSNM practices enhance the resilience of agricultural systems to the challenges posed by climate change, mitigating the adverse effects of extreme weather events and fostering long-term sustainability (Hu et al., 2007; Jat et al., 2021).

The adoption of SSNM practices permeates ecosystems, fostering biodiversity and ecosystem health. By minimizing the ecological impacts of intensive agriculture, SSNM helps maintain the intricate web of life that underpins agricultural productivity and resilience (Chivenge et al., 2022; Mishra and Singh, 2023; Reidsma et al., 2011). Climate change poses unprecedented challenges to global agriculture, jeopardizing food security and livelihoods. SSNM emerges as a potent tool in building resilience against these challenges (Dobermann and White, 1999; Mishra and Mishra, 2023). By enhancing soil health, optimizing nutrient cycling, and minimizing carbon emissions, SSNM contributes to adaptive and mitigative strategies (Chivenge et al., 2021b; Dobermann et al., 2004; Wang et al., 2007).

While SSNM holds immense promise, its widespread adoption is not without challenges. Farmers face barriers such as knowledge gaps, access to resources, and perceived risks associated with changing established practices. Additionally, economic, social, and cultural factors can influence the adoption of SSNM (Mishra et al., 2023; Pasuquin et al., 2014; Sirmour et al., 2017). The future of agriculture is intrinsically linked to the adoption of sustainable practices like SSNM (Asai et al., 2010). As the global population continues to grow, and the impacts of climate change intensify, the imperative for responsible and efficient nutrient management becomes even more pronounced (Tiwari et al., 2023; Xie and Wang, 2015; Zhang et al., 2019).

4.2 OVERVIEW OF SUSTAINABLE SOIL NUTRIENT MANAGEMENT (SSNM)

SSNM is a set of principles and practices aimed at optimizing nutrient use in agriculture while minimizing negative environmental impacts. It emphasizes the efficient utilization of nutrients to ensure healthy plant growth, maintain soil fertility, and safeguard ecosystems for future generations (Ahmad and Malhi, 2018; Brozovic, 2020; Mishra and Mishra, 2023). SSNM takes into account the complex interactions between soil, plants, nutrients, and the environment.

4.2.1 PRINCIPLES OF SSNM

1. **Soil Nutrient Assessment:** SSNM begins with a thorough assessment of soil nutrient levels. Soil testing helps determine the current nutrient status and guides nutrient application decisions. This prevents over- or under-application of fertilizers, reducing nutrient losses and minimizing environmental pollution.
2. **Nutrient Balancing:** SSNM focuses on providing plants with a balanced nutrient supply. The goal is to match nutrient inputs with the actual nutrient requirements of crops at different growth stages. This approach prevents the accumulation of excess nutrients in the soil and reduces the risk of nutrient imbalances that can lead to decreased yields and environmental issues.
3. **Crop-Specific Nutrient Management:** Different crops have varying nutrient needs. SSNM tailors nutrient management plans to the specific requirements of each crop. This includes adjusting nutrient ratios, application rates, and timing to support optimal growth and yield without wastage.
4. **Organic Matter Management:** The addition of organic matter through practices like cover cropping, crop residues, and organic

amendments enhances soil structure, water-holding capacity, and nutrient availability. Organic matter contributes to nutrient recycling and reduces the need for synthetic fertilizers.

5. **Precision Nutrient Application:** SSNM promotes the use of precision agriculture techniques to apply nutrients only where they are needed. Technologies like GPS-guided machinery and sensor-based nutrient application help reduce nutrient wastage by targeting specific areas with nutrient deficiencies.

4.2.2 PRACTICES OF SSNM

1. **Nutrient Source Selection:** SSNM encourages the use of a variety of nutrient sources, including mineral fertilizers, organic amendments, and biofertilizers. The choice of nutrient source depends on factors such as soil type, crop type, and local availability.
2. **Fertilizer Application Timing:** Applying nutrients at the right growth stages of crops improves nutrient uptake efficiency. SSNM considers the critical periods of nutrient demand for different crops and adjusts application schedules accordingly.
3. **Nutrient Application Techniques:** The method of nutrient application plays a role in deciding nutrient use efficiency. SSNM recommends techniques like split application (dividing the total dose into multiple applications) and band placement (placing nutrients closer to the root zone) to ensure maximum uptake.
4. **Monitoring and Adaptation:** Regular monitoring of crop health, soil nutrient levels, and environmental conditions allows for adjustments to nutrient management plans. SSNM emphasizes a dynamic approach, where strategies can be modified based on real-time information.

4.2.3 BENEFITS OF SSNM

1. **Increased Crop Productivity:** SSNM's focus on optimized nutrient use enhances crop growth, leading to improved crop yield and quality.

2. **Soil Health Improvement:** By promoting balanced nutrient application and organic matter management, SSNM enhances soil fertility, structure, and microbial activity.
3. **Environmental Protection:** SSNM minimizes nutrient runoff and leaching, reducing water pollution and its negative impacts on aquatic ecosystems.
4. **Cost Efficiency:** Precision nutrient management and reduced fertilizer wastage lead to a reduction in the cost of cultivation.

Figure 4.1 shows the benefits of using a fertilizer called SSNM. SSNM is shown to have higher NUE (nitrogen use efficiency), potential yield increase, fertilizer use reduction, economic savings, GHG emissions reduction, and soil health improvement compared to conventional fertilizer. By understanding the principles and adopting the practices of SSNM, farmers can contribute to sustainable food production while safeguarding natural resources for future generations (Rejesus et al., 2013; Verma et al., 2020).

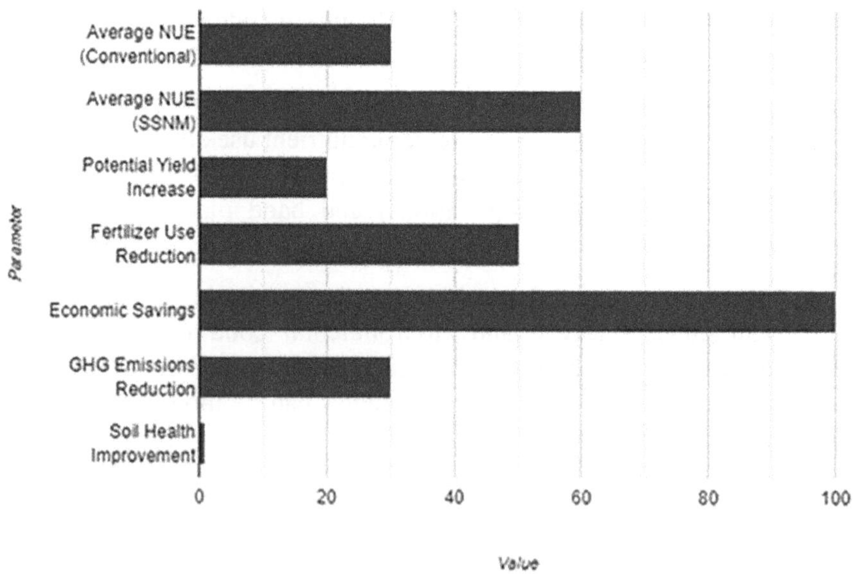

FIGURE 4.1 Benefits of sustainable soil nutrient management.

Source: Adapted from Pampolino et al. (2007).

4.2.4 IMPORTANCE OF TAILORED NUTRIENT MANAGEMENT

1. **Soil Variability:** Soil across different regions and even within the same field exhibits significant variability in its nutrient content, pH levels, and physical properties. Tailoring nutrient management recognizes this variability, ensuring that the right nutrients are applied in the right amounts and at the right time, optimizing their uptake by plants.

2. **Crop-Specific Requirements:** Different crops have varying nutrient demands at different stages of their growth. Tailored nutrient management takes into account the specific nutrient needs of the crop during its various growth phases. By employing this approach, excessive nutrient application is curtailed, minimizing nutrient wastage and mitigating the risk of nutrient runoff into water bodies.

3. **Minimized Environmental Impact:** Excessive use of synthetic fertilizers can lead to nutrient runoff, which can contaminate water bodies, disrupt aquatic ecosystems, and even contribute to issues like algal blooms. By precisely providing nutrients as needed, SSNM helps mitigate these environmental concerns, promoting sustainable farming practices that conserve natural resources.

4. **Cost Efficiency:** Adopting SSNM practices can lead to cost savings for farmers. By avoiding the overuse of fertilizers, farmers can reduce input costs, leading to improved economic viability of their operations.

5. **Soil Health:** Excessive use of synthetic fertilizers can negatively impact soil health by altering microbial populations, reducing soil organic matter, and affecting soil structure. Tailored nutrient management promotes long-term soil health by maintaining nutrient balance and supporting beneficial microbial activity.

SSNM emphasizes the critical importance of understanding the dynamic relationship between soil health, crop requirements, and environmental sustainability (Khurana et al., 2008; Kumar et al., 2023). By customizing nutrient applications based on soil characteristics and crop needs, farmers can achieve optimal yields, preserve natural resources, reduce environmental impacts, and contribute to the long-term viability of agricultural systems (Srivastava et al., 2014).

4.3 ECONOMIC IMPLICATIONS OF SSNM ADOPTION

The economic implications of adopting SSNM practices are crucial for both farmers and policymakers to understand, as they can significantly impact farm profitability, sustainability, and overall agricultural productivity.

4.3.1 COST-BENEFIT ANALYSIS OF ADOPTING SSNM PRACTICES FOR FARMERS

1. **Initial Investment:** The transition from conventional practices to SSNM might require an initial investment in terms of technology adoption, training, and equipment. This could include soil testing kits, precision agriculture tools, and training programs for farmers. Carefully consider the initial expenses in light of the possible long-term advantages.
2. **Reduced Input Costs:** SSNM aims to optimize nutrient application, which can lead to reduced input costs for fertilizers, pesticides, and irrigation. By implementing precise nutrient management practices, farmers can optimize crop growth while minimizing input costs, leading to enhanced profitability.
3. **Yield Enhancement:** By tailoring nutrient application to specific soil and crop conditions, SSNM has the potential to improve crop yields. Increased yield directly contributes to higher revenues for farmers, which can offset the initial investment in adopting SSNM practices.

4.3.2 COMPARISON OF CONVENTIONAL NUTRIENT MANAGEMENT WITH SSNM IN TERMS OF INPUT COSTS

1. **Conventional Approach:** Conventional nutrient management often involves generalized fertilization practices, which can result in over- or under-application of nutrients. This can lead to increased costs due to wastage of fertilizers and potential negative environmental impacts.

2. **SSNM Approach:** SSNM optimizes nutrient application based on precise data about soil nutrient content, crop requirements, and climatic conditions. This precision leads to targeted and efficient use of fertilizers, reducing input costs while maintaining or even improving crop yields (Mishra and Singh, 2023; Yuan et al., 2020; Table 4.1).

TABLE 4.1 Comparison between Conventional Nutrient Management and SSNM

Feature	Conventional Nutrient Management	SSNM
Nutrient application	Blanket application based on average soil fertility.	Tailored application based on site-specific soil fertility and crop requirements.
Input costs	Generally higher due to over application of nutrients.	Generally lower due to more efficient use of nutrients.
Environmental impact	Potential for nutrient runoff and leaching, leading to water pollution.	Reduced risk of nutrient runoff and leaching.
Crop yield	Variable and often suboptimal.	Potential for higher yields due to more precise nutrient application.
Economic returns	Lower due to higher input costs and potentially lower yields.	Higher due to lower input costs and potentially higher yields.

4.3.3 ASSESSING LONG-TERM FINANCIAL BENEFITS

1. **Yield Stability:** One of the significant advantages of SSNM is its potential to stabilize crop yields. By providing crops with the right nutrients at the right time, SSNM minimizes yield loss caused by nutrient deficiencies or toxicity. This stability can help farmers plan better and reduce income variability.

2. **Reduced Input Cost:** Over time, the reduction in input costs due to optimized nutrient management can contribute significantly to overall cost savings. This saving leads to improved profitability and financial resilience for farmers.

3. **Increased Income:** Higher and more stable yields, coupled with reduced input costs, can lead to increased income for farmers. This additional income can be invested in farm infrastructure, diversification of crops, or other income-generating activities.

The economic implications of adopting SSNM practices in agriculture are multifaceted. While there might be initial investments required for technology and training, the long-term benefits are substantial. Through a reduction in input costs, yield enhancement, and increased income, SSNM has the potential to improve farm profitability, sustainability, and the overall economic outlook of agriculture (Apio et al., 2023). Policymakers and agricultural stakeholders should consider these economic implications when promoting and implementing SSNM practices to ensure the well-being of farmers and the agricultural sector as a whole (Khatri-Chhetri et al., 2016).

4.4 MARKET AND TRADE EFFECTS

Embracing SSNM has become a cornerstone of modern agriculture, seeking to balance productivity and sustainability. Consequently, it is crucial to delve into the far-reaching effects of these sustainable practices across the entire agricultural value chain (Karatay and Meyer-Aurich, 2020; Lardo et al., 2020).

4.4.1 EXPLORATION OF MARKET TRENDS INFLUENCED BY SSNM ADOPTION

The widespread integration of SSNM practices within agricultural systems has the potential to reshape market trends significantly. As SSNM optimizes nutrient application according to specific soil and crop requirements, it fosters more consistent yields and enhanced crop quality. Consequently, markets may witness reduced fluctuations in supply, leading to greater stability in prices. Furthermore, the growing environmentally conscious consumer segment is likely to favor products cultivated through sustainable methods, fostering a demand shift towards SSNM-adopted produce. Tracking the correlation between SSNM adoption and market trends offers valuable insights into the efficacy of these practices in aligning with evolving consumer preferences and market dynamics (Pampolino et al., 2012).

4.4.2 IMPACT OF SSNM ON CROP QUALITY AND MARKET VALUE

The tailored approach of SSNM to nutrient management directly influences crop quality and subsequent market value. By precisely supplying nutrients as per crop needs, SSNM not only bolsters yield quantity but also enhances the nutritional content and overall quality of crops (Mishra, 2021). Nutrient balance achieved through SSNM can lead to improved taste, texture, color, and shelf life, all of which are factors influential in determining market value. Farmers implementing SSNM may potentially receive premium prices for their higher-quality produce, thereby incentivizing the adoption of these practices.

4.4.3 CONSIDERATION OF POTENTIAL CHANGES IN INTERNATIONAL TRADE DYNAMICS

The integration of SSNM practices can potentially induce shifts in international trade dynamics. Countries that embrace sustainable agricultural practices, including SSNM, might experience increased competitiveness in global markets. As consumers and regulations worldwide shift towards environmentally friendly products, nations adhering to sustainable practices could gain a comparative advantage in trade. This might lead to a reconfiguration of trade patterns, where countries with robust SSNM adoption could witness an expansion in their export markets, while countries lagging in sustainability might face challenges in maintaining existing trade relationships (Zebarth et al., 2012).

The effects of SSNM adoption extend beyond the fields into the case of market trends, crop quality, market value, and international trade dynamics. The potential of SSNM to stabilize market trends, enhance crop quality, and align with sustainability goals enhances its appeal to both producers and consumers. As global agriculture navigates the complexities of balancing productivity and ecological responsibility, understanding the profound implications of SSNM adoption on market and trade dynamics is paramount for shaping the future of agriculture on a global scale (Sparviero, 2019).

4.5 POLICY AND INSTITUTIONAL SUPPORT

4.5.1 EVALUATION OF GOVERNMENT POLICIES PROMOTING SSNM AND SUSTAINABLE AGRICULTURE

Government policies play a pivotal role in shaping agricultural practices, including SSNM. An in-depth analysis of existing policies is crucial to understanding the level of support and emphasis placed on promoting

SSNM and sustainable agriculture. This evaluation involves studying policy documents, legal frameworks, and strategic plans at various administrative levels, such as national, regional, and local. It also entails assessing the alignment of these policies with international agreements on sustainable development and environmental conservation (Karatay, 2020; Malhi et al., 2022).

4.5.2 ROLE OF AGRICULTURAL EXTENSION SERVICES IN EDUCATING FARMERS ABOUT SSNM

Agricultural extension services act as bridges between researchers, policy-makers, and farmers. Investigating the role of these services in educating farmers about SSNM involves exploring their outreach strategies, training programs, and informational resources. This includes a detailed review of extension materials, workshops, seminars, field demonstrations, and digital platforms employed to disseminate knowledge about SSNM. Understanding how effectively extension services are conveying information, addressing farmers' concerns, and facilitating the adoption of SSNM practices is a critical aspect of this analysis (Rejesus et al., 2014; Shah and Wu, 2019).

4.5.3 INCENTIVE STRUCTURES THAT ENCOURAGE FARMERS TO TRANSITION TO SUSTAINABLE NUTRIENT MANAGEMENT

Adopting SSNM practices necessitates a shift from established routines to innovative methodologies for farmers. Incentive structures are essential tools for encouraging this transition. Investigating these structures involves evaluating both financial and non-financial incentives offered to farmers (D'Souza and Mishra, 2018). Financial incentives might encompass subsidies for purchasing organic fertilizers, grants for soil testing equipment, or compensation for adopting cover cropping. Non-financial incentives could involve recognition programs, certification schemes for sustainable farming, or access to premium markets for sustainably produced crops. This analysis should examine the effectiveness of these incentives in motivating farmers to make the shift to SSNM and assess any challenges or barriers they encounter in doing so (Kumar et al., 2020; Singh and Mishra, 2023).

4.5.4 OVERALL INTEGRATION AND RECOMMENDATIONS

The evaluation of policy and institutional support for SSNM and sustainable agriculture should culminate in an integrated analysis that identifies gaps, strengths, and opportunities within the current framework. This assessment could lead to recommendations for policy adjustments, such as introducing more comprehensive SSNM guidelines into existing agricultural policies or enhancing the scope and effectiveness of agricultural extension services (Srinivasarao et al., 2019). Additionally, based on the effectiveness assessment of incentive structures, suggestions for refining or expanding these mechanisms can be provided. The ultimate goal is to create a well-informed and conducive environment where farmers are empowered and motivated to embrace SSNM practices, fostering both agricultural productivity and environmental preservation (Dhanda et al., 2022).

4.6 ENVIRONMENTAL IMPLICATIONS OF SSNM ADOPTION

4.6.1 ANALYSIS OF REDUCED NUTRIENT RUNOFF AND ITS POSITIVE EFFECTS ON WATER BODIES

Adopting SSNM practices can have a significant impact on reducing nutrient runoff into water bodies. Nutrient runoff, primarily consisting of nitrogen and phosphorus from fertilizers, can lead to harmful algal blooms, oxygen depletion, and overall degradation of aquatic ecosystems. SSNM involves applying fertilizers based on the specific crop nutrient requirements, taking into account soil characteristics and environmental conditions (Betteridge et al., 2008).

When farmers implement SSNM, they meticulously tailor nutrient applications to align with the precise requirements of the crops at different growth stages. This targeted approach minimizes the excessive use of fertilizers that would otherwise be lost during rainfall or irrigation, leading to contamination of rivers, lakes, and oceans. Consequently, the adoption of SSNM can result in reduced nutrient runoff, fostering healthier aquatic ecosystems, improved water quality, and enhanced biodiversity. This positive impact on water bodies can contribute to the sustainability of local ecosystems and the preservation of aquatic life (Olayeni et al., 2021).

4.6.2 MITIGATION OF SOIL DEGRADATION AND EROSION THROUGH IMPROVED NUTRIENT MANAGEMENT

Soil degradation and erosion are critical environmental concerns that jeopardize agricultural productivity and ecosystem health. Excessive and improper fertilizer application can aggravate these problems by disrupting soil structure, diminishing fertility, and amplifying the probability of erosion during rainfall. SSNM addresses this challenge by optimizing fertilizer application based on precise field characteristics and crop requirements (Mishra et al., 2023; Zhang et al., 2023).

By adopting SSNM, farmers can ensure that nutrients are distributed evenly and efficiently throughout the field, preventing over-fertilization in certain areas and under-fertilization in others. This balanced approach enhances soil health and structure, reducing the risk of erosion and promoting sustainable land use. The reduction in soil erosion not only protects agricultural land but also prevents sedimentation in water bodies, which can degrade water quality and aquatic habitats (Kapri and Kesarwani, 2019).

4.6.3 QUANTIFYING GREENHOUSE GAS EMISSIONS REDUCTION FROM ADOPTING SSNM

The adoption of SSNM can also contribute to mitigating greenhouse gas emissions from agriculture. Traditional fertilizer application practices can result in the release of nitrous oxide (N_2O), a potent greenhouse gas that has a significant impact on global warming. Improper fertilization, especially with excess nitrogen, can increase N_2O emissions from soils (Srinivasan, 2006).

SSNM helps minimize these emissions by tailoring nutrient application to actual crop needs, reducing the potential for excess nitrogen to be converted into N_2O. Additionally, SSNM often includes the use of nitrogen inhibitors and controlled-release fertilizers, which further limit nitrogen losses and associated emissions. Through these measures, the adoption of SSNM can contribute to lower agricultural emissions, aligning with climate change mitigation efforts (Mhoja et al., 2021).

The adoption of SSNM practices can lead to several positive environmental outcomes. These benefits include reduced nutrient runoff into water bodies, mitigated soil degradation and erosion, and quantifiable reductions in greenhouse gas emissions. By focusing on precision nutrient application tailored to specific field conditions, SSNM presents a promising pathway toward more sustainable and environmentally friendly agricultural practices.

4.7 BIODIVERSITY AND ECOSYSTEM HEALTH

Biodiversity and ecosystem health are crucial components of sustainable agriculture, and the implementation of SSNM holds significant promise in supporting these objectives. SSNM is an approach that tailors nutrient application to the specific needs of different parts of a field, accounting for factors such as soil type, crop variety, and local conditions. This precision-based approach not only enhances crop productivity but also plays a vital role in promoting biodiversity and ecosystem health within agroecosystems (Mishra, 2022).

4.7.1 SUPPORTING BIODIVERSITY THROUGH SSNM

Traditionally, the widespread use of chemical fertilizers has been linked to negative impacts on biodiversity. These inputs can disrupt natural nutrient cycles, alter soil pH, and harm beneficial soil organisms. In contrast, SSNM emphasizes a more balanced nutrient management strategy. By providing crops with the nutrients they require at appropriate stages of growth, SSNM minimizes nutrient imbalances that can lead to the dominance of certain species over others. This contributes to a more diverse range of plant species thriving in agroecosystems.

4.7.2 BENEFITS OF SSNM PRACTICES ON SOIL HEALTH AND PLANT GROWTH

The adoption of SSNM practices is inherently linked to enhancements in soil structure, microbial diversity, and plant health. With minimized reliance on chemical fertilizers, soil structure is less susceptible to degradation caused by compaction or nutrient imbalances. The judicious application of nutrients also fosters microbial diversity in the soil, as excessive synthetic inputs can hinder the growth of beneficial microorganisms. Improved microbial diversity contributes to nutrient cycling, organic matter decomposition, and enhanced disease resistance. As plants receive the nutrients they require at the appropriate times, their overall health and resilience against pests and diseases are strengthened (Dobermann et al., 2002).

The shift towards SSNM represents a crucial step in supporting biodiversity and ecosystem health in agroecosystems. Through targeted nutrient application, SSNM minimizes the negative impacts of excessive chemical inputs, fostering a diverse range of plant species and promoting the growth of beneficial organisms. This not only aids in maintaining biodiversity but also contributes to the overall sustainability and resilience of agricultural systems (Wan et al., 2018).

4.8 RESILIENCE TO CLIMATE CHANGE

Climate change is a growing threat to global food security. Extreme weather events, such as droughts, floods, and heatwaves, are becoming more frequent and intense, leading to reduced crop yields and increased food insecurity. In the face of these challenges, the need for sustainable agricultural practices has become more critical than ever. SSNM is a promising approach to bolster crop resilience against climate change by optimizing nutrient usage, minimizing runoff, and enhancing soil organic matter. Its effectiveness has been demonstrated worldwide, with rice yield increases of up to 30% in India and fertilizer reductions of up to 20% in China. As climate change intensifies, SSNM is poised to play a pivotal role in safeguarding global food security.

4.8.1 IMPACT OF SSNM ON CROP RESILIENCE

SSNM involves tailoring nutrient application strategies to the unique requirements of individual fields or even smaller management units. Accounting for diverse soil characteristics, topography, and other factors, this approach acknowledges the varying nutrient requirements of different farm areas. SSNM is promoting practices that can help to increase crop resilience to climate change. By providing crops with the precise nutrients they require at different growth stages, SSNM optimizes their physiological processes, making them more robust in withstanding stress induced by erratic weather patterns.

4.8.2 ADAPTATION TO CHANGING PRECIPITATION PATTERNS

One of the significant challenges posed by climate change is the altered precipitation patterns, leading to either intense flooding or prolonged droughts. SSNM offers a dynamic strategy for farmers to adapt to these unpredictable changes (Mishra and Singh, 2023). Through targeted nutrient application, crops can be enriched to better endure waterlogged or arid conditions. For instance, by adjusting nitrogen levels based on the forecasted precipitation, farmers can mitigate the negative effects of waterlogging on root health while simultaneously enhancing the water use efficiency of plants during dry spells (Majumdar et al., 2013).

4.8.3 MITIGATING TEMPERATURE EXTREMES

Extreme weather events, including scorching heat waves and abrupt cold fronts, can wreak havoc on agricultural output. SSNM can be employed as a mechanism to bolster crop resilience to these extremes. The meticulous application of nutrients, particularly micronutrients and biostimulants, can reinforce plants' cellular structures and metabolic processes. This enhanced resilience allows crops to better withstand temperature stress and recover more rapidly when favorable conditions return (Wang et al., 2001).

4.8.4 FOOD SECURITY AMID CLIMATE UNCERTAINTIES

As climate change disrupts traditional growing seasons and agrarian practices, maintaining food security becomes a paramount concern. SSNM is an essential player in the global food system. SSNM enhances crop resilience to unpredictable climatic shifts by promoting efficient nutrient utilization and optimizing crop performance. This approach is particularly crucial in regions already experiencing food insecurity and heightened susceptibility to climate change impacts (Richards et al., 2015).

SSNM holds immense potential as a tool to enhance agricultural resilience in the face of climate change. Its ability to customize nutrient application strategies to unique circumstances enables farmers to foresee and adjust to changing precipitation patterns and extreme temperatures. By enhancing crop resilience, SSNM makes a significant contribution to achieving the goal of maintaining global food security in the face of the uncertainties posed by a changing climate (Srivastava, 2013).

4.9 CHALLENGES AND BARRIERS TO ADOPTION

SSNM holds the promise of revolutionizing agricultural practices by tailoring nutrient application to the specific needs of each field, thereby increasing productivity while minimizing environmental impact (Rahman and Zhang, 2018). However, the widespread adoption of SSNM faces several challenges and barriers that need to be addressed to fully realize its potential.

4.9.1 IDENTIFICATION OF OBSTACLES HINDERING WIDESPREAD SSNM ADOPTION

The implementation of SSNM requires a comprehensive understanding of soil properties, crop requirements, and environmental conditions. This demands substantial investments in terms of technology, knowledge, and data. Key challenges include:

1. **Data Requirements:** Successful SSNM implementation requires accurate and up-to-date data on soil properties, crop genetics, weather patterns, and other relevant factors. Gathering, analyzing, and maintaining this data can be resource-intensive.
2. **Technological Barriers:** Adopting SSNM often necessitates the use of advanced technologies such as precision agriculture tools, remote sensing, and data analytics. Many farmers, particularly in resource-limited regions, may lack access to these technologies or the necessary training to operate them effectively.

4.9.2 SOCIOECONOMIC AND CULTURAL FACTORS INFLUENCING DECISIONS

The decisions of farmers are influenced by a complex interplay of socio-economic and cultural factors that can impact their willingness to adopt SSNM practices:

1. **Risk Tolerance:** Farmers might be hesitant to adopt new practices due to the uncertainty of outcomes. If the benefits of SSNM are not immediately apparent or if initial failures occur, farmers may revert to conventional practices.
2. **Lack of Awareness:** Many farmers might not be aware of the benefits of SSNM or may have misconceptions about its effectiveness. Effective communication and educational programs are necessary to bridge this knowledge gap.
3. **Traditional Practices:** Farming practices are often deeply rooted in tradition and local knowledge. Convincing farmers to switch from familiar practices to SSNM requires demonstrating the clear advantages and economic gains.

4.9.3 STRATEGIES TO ADDRESS CHALLENGES AND PROMOTE WIDER ACCEPTANCE OF SSNM PRACTICES

Overcoming these challenges requires a multi-faceted approach that combines technology, education, policy support, and community engagement:

1. **Capacity Building:** Providing training and workshops to farmers on SSNM principles, technology usage, and data interpretation can enhance their understanding and confidence in adopting these practices.
2. **Financial Incentives:** Governments and agricultural organizations can offer financial incentives such as subsidies for adopting SSNM. These incentives can offset initial implementation costs and encourage wider adoption.
3. **Customized Solutions:** Developing user-friendly decision support systems that offer tailored SSNM recommendations based on local conditions can empower farmers to make informed choices.
4. **Knowledge Exchange:** Facilitating knowledge-sharing platforms where farmers, researchers, and extension agents can exchange experiences and insights can accelerate learning and adoption.

While SSNM holds great promise for sustainable and productive agriculture, its widespread adoption is challenged by technical, economic, and cultural factors. Addressing these challenges requires collaborative efforts from researchers, policymakers, extension services, and farmers themselves. By implementing strategies that integrate technological innovation, education, and supportive policies, we can pave the way for the successful adoption of SSNM practices on a global scale (Huang et al., 2015).

4.10 FUTURE OUTLOOK

In the case of sustainable SSNM, the future holds exciting possibilities shaped by anticipated trends in research, policy, and adoption. The continued convergence of technology, agriculture, and environmental stewardship is expected to drive significant advancements in SSNM practices (Kakraliya et al., 2022; Mishra and Singh, 2023).

4.10.1 ANTICIPATED TRENDS IN SSNM RESEARCH, POLICY, AND ADOPTION

1. **Precision Agriculture Revolution:** SSNM is poised to play a central role in the precision agriculture revolution. Advances in data analytics, remote sensing, and machine learning are expected to lead to more accurate and timely assessments of nutrient needs at a farm-specific level. Researchers and policymakers will collaborate to develop standardized protocols for SSNM integration into precision farming systems.
2. **Climate-Resilient Nutrient Strategies:** With the increasing unpredictability of weather patterns due to climate change, SSNM will likely adapt to offer climate-resilient nutrient management strategies. This could involve dynamic nutrient recommendations that consider weather forecasts and predictive modeling to adjust nutrient application on the fly.
3. **Cross-Disciplinary Collaboration:** Future SSNM research will likely involve closer collaboration between agronomists, soil

scientists, data scientists, and environmental engineers. This multidisciplinary approach will result in more comprehensive and holistic nutrient management solutions that account for ecological, economic, and social factors.

4. **Digital Agricultural Extension Services:** Governments and agricultural organizations may embrace digital extension services, providing farmers with personalized SSNM recommendations through smartphone apps or online platforms. By democratizing access to SSNM knowledge, we can empower farmers with actionable insights, leading to improved agricultural practices and sustainability.

4.10.2 POTENTIAL TECHNOLOGICAL ADVANCEMENTS

1. **IoT-Enabled Soil Sensors:** Miniaturized sensors capable of monitoring soil health and nutrient levels in real time will become more affordable and widespread. These sensors will provide a continuous stream of data to support SSNM decision-making, reducing the need for labor-intensive manual soil testing.
2. **Predictive Modeling with AI:** Advanced AI algorithms will analyze historical data, weather forecasts, and crop growth patterns to predict optimal nutrient application schedules. These models will offer dynamic recommendations that adapt to changing conditions, boosting both productivity and sustainability.
3. **Nanotechnology in Nutrient Delivery:** Nanoparticles could be engineered to encapsulate and slowly release nutrients to plants based on their growth stage and needs. This precise nutrient delivery system could minimize nutrient leaching and runoff.
4. **Blockchain-Backed Traceability:** Blockchain-backed traceability is a system that uses blockchain technology to track the movement of fertilizers from their source to their end use. This can help to ensure that fertilizers are sourced from sustainable sources, produced using environmentally friendly methods, and distributed responsibly (Figure 4.2).

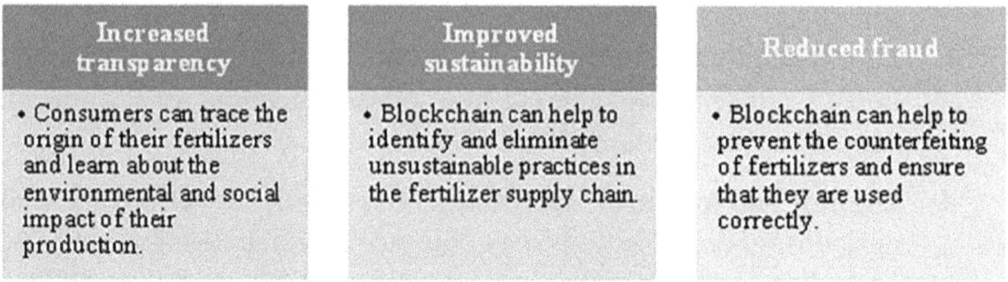

FIGURE 4.2 Benefits of blockchain-backed traceability for fertilizer supply chains.

4. **How Blockchain-Backed Traceability Works:**
   i. **Data Collection:** Fertilizer producers, distributors, and retailers collect data about the origin, production, and distribution of their fertilizers.
   ii. **Data Storage:** The data is stored on a blockchain, which is a distributed ledger that is secure and transparent.
   iii. **Data Access:** Consumers and other stakeholders can access the data on the blockchain to track the movement of fertilizers.
5. **Alignment with Consumer Demands:** Consumers are increasingly demanding eco-friendly and socially responsible products. Blockchain-backed traceability can help to meet these demands by providing consumers with information about the sustainability of their fertilizers.

4.10.3 ENVISIONING THE ROLE OF SSNM IN SUSTAINABLE AND RESILIENT AGRICULTURE

For future agriculture, SSNM will be an indispensable tool for achieving both sustainability and resilience:

1. **Resource Efficiency:** SSNM practices will optimize nutrient usage, minimizing waste and reducing the environmental impact of nutrient runoff into water bodies. This will contribute to cleaner water sources and healthier ecosystems.
2. **Climate Adaptation:** By tailoring nutrient applications to changing climate conditions, SSNM will aid in crop adaptation

and resilience. Nutrient adjustments based on weather forecasts will mitigate the effects of extreme weather events on crop yields.

3. **Yield Stability:** The precision approach of SSNM will lead to more stable and predictable crop yields. This stability will be crucial in ensuring food security, especially in the face of population growth and environmental challenges.

4. **Carbon Sequestration:** SSNM strategies could be designed to enhance soil organic matter and carbon sequestration. Healthy soils will not only improve nutrient retention but also play a pivotal role in mitigating climate change.

Poised at the forefront of agricultural innovation, the future of SSNM is a dynamic tapestry of collaboration, sustainable practices, and cutting-edge technologies. As advancements in technology reshape the world, SSNM will seamlessly adapt, becoming an indispensable force in ensuring global food security, environmental sustainability, and resilient agricultural systems (Medina et al., 2007; Meena et al., 2022).

4.11 CONCLUSION

In conclusion, the chapter discusses the multifaceted dimensions of SSNM and its profound impacts on both economic and environmental fronts. As the global population continues to expand, ensuring food security while safeguarding the delicate balance of our ecosystems has become a paramount concern. SSNM emerges as a promising approach, offering a holistic solution to address these intricate challenges. The economic implications of SSNM adoption are significant and far-reaching. By optimizing nutrient usage, farmers can achieve higher yields, reduced input costs, and improved overall profitability. Moreover, the adoption of SSNM practices can lead to increased market competitiveness and enhanced trade relations, as sustainable practices align with the values of environmentally conscious consumers and international regulatory bodies. The chapter highlights the pivotal role of policy frameworks and institutional support in driving SSNM adoption. Effective policies can incentivize farmers, provide technical guidance, and create an enabling environment for the transition to sustainable practices. Turning towards the environmental dimension, SSNM adoption emerges as a crucial element for promoting biodiversity and ecosystem health. The careful management of soil nutrients minimizes the risk of soil degradation, water pollution, and habitat destruction, thereby fostering diverse and resilient ecosystems. Additionally, the capacity of SSNM to enhance the resilience of agricultural systems against climate change is explored. By nurturing soil health, SSNM mitigates the adverse effects of extreme weather events and contributes to carbon sequestration, playing a pivotal role in climate change adaptation and mitigation.

The chapter does not refrain from acknowledging the challenges and barriers inherent in SSNM adoption. These range from a lack of awareness and technical expertise among farmers to financial constraints and institutional gaps. Nevertheless, these challenges are not insuperable, and with concerted efforts from governments, research institutions, and the private sector, they can be overcome. Looking ahead, the future outlook for SSNM is promising. As sustainable agricultural practices gain momentum on the global stage, SSNM is poised to become a cornerstone of agricultural strategies worldwide. Leveraging technological advancements and data-driven insights, the potential of SSNM can be fully realized, ultimately contributing to a more food-secure, economically viable, and ecologically resilient planet.

KEYWORDS

- **biodiversity conservation**
- **carbon emissions**
- **crop quality**
- **fertilizers**
- **global food security**
- **site-specific nutrient management**
- **soil fertility**
- **sustainable agriculture**
- **sustainable farming practices**

REFERENCES

Ahmad, L., & Mahdi, S. S. (2018). *Satellite Farming* (p. 190). Basel, Switzerland: Springer International Publishing.

Apio, A. T., Thiam, D. R., & Dinar, A. (2023). Farming under drought: An analysis of the factors influencing farmers' multiple adoption of water conservation practices to mitigate farm-level water scarcity. *Journal of Agricultural and Applied Economics, 55*(3), 1–39.

Aryal, J. P., Jat, M. L., Sapkota, T. B., Khatri-Chhetri, A., Kassie, M., Rahut, D. B., & Maharjan, S. (2018). Adoption of multiple climate-smart agricultural practices in the Gangetic plains of Bihar, India. *International Journal of Climate Change Strategies and Management, 10*(3), 407–427.

Aryal, J. P., Rahut, D. B., Maharjan, S., & Erenstein, O. (2018). Factors affecting the adoption of multiple climate-smart agricultural practices in the Indo-Gangetic Plains of India. In *Natural Resources Forum* (Vol. 42, No. 3, pp. 141–158). Oxford, UK: Blackwell Publishing Ltd.

Asai, M., Reidsma, P., & Feng, S. (2010). Impacts of agricultural land-use changes on biodiversity in Taihu Lake Basin, China: A multi-scale cause–effect approach considering multiple land-use functions. *International Journal of Biodiversity Science, Ecosystem Services & Management, 6*(3, 4), 119–130.

Betteridge, K., Schnug, E., & Haneklaus, S. (2008). Will site-specific nutrient management live up to expectations? *Agriculture and Forestry Research, 4*(58), 283–294.

Brozovic, D. (2020). Business model based on strong sustainability: Insights from an empirical study. *Business Strategy and the Environment, 29*(2), 763–778.

Chen, J., Huang, Y., & Tang, Y. (2011). Quantifying economically and ecologically optimum nitrogen rates for rice production in south-eastern China. *Agriculture, Ecosystems & Environment, 142*(3, 4), 195–204.

Chivenge, P., Saito, K., Bunquin, M. A., Sharma, S., & Dobermann, A. (2021). Co-benefits of nutrient management tailored to smallholder agriculture. *Global Food Security, 30*, 100570.

Chivenge, P., Sharma, S., Bunquin, M. A., & Hellin, J. (2021). Improving nitrogen use efficiency—A key for sustainable rice production systems. *Frontiers in Sustainable Food Systems, 5*, 737412.

Chivenge, P., Zingore, S., Ezui, K. S., Njoroge, S., Bunquin, M. A., Dobermann, A., & Saito, K. (2022). Progress in research on site-specific nutrient management for smallholder farmers in sub-Saharan Africa. *Field Crops Research, 281*, 108503.

D'Souza, A., & Mishra, A. K. (2018). Adoption and abandonment of partial conservation technologies in developing economies: The case of South Asia. *Land Use Policy, 70*, 212–223.

Dhanda, S., Yadav, A., Yadav, D. B., & Chauhan, B. S. (2022). Emerging issues and potential opportunities in the rice–wheat cropping system of North-Western India. *Frontiers in Plant Science, 13*, 832683.

Dobermann, A., & White, P. F. (1999). Strategies for nutrient management in irrigated and rainfed lowland rice systems. In *Resource Management in Rice Systems: Nutrients: Papers Presented at the International Workshop on Natural Resource Management in*

Rice Systems: Technology Adaption for Efficient Nutrient Use, Bogor, Indonesia, 2–5 December 1996 (pp. 1–26). Springer Netherlands.

Dobermann, A., Blackmore, S., Cook, S. E., & Adamchuk, V. I. (2004). Precision farming: Challenges and future directions. In *Proceedings of the 4th International Crop Science Congress* (Vol. 26). Australia: Brisbane.

Dobermann, A., Witt, C., Dawe, D., Abdulrachman, S., Gines, H. C., Nagarajan, R., & Adviento, M. A. A. (2002). Site-specific nutrient management for intensive rice cropping systems in Asia. *Field Crops Research, 74*(1), 37–66.

Hengsdijk, H., van den Berg, M., Roetter, R., Guanghuo, W., Wolf, J., Changhe, L., & van Keulen, H. (2005). Consequences of technologies and production diversification for the economic and environmental performance of rice-based farming systems in East and Southeast Asia. *Copyright International Rice Research Institute 2005*, 422.

Hu, R., Cao, J., Huang, J., Peng, S., Huang, J., Zhong, X., & Buresh, R. J. (2007). Farmer participatory testing of standard and modified site-specific nitrogen management for irrigated rice in China. *Agricultural Systems, 94*(2), 331–340.

Huang, J., Huang, Z., Jia, X., Hu, R., & Xiang, C. (2015). Long-term reduction of nitrogen fertilizer uses through knowledge training in rice production in China. *Agricultural Systems, 135*, 105–111.

Jat, H. S., Datta, A., Choudhary, M., Sharma, P. C., & Jat, M. L. (2021). Conservation agriculture: Factors and drivers of adoption and scalable innovative practices in the Indo-Gangetic plains of India–A review. *International Journal of Agricultural Sustainability, 19*(1), 40–55.

Kakraliya, S. K., Jat, H. S., Singh, I., Gora, M. K., Kakraliya, M., Bijarniya, D., & Jat, M. L. (2022). Energy and economic efficiency of climate-smart agriculture practices in a rice–wheat cropping system of India. *Scientific Reports, 12*(1), 8731.

Kapri, M., & Kesarwani, A. (2019). Implementation of improved nitrogen management strategies under intense wheat cultivation. *International Journal of Advanced Research in Agriculture and Allied Sciences, 1*, 32–38.

Karatay, Y. N., & Meyer-Aurich, A. (2020). Profitability and downside risk implications of site-specific nitrogen management with respect to wheat grain quality. *Precision Agriculture, 21*(2), 449–472.

Khatri-Chhetri, A., Aryal, J. P., Sapkota, T. B., & Khurana, R. (2016). Economic benefits of climate-smart agricultural practices to smallholder farmers in the Indo-Gangetic Plains of India. *Current Science, 110*(7), 1251–1256.

Khurana, H. S., Phillips, S. B., Alley, M. M., Dobermann, A., Sidhu, A. S., & Peng, S. (2008). Agronomic and economic evaluation of site-specific nutrient management for irrigated wheat in northwest India. *Nutrient Cycling in Agroecosystems, 82*, 15–31.

Kumar, A., Rana, S. S., & Manuja, S. (2020). Influence of nutrient management practices and varieties on the productivity and economics of maize (*Zea mays* L.) and their residual effect in Gobhi Sarson (*Brassica Napus* L.) under rainfed conditions. *Communications in Soil Science and Plant Analysis, 51*(18), 2323–2330.

Kumar, N., Kushwaha, R. R., Meena, N. R., Mishra, H., & Yadav, A. P. (2023). A study on costs and returns of paddy cultivation in Ambedkar Nagar district of Uttar Pradesh. *International Journal of Statistics and Applied Mathematics, SP-8*(3), 107–111.

Laborte, A. G., Schipper, R. A., Van Ittersum, M. K., Van Den Berg, M. M., Van Keulen, H., Prins, A. G., & Hossain, M. (2009). Farmers' welfare, food production and the

environment: A model-based assessment of the effects of new technologies in the northern Philippines. *NJAS: Wageningen Journal of Life Sciences, 56*(4), 345–373.
Lardo, A., Mancini, D., Paoloni, N., & Russo, G. (2020). The perspective of capability providers in creating a sustainable I4.0 environment. *Management Decision, 58*(8), 1759–1777.
Majumdar, K., Jat, M. L., Pampolino, M., Satyanarayana, T., Dutta, S., & Kumar, A. (2013). Nutrient management in wheat: Current scenario, improved strategies, and future research needs in India. *Journal of Wheat Research, 4*(1).
Malhi, G. S., Kaur, M., Singh, A., Singh, V. K., Saini, S. P., & Jatav, H. S. (2022). Agronomic and economic assessment of site-specific nutrient management in crop production. In *Ecosystem Services: Types, Management and Benefits* (pp. 241–257). Nova Science Publishers, USA.
Mariano, M. J., Villano, R., & Fleming, E. (2012). Factors influencing farmers' adoption of modern rice technologies and good management practices in the Philippines. *Agricultural Systems, 110*, 41–53.
Medina, C. M., Labios, J. D., Labios, R. V., Ocampo, A. M., Gines, H. C., Tumamang, S. C., & Pasuguin, J. M. (2007). Socio-economic of site-specific nutrient management for maize in the Philippines. *Philippine Journal of Crop Science* (Philippines).
Meena, B. R., Jatav, H. S., Dudwal, B. L., Kumawat, P., Meena, S. S., Singh, V. K., & Sathyanarayana, E. (2022). Fertilizer recommendations by using different geospatial technologies in precision farming or nanotechnology. *Ecosystem Services, 14*, 241–257.
Mhoja, L. S., Mtae, H. G., & Haule, C. (2021). Effect of advanced rice production technologies on household economic development in Kilombero Valley, Tanzania. *East African Journal of Education and Social Sciences (EAJESS), 2*(2), 222–230.
Mishra, D., Singh, K. K., Mishra, H., & Srivastava, A. B. (2023). Resource use efficiency (RUE) of lentil cultivation in Sultanpur district of Uttar Pradesh. *Environment and Ecology, 41*(2B), 1209–1216.
Mishra, H. (2021). A new era for Indian farming. *Times of Agriculture, 17*, 80, 81.
Mishra, H. (2022). Production and marketing of cucumber and bitter gourd in Sultanpur district of Uttar Pradesh. *M.Sc. Thesis*. Acharya Narendra Deva University of Agriculture and Technology, Ayodhya, U.P. pp. 152.
Mishra, H., & Mishra, D. (2023). Artificial intelligence and machine learning in agriculture: Transforming farming systems. In *Research Trends in Agriculture Science* (Volume I, pp. 1–16). Bhumi Publishing.
Mishra, H., & Mishra, D. (2023). From comparative advantage to protectionism: Economic effects of trade wars on agricultural markets. *Agriallis, 5*(7), 29–36. www.agriallis.com (accessed on 16 January 2025).
Mishra, H., & Singh, M. (2023). Agritech in India: An overview. *Agriallis*, 28–38.
Mishra, H., & Singh, M. (2023). Market liberalization and agricultural sector transformation: Paving the path to a dynamic future. *Times of Agriculture, 3*(7), 67–68.
Mishra, H., & Singh, M. (2023). Revolutionizing agriculture: Empowering farmers through technological adoption for increased productivity and profitability. *The Agriculture Magazine, 2*(8), 17–22. www.theagricultureonline.com (accessed on 16 January 2025).
Mishra, H., & Singh, M. (2023). The significance of wildlife sanctuaries for Indian agriculture. *Indian Farmer, 10*(05), 243–246.

Mishra, H., Neerugatti, M. P., Gautam, S., & Mishra, D. (2023). Economic analysis of cucumber market performance and their constraints in Sultanpur district of Uttar Pradesh. *Asian Journal of Agricultural Extension, Economics & Sociology, 41*(4), 82–95.

Olayeni, A., Ogbo, A., Okwo, H., Chukwu, B., Ifediora, C., & Ezenwakwelu, C. (2021). Green strategy effect on financial and environmental performance: A mediation analysis of product quality. *Sustainability, 13*(4), 2115.

Pampolino, M. F., Manguiat, I. J., Ramanathan, S., Gines, H. C., Tan, P. S., Chi, T. T. N., & Buresh, R. J. (2007). Environmental impact and economic benefits of site-specific nutrient management (SSNM) in irrigated rice systems. *Agricultural Systems, 93*(1–3), 1–24.

Pampolino, M. F., Witt, C., Pasuquin, J. M., Johnston, A., & Fisher, M. J. (2012). Development approach and evaluation of the Nutrient Expert software for nutrient management in cereal crops. *Computers and Electronics in Agriculture, 88*, 103–110.

Parihar, C. M., Jat, S. L., Singh, A. K., Majumdar, K., Jat, M. L., Saharawat, Y. S., & Kuri, B. R. (2017). Bio-energy, water-use efficiency, and economics of maize-wheat-mungbean system under precision-conservation agriculture in semi-arid agro-ecosystem. *Energy, 119*, 245–256.

Pasuquin, J. M., Pampolino, M. F., Witt, C., Dobermann, A., Oberthür, T., Fisher, M. J., & Inubushi, K. (2014). Closing yield gaps in maize production in Southeast Asia through site-specific nutrient management. *Field Crops Research, 156*, 219–230.

Rahman, K. A., & Zhang, D. (2018). Effects of fertilizer broadcasting on the excessive use of inorganic fertilizers and environmental sustainability. *Sustainability, 10*(3), 759.

Ranjan, S., Sow, S., Kumar, S., Ghosh, M., Roy, D. K., Dutta, S. K., & Pramanick, B. (2022). Enhancing nutrient use efficiency and productivity of cereals through site-specific nutrient management. *Journal of Cereal Research, 14*(3), 229–242.

Reidsma, P., Feng, S., van Loon, M., Luo, X., Kang, C., Lubbers, M., & Qu, F. (2012). Integrated assessment of agricultural land use policies on nutrient pollution and sustainable development in Taihu Basin, China. *Environmental Science & Policy, 18*, 66–76.

Reidsma, P., König, H., Feng, S., Bezlepkina, I., Nesheim, I., Bonin, M., & Brouwer, F. (2011). Methods and tools for integrated assessment of land use policies on sustainable development in developing countries. *Land Use Policy, 28*(3), 604–617.

Rejesus, R. M., Martin, A. M., & Gypmantasiri, P. (2013). Meta-impact assessment of the irrigated rice research consortium. International Rice Research Institute (IRRI).

Rejesus, R. M., Martin, A. M., & Gypmantasiri, P. (2014). Enhancing the impact of natural resource management research: Lessons from a meta-impact assessment of the Irrigated Rice Research Consortium. *Global Food Security, 3*(1), 41–48.

Richards, M. B., Butterbach-Bahl, K., Jat, M. L., Ortiz Monasterio, I., Sapkota, T. B., & Lipinski, B. (2015). Site-specific nutrient management: Implementation guidance for policymakers and investors. *CSA Practice Brief*, 1–10.

Rodriguez, D. G. P., & Nga, N. T. D. (2012). Impacts of site-specific nutrient management in irrigated rice farms in the Red River Delta, Northern Vietnam. https://www.researchgate.net/publication/238048232_Impacts_of_Site-Specific_Nutrient_Management_in_Irrigated_Rice_Farms_in_the_Red_River_Delta_Northern_Vietnam (accessed on 16 January 2025).

Sarkar, D., Meena, V. S., Haldar, A., & Rakshit, A. (2017). Site-specific nutrient management (SSNM): A unique approach towards maintaining soil health. In *Adaptive Soil Management: From Theory to Practices* (pp. 69–88).

Senyolo, M. P., Long, T. B., Blok, V., & Omta, O. (2018). How the characteristics of innovations impact their adoption: An exploration of climate-smart agricultural innovations in South Africa. *Journal of Cleaner Production, 172*, 3825–3840.

Shah, F., & Wu, W. (2019). Soil and crop management strategies to ensure higher crop productivity within sustainable environments. *Sustainability, 11*(5), 1485.

Sida, T. S., Gameda, S., Chamberlin, J., Andersson, J. A., Getnet, M., Woltering, L., & Craufurd, P. (2023). Failure to scale in digital agronomy: An analysis of site-specific nutrient management decision-support tools in developing countries. *Computers and Electronics in Agriculture, 212*, 108060.

Singh, M., & Mishra, H. (2023). Medicinal plant farming to improve farmers' economy. *Agriallis*, 30–36.

Singh, M., & Mishra, H. (2023). Trade development and value chains in agriculture: Unleashing the potential for sustainable growth. *Times of Agriculture*, 56, 57.

Sirmour, A., Pandey, M. K., & Verma, A. (2017). Adoption of new resource conservation technology for sustainable production. *Journal of Pharmacognosy and Phytochemistry, 6*(6S), 655–661.

Sparviero, S. (2019). The case for a socially oriented business model canvas: The social enterprise model canvas. *Journal of Social Entrepreneurship, 10*(2), 232–251.

Srinivasan, A. (2006). *Handbook of Precision Agriculture: Principles and Applications*. CRC Press.

Srinivasarao, C., Kundu, S., Lakshmi, C. S., Rani, Y. S., Nataraj, K. C., Gangaiah, B., & Manasa, R. (2019). Soil health issues for sustainability of South Asian agriculture. *EC Agriculture, 5*(6), 310–326.

Srivastava, A. K. (2013). Site-specific nutrient management in citrus. *Scientific Journal of Agricultural, 2*(2), 53–67.

Srivastava, A. K., Das, S. N., Malhotra, S. K., & Majumdar, K. (2014). SSNM-based rationale of fertilizer uses in perennial crops: A review. *Indian Journal of Agricultural Sciences, 84*(1), 3–17.

Tiwari, A. K., Mishra, H., Nishad, D. C., & Pandey, A. (2023). Sustainable water management in agriculture: irrigation techniques and water conservation (pp. 53–68). In *Research Trends in Agriculture Science* (Volume II). Bhumi Publishing.

Verma, P., Chauhan, A., & Ladon, T. (2020). Site-specific nutrient management: A review. *Journal of Pharmacognosy and Phytochemistry, 9*(5S), 233–236.

Wan, Z., El Makhloufi, A., Chen, Y., & Tang, J. (2018). Decarbonizing the international shipping industry: Solutions and policy recommendations. *Marine Pollution Bulletin, 126*, 428–435.

Wang, G., Dobermann, A., Witt, C., Sun, Q., & Fu, R. (2001). Performance of site-specific nutrient management for irrigated rice in southeast China. *Agronomy Journal, 93*(4), 869–878.

Wang, G., Zhang, Q. C., Witt, C., & Buresh, R. J. (2007). Opportunities for yield increases and environmental benefits through site-specific nutrient management in rice systems of Zhejiang province, China. *Agricultural Systems, 94*(3), 801–806.

Xie, H., & Wang, W. (2015). Exploring the spatial-temporal disparities of urban land use economic efficiency in China and its influencing factors under environmental constraints based on a sequential slacks-based model. *Sustainability, 7*(8), 10171–10190.

Yamano, T., Arouna, A., Labarta, R. A., Huelgas, Z. M., & Mohanty, S. (2016). Adoption and impacts of international rice research technologies. *Global Food Security, 8*, 1–8.

Yi, B., Zhang, Q., Gu, C., Li, J., Abbas, T., & Di, H. (2018). Effects of different fertilization regimes on nitrogen and phosphorus losses by surface runoff and bacterial community in a vegetable soil. *Journal of Soils and Sediments, 18*, 3186–3196.

Yuan, H., Feng, Y., Lee, J., Liu, H., & Li, R. (2020). The spatial threshold effect and its regional boundary of financial agglomeration on green development: A case study in China. *Journal of Cleaner Production, 244*, 118670.

Zebarth, B. J., Bélanger, G., Cambouris, A. N., & Ziadi, N. (2012). Nitrogen fertilization strategies in relation to potato tuber yield, quality, and crop N recovery. In *Sustainable Potato Production: Global Case Studies* (pp. 165–186).

Zhang, K., Liang, X., Zhang, Y., Liu, X., Cao, Q., Zhu, Y., & Tian, Y. (2023). Unveiling the environmental and socioeconomic benefits of precision nitrogen management for paddy fields in subtropical China. *European Journal of Agronomy, 142*, 126663.

Zhang, Q., Yi, B., Gu, C., & Di, H. (2019). Effects of different fertilization regimes on nitrogen and phosphorus losses by surface runoff and bacteria community in a vegetable soil. In *SSSA International Soils Meeting* (2019). ASA-CSSA-SSSA.

CHAPTER 5

Millets: A Promising Small-Grained Crop for Ensuring Nutritional Food Security in the Era of Climate Change

PRIYANKA SHARMA,[1,*] MONISH ROY,[2] SHARMILA DUTTA DEKA,[3] KISHORE KUMAR SHARMA,[4] PURNA KANTA BARUA,[4] SUMIT SOW,[5] HARUN GITARI,[6] and SHIVANI RANJAN[5]

[1]AICRP on Seed (Crops), Department of Plant Breeding of Genetics, Faculty of Agriculture, Assam Agricultural University, Jorhat, Assam, India

[2]Department of Seed Science and Technology, Faculty of Agriculture, Uttar Banga Krishi Vishwavidyalaya, Pundibari, Cooch Behar, West Bengal, India

[3]AICRP on Crops, Department of Plant Breeding and Genetics, Faculty of Agriculture, Assam Agricultural University, Jorhat, Assam, India

[4]Department of Plant Breeding and Genetics, Faculty of Agriculture, Assam Agricultural University, Jorhat, Assam, India

[5]Department of Agronomy, Dr. Rajendra Prasad Central Agricultural University, Pusa, Samastipur, Bihar, India

[6]Department of Agricultural Sciences and Technology, School of Agriculture and Environmental Sciences, Kenyatta University, Nairobi, Kenya

*Corresponding author

Climate-Resilient Agriculture for Sustainable Crop Production: New Adaptation and Mitigation Strategies. Sanjay Kumar, Shivani Ranjan, Sumit Sow, Smruti Ranjan Padhan, and Dhirendra Kumar Roy (Eds.)

DOI: 10.1201/9781998511150-5

ABSTRACT

Poverty and hunger, particularly among populations whose diets consist mainly of cereals and lack micronutrients, pose a serious global threat. Millets possess unique nutritional properties, including complex carbohydrates, dietary fiber, phenolic compounds, and phytochemicals with therapeutic benefits. They also naturally contain essential nutrients such as iron, zinc, and calcium, making them a viable solution to malnutrition. Climate change has become a major factor limiting agricultural yields, especially in grain production. In this context, millets are increasingly recognized for their climate resilience compared to other grains. The adverse effects of climate change, though variable across regions, are particularly severe in developing countries like India, where food insecurity is a major concern. Millets contribute to climate change mitigation due to their lower carbon footprint—substantially less than that of wheat and rice, which emit 3968 and 3401 kg of CO_2 equivalents per hectare, respectively. Known as "nutri-cereals" or "smart foods," millets are well adapted to a range of climatic conditions and thrive in resource-constrained environments with limited water. As such, they support sustainable food systems and offer a practical alternative to conventional cereals. Promoting millet cultivation is essential to addressing hunger, improving nutrition, and enhancing food security in a changing climate.

5.1 INTRODUCTION

5.1.1 MILLETS: THE FAMINE SUPER CROP

Millets are a type of small-grained cereal food crop that is extremely nutrient-rich and can be grown on marginal soils with little to no inputs of pesticides and fertilizers. These crops significantly increase food security. The majority of millet crops are indigenous to India, and they are referred to as "nutri-cereals" since they are abundant in the majority of nutrients that the body needs in order to function effectively. Millets are crops that require rainfall to adapt well in regions with low precipitation, giving them a bigger role in sustaining agriculture and ensuring global food security. The millets are divided into major millets and minor millets based on the area under cultivation and the size of the grains. The two most common millets are pearl millet (*Pennisetum glaucum*) and sorghum (*Sorghum bicolor*). Other types of minor millets include small millet, commonly known as finger millet (*Eleusine coracana*), kodo millet (*Paspalum scrobiculatum*), barnyard millet (*Echinochloa frumentacea*), proso millet (*Panicum miliaceum*), and browntop millet (*Urochloa ramosa*). These crops are also known as famine crops because they provide yield under unfavorable climatic conditions. Because they were the last to be considered for cultivation, these crops were previously referred to as orphan crops due to poor market demand and lower earnings than other crops. However, these underutilized crops are crucial for the livelihood, food, and nutritional security of the underprivileged around the world, as well as for diversifying our food supply. The details regarding the different kinds of millets are given below:

1. **Sorghum:** It is commonly known as jowar, is the fifth-largest cereal food crop in the world, in terms of area and production, compared to other major cereal crops. The stover that is left over after the grains are collected is extremely valuable as nutrient-rich animal feed, even though sorghum grains are mostly consumed in India as food. The crop may be produced in a variety of environmental conditions and is naturally drought-tolerant. Since the seed and biomass are both highly valued for use in human and animal nutrition, sorghum is a dual-use crop in the semi-arid region. Sorghum is a fodder crop that is grown specifically for animal nutrition and is preferred over other grain fodders, especially in the North Indian states. It is widely grown throughout northwestern, western, and central regions, as well as the southern peninsula and a few states in the northeast, with Maharashtra and Karnataka having the largest acreages.

2. **Pearl Millet:** The sixth-largest crop by area and production, pearl millet, sometimes referred to as bajra, has the highest potential for drought resistance of all the millets. More than 90% of the pearl millet produced in the country is produced in the northern Indian states, specifically in Rajasthan, Maharashtra, Gujarat, Uttar Pradesh, and Haryana. It is a tall, upright plant that grows up to 15 feet in height. The plant produces a brownish flower with a thick panicle resembling a spike. With the exception of sorghum, which can be virtually white, pale yellow, brown, grey, slate blue, or purple in color, pearl millet contains the largest kernels of any type of millet, measuring 3–4 mm in length. Due to its

high mineral and protein content, pearl millet has several health benefits.

3. **Finger Millet:** It is a staple grain, particularly for rural people in South and East India as well as Central Africa. Ragi, also known as finger millet, can be produced in a variety of climatic conditions, including those at sea level and in the Himalayan foothills, but it grows best on loamy, well-drained soils. Finger millet is a tiny, rapidly tillering plant with characteristic terminal inflorescences that resemble fingers. Mature plants have tiny, mustard-like seeds that can range in color from white to light brown to dark brown and resembles a height of 30 cm–150 cm. Depending on the cultivar and growth conditions, the crop matures in about 3–5 months. The crop has a deep root structure and is adapted to reasonably regular rainfall conditions. The grain has outstanding malting qualities and is particularly nutrient-rich. As the grains may be stored for up to 45–50 years without significantly deteriorating due to storage insect pests, finger millet is an advantageous crop as a source of food during famine.

4. **Foxtail Millet:** The third-largest millet crop, foxtail millet is grown for food as well as for forage in semi-arid tropical regions of Asia, Europe, North America, Australia, and North Africa. Although China a country where the majority of the crop is grown, Italian millet is also cultivated in Russia, Japan, and India. It develops into a thin, upright, leafy stem that can grow upto a height of 1–5 feet. The seeds are produced in a compact panicle that resembles a large, yellow or green foxtail. The structure of the grain is quite similar to that of paddy rice. In order to use them as food, the outer husk must be removed. It is typically grown on black cotton soils, with sorghum serving as an alternative crop when rainfall is low, and can endure drought, developing at higher elevations of above 600 feet and withstanding higher altitudes. It also thrives in soils that are clayey, loamy, or alluvial.

5. **Proso Millet:** A short-season grain called proso millet is grown in arid parts of Asia, Africa, Europe, Australia, and North America. Proso millet is a highly drought-resistant crop, growing in areas with limited water supplies and extended dry spells. Millet is a short-season crop, maturing 60–75 days after sowing. It is frequently grown as a late-sown summer crop. Commonly, it is

also known as broom corn millet, which has got its name from the way that its small panicle has a droopy tip that resembles an old broom. Its round seeds have a smooth, shiny shell that is about 1/8 inch across. The color of the seeds might be white, brown, orange-red, or yellow.

6. **Barnyard Millet:** In India, China, Japan, and Korea, barnyard millet is cultivated for food and fodder. Even though it takes three to four months to grow, it flourishes in all types of weather and higher elevations. Barnyard millet is an upright plant, 60–130 cm tall, with brownish-purple spikelets. In Eastern Asia, particularly in the hilly districts of Uttarakhand, India, some sections of Africa, and the Eastern United States, it is produced for both grain and feed.
7. **Kodo Millet:** Even though kodo millet is rarely grown anywhere else, it is commonly planted in India on the poorest soils. It may reportedly thrive on soils endowed with stones and gravels, where other crops cannot grow. It has a maturation period of about four to six months, compared to two to four months for the other millets. The grain is encircled by difficult-to-remove hard, corneous husks.
8. **Little Millet:** Little millet looks similar to proso, but it is shorter, has smaller panicles and seeds, and is cultivated on a small scale in underdeveloped areas. Little millet grows quickly and can resist both dry conditions and standing water.
9. **Brown Top Millet:** Although it is a widespread weed throughout the country, brown top millet, an indigenous crop of India, is only grown in a small region in the states of Karnataka and Andhra Pradesh. It is the most affordable crop to plant, requires little weeding, and has no major pests or diseases. It matures in 60–80 days and thrives in sandy loam soils that are less productive. It grows more quickly and has finer stems in comparison to pearl millet.

5.2 DIVERSITY OF RACE

Races and subraces of little millets are distinguished based on comparisons between the morphologies of accessions and show substantial intraspecific variation. There are two subspecies of *Eleusine coracana*: the cultivated *coracana* and the wild *africana*. The *spontanea* and the *africana* are the two wild races that make up the subspecies *africana*, whereas the *elongata*, *plana*, *compacta*, and *vulgaris* are the four domesticated races of the subspecies *coracana*. These cultivated races are further divided into the races *vulgaris*, *liliacea*, *stellata*, *incurvata*, and *digitata*; *elongata*, *laxa*, and *sparsa*; and *plana*, *seriata*, *confundere*, and *grandigluma*. According to Rao and de Wet (1997), there are no subraces within the *compacta* race.

Based on the comparative morphology of the foxtail millet accessions, Rao et al. (1987) hypothesized three races of foxtail millet: *Indica*, *Maxima*, and *Moharia*. The *aristata*, *fusiformis*, and *glabra* of the *Moharia*; the *compacta*, *spongiosa*, and *assamense* of the *Maxima*; and the *erecta*, *glabra*, *nana*, and *profusa* of the *Indica* are the ten subraces of these races.

Proso millet can be divided into five races, according to de Wet (1986): *miliaceum*, *patentissimum*, *contractum*, *compactum*, and *ovatum*. De Wet et al. (1983a) opined that small millet has two species, *nana* and *robusta*, each of which has two subspecies, namely *laxa* and *erecta* in *nana* and *laxa* and *compacta* in *robusta*.

The species *Echinochloa crus-galli* is divided into two subspecies: *crus-galli* and *utilis*. Additionally, subsp. *utilis* includes two races, *utilis* and *intermedia*, while subsp. *crus-galli* comprises the *crus-galli* and *macrocarpa* races. The species *Echinochloa colona* has two subspecies: *colona* and *frumentacea*. According to de Wet et al. (1983b), subsp. *frumentacea* includes four races: *stolonifera*, *intermedia*, *robusta*, and *laxa*, whereas subsp. *colona* has no recognized races.

5.3 A HISTORICAL VIEW ON MILLETS WORLDWIDE

Native populations in Asia and Africa domesticated millet crops, which were later spread to new parts of the world. Additional domestication in other regions produced secondary areas with diversity, adaptation, and a variety of potential applications. Below is a brief discussion of the origin of each millet:

1. **Sorghum (*Sorghum bicolor*):** Commonly known as a C_4 plant, is one of the crops with the lowest water and sunlight energy requirements for the production of food and biomass. The continent of India is the secondary origin of sorghum, with evidence of early cereal production there extending back 4500 years. For more than

500 million people, sorghum is a basic grain, and because of its diverse applications, it also provides an opportunity for income generation for farming communities worldwide.

2. **Pearl Millet:** Scientifically known as *Pennisetum glaucum,* pearl millet originated in West Africa and is the crop which is widely cultivated in India after major cereals. Around 4500 BC, pearl millet was domesticated in northern and central Sahelian Africa. Pearl millet provides food and income for 90 million people living in poverty. They usually live in the arid parts of Africa and Asia, where most other crops cannot thrive, and the indigenous farming populations have no other way to ensure their food supply.

3. **Finger Millet:** Particularly for rural people in East and Central Africa and Southern India, *Eleusine coracana*, also referred to as finger millet, is a staple food. It appears that the highlands of Ethiopia or western Tanzania are the regions where finger millet was first occured. It is a domesticated grain with African origins that made its way to Asia in ancient times. Upland races appear to represent a secondary adaptation in Asia, where they are particularly common from India to Nepal and southern China in the Himalayas. Finger millet has a substantial impact on the nutritional requirements and earnings of several rural communities in South Asia, Eastern Africa, and Southern Africa.

4. **Barnyard Millet:** In Eastern Asia, sections of Africa, and India, particularly in the mountainous regions of Uttarakhand, barnyard millet is produced for both grain and fodder. It has been a valuable grazing crop in the eastern USA. The history of barnyard millets is mostly unknownand therefore, majority of the Scientists agree that these evolved from Central Asia, while *Echinochloa crus-galli* was cultivated in Southeast Asian countries, and *Echinochloa frumentacea* was first domesticated in India.

5. **Proso Millet:** A short-season crop (botanically known as *Panicum miliaceum*) is grown in arid regions of Asia, Africa, Europe, Australia, and North America. Because variation grows towards Mongolia, China, and Eastern Asia, the data suggests an origin in Central or Eastern Asia. It has been proposed that the Yellow River valley in China is the source.

6. **Foxtail Millet *(Setaria italica):*** It has a grain structure that is strikingly similar to paddy. Despite having moderate drought tolerance, foxtail millet can occasionally evade droughts due to its early maturity. It can be used as a short-duration catch crop due to its rapid growth. It can endure a range of climatic conditions, soil types, and elevations but cannot withstand standing water.
7. **Little Millet:** The origin of little millet is not widely known, with the exception of its likely Indian endemicity and the fact that it has names in all of India's regional vernacular languages. The lack of variation and the absence of comparable wild species outside of India point towards its Indian origin. It was also cultivated in neighbouring countries like Sri Lanka and India.
8. **Kodo Millet** (*Paspalum scrobiculatum*): It is a tropical African species that has been domesticated in India for approximately 3000 years. It is also known as Native Paspalum, Indian Crown Grass, cow grass, rice grass, and ditch millet. In contrast to *Paspalum scrobiculatum var. commersonii*, the wild variety native to Africa, *Paspalum scrobiculatum var. scrobiculatum* is cultivated as an important crop in India.
9. **Brown Top Millet:** Indian origin brown top millet (*Urochloa ramose*) is cultivated only in a small portion of Andhra Pradesh and Karnataka, while it is found as a weed in all the other states of India.

5.4 GLOBAL STATUS OF GENETIC RESOURCES

The scientific community has access to several small millet germplasm accessions. Over 46,000 foxtail millet, 37,000 finger millet, 29,000 proso millet accessions, 8000 each of barnyard and kodo millet, and 3000 accessions of tiny millet germplasm accessions have been kept *ex situ* in global gene banks. India has the largest accessions of finger millet germplasm, and many African countries, particularly Kenya, Ethiopia, Uganda, and Zambia (Chandrashekar, 2010). The main gene banks in China, India, France, and Japan have the greatest foxtail millet germplasm accessions collections (Vetriventhan et al., 2015). China, India, the Russian Federation, and Ukraine were the countries where the majority of the germplasm accessions for proso millet were collected. Japan and India both collected barnyard millet. As per the researchers report, minimum quantity of kodo millet was harvested in India and the USA (Vetriventhan et al., 2015). Germplasm is a resource for creating high-yielding cultivars that are resistant to biotic and abiotic stress. Germplasm accessions can be used in breeding operations to enhance small subsets of a few 100 germplasm lines that cover the full range of millets. Similar to the core and mini-core collections, these diversity sets can be usefully studied to identify germplasm trait-specific sources for their improved use in developing high-yielding cultivars with a range of adaptations.

5.5 INDIAN AND GLOBAL MILLETS PRODUCTION SCENARIO

Since India is considered as largest producer of millets globally, accounting for a yield of 1,247 kg/ha and cultivation spread across approximately 21 states, the total millet production amounts to 15.53 million tons annually. Sorghum is the fourth-largest food crop in India in terms of acreage (3.84 million hectares) and production (4.31 million metric tons [MT]), following rice, wheat, and maize. Although millets are grown on more than 50% of the cultivated area, the area covered for *bajra* (pearl millet) production is only 7.05 million hectares (ha).

India is the largest producer of barnyard millet (99.9%), finger millet (53.3%), kodo millet (100%), little millet (100%), and pearl millet (44.5%), covering an area of 8.87 million hectares. The cultivation of millets has expanded recently in the states of Gujarat and Madhya Pradesh. However, Gujarat (1,762.05 kg/ha), Madhya Pradesh (1,729.70 kg/ha), Haryana (1,906.78 kg/ha), Tamil Nadu (2,153.22 kg/ha), and Andhra Pradesh (2,626.58 kg/ha) contributed the highest yields (Sood, 2022), demonstrating higher output levels compared to other states like Gujarat and Andhra Pradesh.

According to the Food and Agriculture Organization (FAO), 74.00 million hectares were used to produce 89.17 million MT of millets globally in 2020. More than 90% of the world's millet production comes from the two main millet crops, sorghum and pearl millet, while only a small proportion is contributed by finger millet, foxtail millet, proso millet, barnyard millet, small millet, and kodo millet. Additionally, foxtail millet is the most productive of all millets, yielding approximately 2,166 kg/ha. According to Sood (2022), finger millet (1,623 kg/ha), proso millet (1,535 kg/ha), sorghum (1,426 kg/ha), barnyard millet (1,034 kg/ha), pearl millet

(850 kg/ha), little millet (469 kg/ha), and kodo millet (419 kg/ha) are the main minor millets that have contributed primarily in terms of productivity.

Between 2010 and 2020, sorghum production fluctuated between 60.18 million and 58.70 million MT, while the acreage remained relatively stable between 42.16 million and 40.98 million hectares (Sood, 2022). Although production declined over this decade, the area planted with other millets also decreased from 36 million hectares in 2010 to 33.02 million hectares in 2020 (Das et al., 2019), leading to a reduction in production from 32.79 million MT to 30.46 million MT. However, the decreasing area under millet cultivation is a serious concern that can be addressed through policy measures such as promoting millet cultivation and establishing millet processing facilities.

5.6 REVIVING MILLETS FOR BENEFITS TO HEALTH AND DIET

Millet grains are a better option than cereal grains because they contain significantly higher amounts of B complex vitamins, iron, manganese, zinc, calcium, potassium, magnesium, and other minerals. The nutrients glucose, protein, dietary fiber, and good quality fat are found in abundance in them as well (Rao et al., 2017). Additionally, millets include beneficial phytochemicals such as ferulic acid, caffeic acid, and quercetin, as well as lignans, glucans, inulin, sterols, and feraxans. According to studies, polyphenols are essential for preventing diseases including diabetes, high blood pressure, Parkinson's disease, inflammatory illnesses, cardiovascular disease, and cancer (Pei et al., 2022). They also have anti-carcinogenic, anti-inflammatory, antiviral, and neuroprotective properties. Due to their phytochemical content, millets are also thought to offer antibacterial and DNA damage protective properties (Aknabi et al., 2019). According to the IBEF, the millet grain has a very high content of non-starch polysaccharides and dietary fiber, both of which aid in controlling weight. As millets release glucose gradually, they are a great option for persons suffering from diabetes (Dayakar et al., 2017).

Millet is a superior source of fiber and a carbohydrate that digests slowly in the stomach, which is home to billions of bacteria including *Lactobacillus acidophilus, rhamnosus GG, Actinobacteria*, and *Bifido* species. Excellent sources of dietary fiber are millets because the non-starch polysaccharides in them degrade resistant starch into short-chain fatty acids (Bajaj et al., 2021). Due to the growth of gram-negative bacteria that is promoted during the fermentation of millets using a range of cultures, millets are a powerful probiotic meal in the stomach (Sood, 2022). The high fiber content of millets contributes to their lower glycemic index. Even if the degree of food processing, food matrix, ingredient makeup, food shape (grain or flour form), and so forth, affect the glycemic characteristics of various foods, the majority of millet dishes can be prepared to have a low glycemic index (Geetha et al., 2020). Comparative data on the nutritional properties of millets and rice, including wheat, are presented in Figure 5.1. Key values include the following: calcium content of 36.4 mg in ragi (i); fat content of 5.43 g in pearl millet (ii); protein content of 12.50 g in proso millet (iii); dietary fiber of 11.49 g in pearl millet (iv); magnesium content of 153 mg in proso millet (v); zinc content of 3 mg in barnyard millet (vi); iron content of 5 mg in barnyard millet and 6.42 mg in pearl millet (vii); and folic acid content of 39.49 µg in kodo millet and 39.42 µg in sorghum.

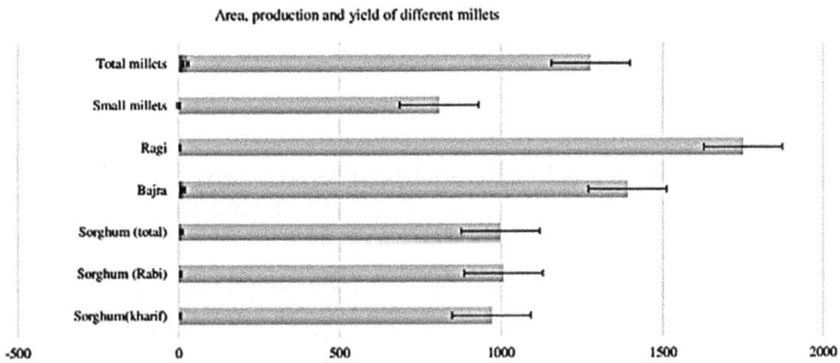

	Sorghum(kharif)	Sorghum (Rabi)	Sorghum (total)	Bajra	Ragi	Small millets	Total millets
■ Area (m ha)	1.76	3.07	4.83	7.55	1.01	0.459	13.83
■ Production (m tons) *	1.58	2.73	4.31	9.22	1.67	0.33	15.53
■ Yield (kg per ha)	967	1002	989	1374	1747	809	1248

■ Area (m ha) ■ Production (m tons) * ■ Yield (kg per ha)

FIGURE 5.1 Graph indicating the amount of land cultivated, millets produced, and their yield.

Source: Modified from Final Estimates, DES, GOI (2021–2022).

Millets also aid in the body's detoxification process by eradicating dangerous pollutants, free radicals, and the enzymes in the organs that neutralize them. They can help prevent many diseases like cancer and heart disease. Numerous nutrients, including trace minerals like zinc, copper, selenium, and iron, as well as vitamins like A, B6, B12, C, D, and E, have been shown to play important roles in boosting the human immune system and decrease the risk of infections. By acting as a vasodilator and lowering blood pressure, millet's high magnesium and potassium content reduces the risk of heart disease and stroke, especially in persons with atherosclerosis. One of the finest ways to protect cardiovascular health is to improve the circulatory system. Millets contain significant amounts of lignans, which the body can absorb and convert to animal lignans. Several chronic diseases, such as cancer and heart disease, have been shown to be mitigated by animal lignans (Peterson et al., 2010). Consuming millet products can reduce insulin resistance by 43% and lower fasting glucose by 32% (Muthamilarasan and Prasad, 2021). Magnesium is also regarded as an essential element for increasing the activation of the body's insulin and glucose receptors, which aids in the management of diabetes. In populations divided between diets containing or excluding magnesium, a 30% decline in diabetes has been observed (Carlson et al., 2018). Due to their high dietary fiber content, which helps in increasing the benefits of "good cholesterol" (HDL) while removing detrimental "bad cholesterol" (LDL) from the body, they are also very helpful for lowering cholesterol. According to Rodriguez et al. (2020), the high fiber and mineral content of millet develops issues issues such as constipation, gastritis, bloating, and cramps. Regulation of the gastrointestinal system improves nutrient retention while also helping in kidney and liver function optimization, which impacts the immune system due to the close relationship between these organs and the body's metabolic processes (Annor et al., 2017). Recent research has shown that eating a diet rich in dietary fiber can significantly lower the risk of breast cancer (Carlson et al., 2018). For resource-constrained farmers in underdeveloped nations, millets are essential for ensuring access to food, nourishment, and revenue generation. Very little scientific research has been conducted on the development of genetic resources and breeding to enhance yield under stress conditions, and other "omics" resources have also been discussed (Bandyopadhyay et al., 2017; Kumar et al., 2018). To assure the amount and quality of food required to feed more than 7 billion people, several multifaceted approaches are required. Many of them have previously experienced hardship as a result of health and weather shocks (Bisoffi et al., 2021; Fróna et al., 2019). For example, the Indian government previously prioritized improving farmers' incomes over agricultural yield and ensuring nutritional security in addition to food security. The government already has changing consumer behavior with appropriate programs and incentives on its agenda (Bisoffi et al., 2021).

5.7 CLIMATE CHANGE RESILIENCE

Climate change and the increasing number of people on the globe are two major causes of suffering (Fróna et al., 2019). The interactions between meteorological, hydrological, ecological, and agricultural systems may change as a result of climate change, and this could have a negative impact on their viability, according to Rosenzweig et al. (2001); Obwocha et al. (2022). Global agricultural production is predicted to decline as a result of water stress brought on by increasing temperatures and higher evaporation caused by climate change, as well as changes in the intensity and distribution of rainfall patterns (Fujimori et al., 2019). Food production is severely impacted by climate change and crop production, especially for essential cereals like rice, wheat, and maize, which directly affects both the amount and quality of these resources, particularly rice, wheat, and maize. The threat of climate change and global warming persists even though some of these causes of hunger can be addressed. As a result, the number of people who suffered due to hunger and malnutrition has marginally decreased from nearly 1 billion in 1990–1992 to 850 million in 2010–2012 (FAO, 2012). Due to the declining food production and the additional strain of feeding a population that will reflect more than 9 billion by 2050, forecasts state that 2–3 billion population will face hunger in the near future (Wheeler and Von, 2013). Climate change and increasing global average temperatures are considered to have a direct impact on crop yields, agricultural output, and overall food system sustainability. Some estimates claim that while certain locations may experience improved production as a result of climate change, this won't be enough to feed the growing world population. Furthermore, the majority of scientists agree that the current rates of global warming and greenhouse gas emissions would significantly reduce crop productivity. Therefore, lowering greenhouse gas emissions is crucial for maintaining global temperature regulation and food security. Higher emissions are often the outcome of intensive agricultural practices, which are used throughout the world (Olesen and Bindi, 2002). Natural and man-made disasters have detrimental effects on agriculture and the food system collectively, putting the ability to satisfy the increasing demand of nutritional and energy needs at risk (Mishra et al., 2021). It is even more crucial to adapt to climate-resilient agricultural practices to deal with natural and human calamities (Bisoffi et al., 2021). Due to their potential to increase nutrient rich food sources and diversify crop production, "famine crops" are one potential solution to these issues (Ye and Fan, 2021). Since they are non-commodity crops that are not widely commercialized and are cultivated in confined areas, many edible species of "famine crops" receive less importance (Ribaut and Ragot, 2019). Under the scenario of a changing climate, famine crops outperform major crops as a result of their unique qualities, which include diversified adaptation on marginal soils, low water needs, tolerance to biotic and abiotic stressors, and greater nutritional content compared to other key grain staples (Varshney, 2021). Sorghum and millet are often regarded as important climate-resilient crops by agronomists. This assertion is typically based on the fact that these crops are more productive in tropical regions than other cereals (such as soybean, rice, and maize) because of their drought-tolerant characteristics (Nciizah et al., 2021). The fact that these crops may be kept in storage for a considerable amount of time without deteriorating strengthens the case for their climate resilience (Mukarumbwa and Mushunje, 2010). In addition, it has been demonstrated that they can withstand high temperatures, making them climate-resilient crops for semiarid and arid regions with little rainfall (Dhankher and Foyer, 2018). Developing resilient agricultural systems is a low-cost and effective way to tackle the rising risk of droughts and other climate disasters (McDermid et al., 2014). To reduce yield losses brought on by climate change, many adaptation measures have been developed (Challinor et al., 2014). The core of this strategy is the identification and mainstreaming of crops that are currently underutilized or neglected (Wimalasiri et al., 2022). Underutilized crops, especially millets, can contribute to the diversification and resilience of agroecosystems, which may help ensure the security of the global food supply. Millets are often grown in low-input systems and are adaptable to marginal locations. Therefore, investments in research should support the development of evidence bases for these crops (Mabhaudhi et al., 2019). A lot of effort needs to be done to predict the appropriateness and responsiveness of neglected and underutilized crops to climate and economic situations, yet yield forecasting is gradually becomming popular (Jahansiri et al., 2022). Crop simulation modeling permits the evaluation of the potential of crops and crop varieties in diverse geographic areas under varying climates without the need for extensive agronomic studies, which can be expensive and time-consuming. In order to accurately simulate the production of undervalued and neglected crops, cutting-edge crop modeling methodologies and supporting datasets have been developed (Wimalasiri et al., 2021).

5.8 DROUGHT RESISTANCE

Millets are a dependable option in areas with unpredictable rainfall patterns or water restrictions since they are highly resistant to water scarcity and can endure extended droughts. Abiotic stresses, such as drought, salt, flooding, and severe temperatures (cold, frost, and heat), are the primary variables that restrict production in semi-arid and arid environments. Millets are more resistant to environmental stresses than major grains because of differences in their morphophysiological, molecular, and biochemical properties. Drought stress will continue to be a severe abiotic limitation on food production, according to climate change models (Simmons et al., 2020). Millets are the best alternative in semi-arid and dry regions since they can naturally resist droughts. Drought has been demonstrated to negatively affect the physiological growth, yield, membrane integrity, pigment, osmotic adjustment, water relations, and photosynthetic activity of millets (Veach et al., 2020). They also possess a highly beneficial C_4 photosynthetic system, which gives them the ability to survive both high and low temperatures. Plants can keep their stomata closed all day long to minimize water loss due to the C_4 mechanism, which increases the concentration of CO_2 in the bundle sheath and stimulates photorespiration (Lundgren et al., 2014). According to Wang et al. (2012), when growing in warm regions, the efficiency of millets in water and nitrogen consumption is much higher than in C_3 plants. Millets also gain secondary advantages from C_4 photosynthesis, improvements in ecological performance at warm temperatures, more adaptable biomass allocation patterns, and a decline in hydraulic conductivity per unit leaf area (Lundgren et al., 2014). In the context of a changing environment, these examples present opportunities for the promotion of these crops to a higher degree of production through a number of metabolic processes, such as the control of reactive oxygen species (ROS), the activation of ROS-scavenging enzymes (catalase and superoxide), and the generation of proteins related to stress.

Climate change is expected to cause a 15%–20% global decline in the output of major grains over the next few decades, according to CGIAR forecasts. To feed the 9 billion people by the year 2050, global agriculture needs to increase productivity by 70%. In that context, it is significant to note that millets have demonstrated an outstanding growth in productivity over the last five decades, in contrast to a trend of variable productivity for key food crops. As agriculture in India is heavily dependent on the monsoon, millets, also known as "famine reserves," are vital. This is because they can be stored for an extended period of time and are easy to do so. They are best suited for mixed and intercropping, providing farmers with secure access to food and a means of subsistence.

5.9 BENEFITS OF BIODIVERSITY FOR THE ENVIRONMENT

Through a number of processes, millet farming can significantly contribute to biodiversity conservation. Millets are a group of small-seeded, drought-tolerant cereal plants that have been cultivated for thousands of years and have a number of advantageous effects on the environment. Here are some ways that millet farming can help to conserve biodiversity.

5.9.1 CROP DIVERSITY

Millets help in enhancing crop diversity by reducing dependency on a small number of important staple crops. Farmers may build a more resilient and diverse agricultural landscape that is less vulnerable to pests, diseases, and the effects of climate change by cultivating a diversity of millet species.

5.9.2 PRESERVING TRADITIONAL KNOWLEDGE

Traditional agricultural methods and information that have been passed down through the years are frequently used when cultivating millets, preserving traditional wisdom. Farmers preserve local knowledge of agroecological systems and seed-saving methods, which are frequently well-adapted to particular ecological circumstances, by upholding these practices.

5.9.3 LESS WATER INTENSIVE

Millets are typically more drought-resistant and use less water than certain other main cereal crops, such as rice and wheat. Farmers can reduce water use, save water supplies, and contribute to the preservation of aquatic habitats by utilizing less water in millet production areas that could otherwise be depleted by growing millets in water-scarce areas.

5.9.4 HABITAT CONSERVATION

Millets frequently grow well in marginal lands and infertile soils, which is suitable for habitat conservation. Farmers can reduce the need to expand into natural habitats by growing millets on these sites, assisting in the effort to prevent the conversion of biodiverse places like forests and wetlands to agriculture. These crops also have the benefit of being "good for the planet" since they can survive in barren, infertile soils that cannot be adapted by other cereal crops (environmentally sustainable), in addition to being nutritious, healthy, and climate-resilient, which is considered to be good for the individual and farmer (Kane-Potaka et al., 2021).

5.9.5 SUPPORTING POLLINATORS

Millets can serve as important sources of feed for pollinators like bees and butterflies. Millet production indirectly promotes the conservation of pollinator populations, which is vital for preserving biodiversity in agricultural environments, by offering an additional food supply for these significant insects.

5.9.6 REDUCED CHEMICAL INPUTS

Millets require less chemical input, such as insecticides and herbicides, because they are typically more resistant to disease and pest attacks. By using fewer agrochemicals, it will be possible to avoid harming non-target species like beneficial insects, birds, and soil microbes.

5.9.7 GENETIC DIVERSITY

Millets themselves have a significant genetic variety due to their many kinds and landraces. Farmers can aid in preserving and enhancing genetic diversity, which is necessary for developing more robust crop types that can withstand changing environmental conditions, by encouraging the cultivation of a variety of millet kinds.

5.9.8 CONSERVING WILD RELATIVES

Wild relatives of millet are frequently found growing in barren lands. Farmers indirectly support the preservation of these wild relatives, which may serve as a priceless genetic resource for future crop enhancement initiatives, by continuing to cultivate millet. In general, encouraging millet production can enable resilient and biodiverse agricultural systems that protect natural ecosystems while also supporting human livelihoods. It is crucial to make sure that the farming practices are carried out with consideration for the regional ecological contexts, with respect to indigenous knowledge, and use of sustainable agricultural methods.

5.10 BUILDING RESILIENCE AGAINST DISEASE AND PEST INFESTATION IN MILLETS

The cultivation of millet is severely constrained by insect pests, which results in significant production losses and hinders the food security of vulnerable people. Stem borers, shoot flies, ear-head bugs, and aphids are a few typical insect pests that damage millets. Millet crops can also suffer from diseases like blast, rust, smut, and downy mildew. Chemical pesticides have been commonly used as traditional pest management techniques, but they come with their own set of environmental and health risks. Along with the previously mentioned benefits, Chazovachii et al. (2010) found that compared to other grains, millet and sorghum are less vulnerable to pests and illnesses during pre-harvest and post-harvest storage. Additionally, they mature apparently quickly and require the least amount of input, according to Saleh et al. (2013). Farmers and agricultural researchers are concentrating more on developing and promoting millet varieties with improved insect pest and disease resistance in an effort to handle these issues sustainably. Scientists are working to develop millet varieties that can withstand and recover from pest and disease attacks using both conventional breeding techniques and cutting-edge biotechnological approaches, thereby reducing the dependency on chemical inputs and minimizing the environmental impact. Identifying naturally occurring resistance genes within millet germplasm or closely related species is one of the main techniques used in breeding for resistance. Researchers can improve the capacity of the crop to reduce pest and disease issues naturally, without the need for outside intervention. This can be done by introducing these genes into sensitive species. Furthermore, it is critical to investigate alternate methods for sustainable pest management as climate change continues to affect pest populations and the spread of disease. Integrated pest management strategies provide a thorough and environmentally beneficial way to conserve millet crops by integrating several pest management techniques such as cultural practices, biological control agents, and natural enemies with resistant cultivars. Scientists and researchers can strengthen millet crops against the risks of pests and diseases by utilizing the inherent defense mechanisms present in millets and embracing novel techniques, providing a more secure and resilient future for millet-based agriculture.

5.11 IMPROVEMENT OF MILLET THROUGH BIOTECHNOLOGICAL METHODS INCLUDING TISSUE CULTURE AND GENETIC ENGINEERING

Plant biotechnology approaches have become an essential complement to conventional breeding methods for integrating new genes into current crop types and accelerating genetic development. The majority of cereal and grass species have had success with *in vitro* cell culture establishment and plant regeneration, which has shedlighted interest in other, less-studied crop species, particularly millets, which are crucial for resolving the food and forage issues for several countries (Yemets et al., 2013). The genetic quality of millets could be improved using both traditional methods and cutting-edge technologies like genetic engineering and transgenics. The principal advantages of regeneration and transformation in millet improvement include: (i) the replication of identical copies of pest- and disease-free plants; and (ii) the regeneration of the entire plant from changed tissues with desirable features. Therefore, in order to improve transformation efficiency for various millet kinds, the regeneration process needs to be optimized (Wuthrich and Tadele, 2012). The first millet regeneration research for proso, finger, pearl, and kodo millets was carried out in the 1970s (Rangan, 1973, 1976) through somatic embryogenesis or organogenesis to regenerate plants. Millets seldom undergo organogenesis; only finger millet and pearl millet have done so to date (Jha et al., 2009). Because an effective regeneration technique is required for optimal transformation in cereal crops, *in vitro* regeneration is a crucial part of the transformation process. Pathogen resistance was one of the features that several of these modifications targeted (O'Kennedy et al., 2011a). The right kind of explant and the right medium composition must be optimized in order to establish an effective regeneration system. Compared to other significant crops like wheat and rice, the regeneration of the millet plant has not made much progress. When selecting the right explant, it is important to take the explant's capacity for regeneration into account because different genotypes of the same species exhibit significant variation in this capacity. Significant efforts are needed to be undertaken to research suitable regeneration strategies (Wuthrich and Tadele, 2012).

5.12 INITIATIVE IN ENCOURAGING MILLET CULTIVATION

Over the past few years, the Indian government has been shedlighted an initiative to promote millet both domestically and internationally. India has established itself as a major producer of millets by encouraging and developing millet farming. The Indian government has been on a mission to promote the branding and production of millets since 2018 in order to solve systemic environmental concerns as well as challenges of nutritional security. With 18% of the world's millets produced and about 20% of the land under cultivation, India can proudly claim to be the world's greatest producer of millets. Several initiatives are currently being carried out under the National Food Security Program to increase millet production in India. The initiatives are comprised in developing the Millet Mission and centers of excellence in several Indian states. In order to increase food security in India, the World Food Program (WFP) and Niti Aayog signed a Statement of Intent in 2021 that emphasized the development of climate-resilient agriculture. The plan would also consider ways to use

India's millet expertise to help other developing countries through the development of knowledge management platforms and the promotion of information exchange.

5.13 TECHNIQUES TO INCREASE THE NATION'S FOOD SECURITY AND NUTRITION

In particular, the economic aspect of farmers' crop choices, which probably contributes to the historical trend towards the cultivation of rice and wheat, needs to be heavily pushed in order to establish the millet ecosystem. This is because the government's MSP and PDS for the National Food Security Programs heavily favor wheat and paddy and do not provide farmers with any incentives to produce other crops. Although millet has long been consumed historically, there has been a recent decline in millet consumption as a result of challenges regarding price, flavor, perception, and availability. For the crop to last longer, it is necessary to increase awareness of its nutritional benefits and build better storage facilities. India's G20 presidency is the perfect time to capitalize on the Year of Millets campaign for global awareness because it has built channels for popularization and knowledge dissemination for the global market. There are scientific ways to get through current obstacles, including the presence of antinutritional ingredients and the low sensory acceptance of millet-based goods. Farmers will be encouraged to cultivate millets since it will open up new options for them and give them renewed energy if millets are used in processed or packaged goods. The current nutrient deficits of protein, calcium, and iron in under developed will be reduced by the addition of millet-based foods in international, national, and state-level feeding program. This will lead to a paradigm shift from food security to nutritional security, with climate change serving as the pivot. It is time to unleash the full potential of millets by educating consumers about their nutritional advantages and influencing their purchasing behavior, a task that can be given to civil societies because they have a large grassroots audience. Therefore, it is imperative to preserve these underutilized crops in order to maintain both agro-diversity and provide food security. To meet the objectives of sustainable development and biodiversity conservation, traditional farming practices must be strengthened (Kumar, 2023).

5.14 CONCLUSION AND FUTURE PROSPECTIVES

The next crop that might withstand climate change is millets, which historically made up the majority of people's diets in less irrigated regions of Asia, particularly India. The Year of Millets campaign has established avenues for popularization and information dissemination for the global market, making the G20 presidency of India the ideal opportunity to leverage its efforts to raise awareness of millets worldwide. This will increase demand for this astonishing super nutritious crop and improve the availability of strong market connections. The addition of millet-based foods in international, national, and state-level feeding programs will address the current nutrient deficiencies of protein, calcium, and iron in poor countries. Therefore, in order to sustain agro-diversity and ensure food security in addition to restoring ecological harmony and public well-being, millet farming must be aggressively promoted while exploiting the "super grain" as a source of wholesome food. Nutri-cereal growers and purchasers should receive incentives, and through increasing consumer knowledge, incentives should also enhance domestic consumption.

KEYWORDS

- **barnyard millet**
- **carbohydrates**
- **carbon dioxide**
- **climate change**
- **drought resistant**
- **famine crops**
- **finger millet**
- **kodo millet**
- **malnutrition**
- **micronutrients**
- **phenolic compounds**
- **phytochemicals**
- **sustainable food systems**

REFERENCES

Akanbi, O. T., Timilsena, Y., & Dhital, S. (2019). *Bioactives from Millet: Properties and Effects of Processing on Bioavailability*. Springer Nature Singapore Pvt. Ltd.

Annor, G. A., Tyl, C., Marcone, M., Ragaee, S., & Marti, A. (2017). Why do millets have slower starch and protein digestibility than other cereals? *Trends in Food Science and Technology, 66*, 73–83.

Bajaj, A., Sangwan, V., Kawatra, A., & Rani, V. (2021). Non-starch polysaccharides (dietary fiber) of bio-fortified pearl millet varieties as influenced by malting. *The Indian Journal of Nutrition and Dietetics, 58*(1), 26111.

Bandyopadhyay, T., Muthamilarasan, M., & Prasad, M. (2017). Millets for next generation climate-smart agriculture. *Frontiers in Plant Science, 8*, 1266.

Bisoffi, S., Ahrné, L., Aschemann-Witzel, J., Báldi, A., Cuhls, K., & DeClerck, F. (2021). COVID-19 and sustainable food systems: What should we learn before the next emergency. *Frontiers in Sustainable Food Systems, 5*, 650–987.

Carlson, J. L., Erickson, J. M., Lloyd, B. B., & Slavin, J. L. (2018). Health effects and sources of prebiotic dietary fiber. *Current Developments in Nutrition, 2*(3), nzy022.

Challinor, A. J., Watson, J., Lobell, D. B., Howden, S. M., Smith, D. R., & Chhetri, N. A. (2014). A meta-analysis of crop yield under climate change and adaptation. *Nature Climate Change, 4*(4), 287–291.

Chandrashekar, A. (2010). Finger millet: *Eleusine coracana*. In *Advances in Food and Nutrition Research, 59*, 215–262.

Das, B., Satyapriya, Singh, P., Bhowmik, A., & Ray, P. (2019). Growth and instability in area, production, productivity and consumption of millets in India: An analysis. *Indian Journal of Extension Education, 55*(4), 158–161.

Dayakar Rao, B., Bhaskarachary, K., Arlene Christina, G. D., Sudha Devi, G., & Vilas, A. T. (2017). Nutritional and health benefits of millets. ICAR-Indian Institute of Millets Research (IIMR), Rajendranagar, Hyderabad, pp. 112.

de Wet, J. M. J. (1986). Origin, evolution and systematics of minor cereals. In A. Seetharam, K. W. Riley, & G. Harinarayana (Eds.), *Small Millets in Global Agriculture* (pp. 19–30). Oxford & IBH Publishing Co. Pvt. Ltd.

de Wet, J. M. J., Prasada Rao, K. E., Mengesha, M. H., & Brink, D. E. (1983b). Domestication of sawa millet (*Echinochloa colona*). *Economic Botany, 37*(3), 283–291.

de Wet, J. M. J., Prasada Rao, K. E., Mengesha, M. H., & Brink, D. E. (1983a). Diversity in kodo millet (*Paspalum scrobiculatum*). *Economic Botany, 37*, 159–163.

Dhankher, O. P., & Foyer, C. H. (2018). Climate resilient crops for improving global food security and safety. *Plant Cell and Environment, 41*(5), 877–884.

Food and Agriculture Organization of the United Nations. (2012). *Technical Note: FAO Methodology to Estimate the Prevalence of Undernourishment*. Food and Agricultural Organization.

Fróna, D., Szenderák, J., & Harangi-Rákos, M. (2019). The challenge of feeding the world. *Sustainability, 11*(20), 5816.

Fujimori, S., Hasegawa, T., Krey, V., Riahi, K., Bertram, C., Bodirsky, B. L., et al. (2019). A multi-model assessment of food security implications of climate change mitigation. *Nature Sustainability, 2*(5), 386–396.

Geetha, K., Yankanchi, G. M., Hulamani, S., & Hiremath, N. (2020). Glycemic index of millet-based food mix and its effect on pre-diabetic subjects. *Journal of Food Science & Technology, 57*(7), 2732–2738.

Jahanshiri, E., Mohd Nizar, N. M., Tengku Mohd Suhairi, T. A. S., Gregory, P. J., Mohamed, A. S., & Wimalasiri, E. M. (2022). A land evaluation framework for agricultural diversification. *Sustainability, 12*, 3110.

Jha, P., Yadav, C. B., Anjaiah, V., & Bhat, V. (2009). *In vitro* plant regeneration through somatic embryogenesis and direct shoot organogenesis in *Pennisetum glaucum* (L.) R. Br. *In Vitro Cellular & Developmental Biology–Plant, 45*, 145–154.

Kane-Potaka, J., Anitha, S., Tsusaka, T. W., Botha, R., Budumuru, M., et al. (2021). Assessing millets and sorghum consumption behavior in urban India: A large-scale survey. *Frontiers in Sustainable Food Systems, 5*, 680–777.

Kang, Y., Khan, S., & Ma, X. (2009). Climate change impacts on crop yield, crop water productivity and food security—A review. *Progress in Natural Science, 19*(12), 1665–1674.

Kumar, A. (2023). Potential prospects for G20, millets: An environmentally sustainable superfood. *Focus Global Reporter*.

Kumar, A., Tomer, V., Kaur, A., Kumar, V., & Gupta, K. (2018). Millets: A solution to agrarian and nutritional challenges. *Agriculture and Food Security, 7*(1), 31.

Lundgren, M. R., Osborne, C. P., & Christin, P. A. (2014). Deconstructing Kranz anatomy to understand C4 evolution. *Journal of Experimental Botany, 65*(13), 3357–3369.

Mabhaudhi, T., Chibarabada, T. P., Chimonyo, V. G. P., Murugani, V. G., Pereira, L. M., Sobratee, N., Govender, L., Slotow, R., & Modi, A. (2019). Mainstreaming underutilized indigenous and traditional crops into food systems: A South African perspective. *Sustainability, 11*(1), 172.

McDermid, S. P., Ruane, A. C., Rosenzweig, C., Hudson, N. I., Morales, M. D., & Agalawatte, P. (2014). The AgMIP coordinated climate-crop modeling project (C3MP): Methods and protocols. In D. Fuccilo, L. Sears, & P. Stapleton (Eds.), *Handbook of Climate Change and Agro-Ecosystems, 3* (pp. 191–220). Imperial College Press.

Mishra, A., Bruno, E., & Zilberman, D. (2021). Compound natural and human disasters: Managing drought and COVID-19 to sustain global agriculture and food sectors. *Science of the Total Environment, 754*, 142–210.

Mukarumbwa, P., & Mushunje, A. (2010). Potential of sorghum and finger millet to enhance household food security in Zimbabwe's semi-arid regions: A review. *Third Conference/AEASA 48th Conference*, Cape Town, South Africa, African Association of Agricultural Economists (AAAE).

Muthamilarasan, M., & Prasad, M. (2021). Small millets for enduring food security amidst pandemics. *Trends in Plant Science, 26*(1), 33–40.

Nciizah, T., Nciizah, E., Mubekaphi, C., & Nciizah, A. D. (2021). Role of small grains in adapting to climate change: Zvishavane District, Zimbabwe. In W. Leal Filho, N. Oguge, D. Ayal, L. Adeleke, & I. da Silva (Eds.), *African Handbook of Climate Change Adaptation* (pp. 581–599). Springer.

Obwocha, E. B., Ramisch, J. J., Duguma, L., & Orero, L. (2022). The relationship between climate change, variability, and food security: Understanding the impacts and building resilient food systems in West Pokot County, Kenya. *Sustainability, 14*(2), 765.

O'Kennedy, M. M., Crampton, B. G., Lorito, M., Chakauya, E., Breese, W. A., Burger, J. T., & Botha, F. C. (2011). Expression of a beta-1,3-glucanase from a biocontrol fungus in transgenic pearl millet. *South African Journal of Botany, 77*(2), 335–345.

Olesen, J. E., & Bindi, M. (2002). Consequences of climate change for European agricultural productivity, land use and policy. *European Journal of Agronomy, 16*(4), 239–262.

Pei, J., Umapathy, V. R., Vengadassalapathy, S., Hussain, S. F. J., Rajagopal, P., Jayaraman, S., Veeraraghavan, V. P., Palanisamy, C. P., & Gopinath, K. A. (2022). Review of the potential consequences of pearl millet (*Pennisetum glaucum*) for diabetes mellitus and other biomedical applications. *Nutrients, 14*, 2932.

Peterson, J., Dwyer, J., Adlercreutz, H., Scalbert, A., Jacques, P., & McCullough, M. L. (2010). Dietary lignans: Physiology and potential for cardiovascular disease risk reduction. *Nutrition Reviews, 68*(10), 571–603.

Prasada Rao, K. E., & de Wet, J. M. J. (1997). Small millets. In D. Fuccilo, L. Sears, & P. Stapleton (Eds.), *Biodiversity in Trust* (pp. 259–272). Cambridge University Press.

Rangan, T. S. (1973). Morphogenic investigations on tissue cultures of *Panicum miliaceum. Zeitschrift für Pflanzenphysiologie, 72*, 456–459.

Rangan, T. S. (1976). Growth and plantlet regeneration in tissue cultures of some Indian millets: *Paspalum scrobiculatum* L., *Eleusine coracana* Gaertn., and *Pennisetum typhoideum. Zeitschrift für Pflanzenphysiologie, 78*(3), 208–216.

Rao, K. E. P., de Wet, J. M. J., Brink, D. K., & Mengesha, M. H. (1987). Intraspecific variation and systematics of cultivated *Setaria italica*, foxtail millet (Poaceae). *Economic Botany, 41*(1), 108–116.

Ribaut, J. M., & Ragot, M. (2019). Modernising breeding for orphan crops: Tools, methodologies and beyond. *Planta, 250*, 971–977.

Rodríguez, J. P., Rahman, H., Thushar, S., & Singh, R. K. (2020). Healthy and resilient cereals and pseudo-cereals for marginal agriculture: Molecular advances for improving nutrient bioavailability. *Frontiers in Genetics, 11*, 49.

Rosenzweig, C., Iglesias, A., Yang, X. B., Epstein, P., & Chivian, E. (2001). Climate change and extreme weather events-implications for food production, plant diseases, and pests. *Global Change and Human Health, 2*(2), 90–104.

Simmons, T., Styer, A. B., Pierroz, G., Gonçalves, A. P., Pasricha, R., et al. (2020). Drought drives spatial variation in the millet root microbiome. *Frontiers in Plant Science, 11*, 599.

Sood, D. (2022). *Millets: The Future Superfood for India.* The Associated Chambers of Commerce and Industry of India, pp. 1–44.

Varshney, R. K. (2021). Mighty millets: Super grains of power. *India Perspectives, 2*, 74–81.

Veach, A. M., Chen, H., Yang, Z. K., Labbe, A. D., Engle, N. L., Tschaplinski, T. J., Schadt, C. W., & Cregger, M. A. (2020). Plant hosts modify belowground microbial community response to extreme drought. *Systems, 5*(3).

Vetriventhan, M., Upadhyaya, H. D., Dwivedi, S. L., Pattanashetti, S. K., & Singh, S. K. (2015). Finger and foxtail millets. In M. Singh & H. D. Upadhyaya (Eds.), *Genetic and Genomic Resources for Grain Cereals Improvement* (pp. 291–319). Academic Press, Elsevier.

Wang, C., Guo, L., Li, Y., & Wang, Z. (2012). Systematic comparison of C3 and C4 plants based on metabolic network analysis. *BMC Systems Biology, 6*(2), 1–14.
Wheeler, T., & Von Braun, J. (2013). Climate change impacts on global food security. *Science, 341*(6145), 508–513.
Wimalasiri, E. M., Jahanshiri, E., Chimonyo, V., Azam-Ali, S. N., & Gregory, P. J. (2021). Crop model ideotyping for agricultural diversification. *Methods X, 8*, 101–420.
Wimalasiri, E. M., Jahanshiri, E., Perego, A., & Azam-Ali, S. N. (2022). A novel crop shortlisting method for sustainable agricultural diversification across Italy. *Agronomy, 12*(7), 16–36.
Wuthrich, S. P., & Tadele, Z. (2012). Millet improvement through regeneration and transformation. *Biotechnology and Molecular Biology Reviews, 7*(2), 48–61.
Ye, C. Y., & Fan, L. (2021). Orphan crops and their wild relatives in the genomic era. *Molecular Plant, 14*(1), 27–39.
Yemets, A. I., Bayer, G. A., & Blume, Y. B. (2013). An effective procedure for *in vitro* culture of *Eleusine coracana* (L.) Gaertn. and its application. *ISRN Botany, 21*, 503–510.

CHAPTER 6

Climate Change-Driven Disease Management Strategies

VICTOR V. NGAIZA,[1] EMMANUEL Z. NUNGULA,[2,*] LUCIANA RAPHAEL CHAPPA,[3] SHIVANI RANJAN,[4] SUMIT SOW,[4] JERRY C. DLAMINI,[5] ABHISHEK RAJ,[6] RIZIKI MWADALU,[7] and HARUN GITARI[8]

[1]*School of Life Sciences and Bioengineering, The Nelson Mandela Institution of Sciences and Technology, Arusha, Tanzania*

[2]*Department of Environment and Sustainable Development, Mzumbe University, Morogoro, Tanzania*

[3]*Mwalimu Nyerere University of Agriculture and Technology (MNUAT); Department of Agricultural sciences. Butiama-Mara Tanzania*

[4]*Department of Agronomy, Dr. Rajendra Prasad Central Agricultural University, Pusa, Samastipur, Bihar, India*

[5]*Unit for Environmental Sciences and Management, Faculty of Natural and Agricultural Sciences, North-West University, Potchefstroom, South Africa*

[6]*Pandit Deendayal Upadhyay College of Horticulture and Forestry, Dr. Rajendra Prasad Central Agriculture University, Samastipur, Bihar, India*

[7]*Central Highlands Eco-Region Research Program, Kenya Forestry Research Institute, Nairobi, Kenya*

[8]*Department of Agricultural Sciences and Technology, School of Agriculture and Environmental Sciences, Kenyatta University, Nairobi, Kenya*

*Corresponding author

Climate-Resilient Agriculture for Sustainable Crop Production: New Adaptation and Mitigation Strategies. Sanjay Kumar, Shivani Ranjan, Sumit Sow, Smruti Ranjan Padhan, and Dhirendra Kumar Roy (Eds.)

DOI: 10.1201/9781998511150-6

ABSTRACT

Plant disease outbreaks pose a significant loss of primary productivity and biodiversity, leading to increased risks to global food security. Disease outbreaks increase as a response to climate change, as such changes affect host-pathogen interactions, pathogen evolution, and the establishment of new pathogenic strains. It is predicted that the changes in disease pressure brought about by known and developing pathogens due to climate change will cancel out any possible yield gains over the next 50 years. This chapter aims to explore the effect of climate change on plant diseases and their management by characterizing the change in each climatic factor (temperature, carbon dioxide level, and humidity change) and its associated effect on plant diseases. The incidence, prevalence, severity, and epidemiology of plant diseases are all impacted by climate change. When the virulence and pathogenicity of plant disease-causing pathogens are automatically altered, it renders disease management more difficult. Lastly, the chapter sheds light on strategies for managing diseases in light of climate change, which requires every player in the industry to play a role, including plant breeders, policymakers, extension officers, researchers, forecasters, and farmers at large.

6.1 INTRODUCTION

Climate is a long-term pattern in the weather that is typically averaged over 30 years (Bootsma, 1994). Technically speaking, it is the average variability of climate variables over timescales spanning millions of years to months (Andresen et al., 2012; Mote et al., 2013). Among the meteorological variables that are commonly measured are temperature, humidity, air pressure, wind, and precipitation. More generally, the climate system is made up of the conditions of the hydrosphere, cryosphere, lithosphere, biosphere, and atmosphere, as well as their interconnections (Crickenberger and Wethey, 2018; Peixóto and Oort, 1984). The Earth's surface is, at present, 1.1°C warmer on average than it was in the late 1800s (before the industrial revolution) and warmer than it has been in the previous 1,00,000 years. The four decades that have preceded it since 1850 have all been warmer than the previous ones, with the most recent decade being the warmest on record, 2011–2020. These temperature fluctuations may be caused by natural events like volcanic eruptions or variations in the sun's activity (Matsaba, 2021; Mohajan, 2013). But the main concern is the anthropogenic activity causing climate change and, in turn, alteration in atmospheric activity.

Melillo et al. (2014) report that human emissions of greenhouse gases (GHGs) have been the primary source of the global warming that has been observed over the past 50 years. Carbon dioxide and methane are the primary GHGs responsible for climate change. These emanate from using coal to heat a building or gasoline to operate a vehicle, for instance. The concentration of carbon dioxide in the atmospheric pool can be further aggravated by clearing land and destroying trees. Emissions of methane are mostly produced by the oil and gas industry and agriculture. According to Chen and Zhang (2010), the primary sectors responsible for greenhouse gas emissions include energy, industry, transport, buildings, agricultural systems, and land use. Agricultural systems that are already stressed and at high risk could be further impacted by climate change, which is considered a silent enemy because its impacts take time to manifest (Ljuša and Sitaula, 2016). In Limpopo, South Africa, Maponya and Mpandeli (2012) noted that the province's agricultural output is highly dependent on rainfall during the rainy period, making it particularly vulnerable to climate unpredictability and adaptation. Since the agriculture industry is one of the most susceptible sectors to the dangers and effects of global climate change, agricultural productivity will impact agro-industrial output (Parry et al., 1999).

Gowda et al. (2018) claim that climate change might improve or worsen growth conditions for crops in various areas. For instance, extended growing seasons are a result of changes in nearly every state's rainfall, temperature, and number of frost-free days. Changes in precipitation and temperature will also probably increase the incidence and range of diseases, weeds, and insects, which may increase the requirement for weed and insect pest control (Gowda et al., 2018; Walsh et al., 2020; Ziska et al., 2016). Such risks include being subjected to heat and other extreme weather, increased exposure to pesticides due to increased pest presence, disease-carrying pests such as mosquitoes and ticks, and worsened air quality. In addition, animals nurtured for meat, milk, and eggs may experience adverse health impacts from high temperatures and humidity (SDA, 2022). Climate change tends to increase plant disease outbreaks and worsen plant disease conditions; this is due to the evolution of aggressive and resistant pathogenic races, and it also alters host-pathogen interactions. On the other hand, climate change tends to weaken the host plant's defense, thereby making it susceptible to infections (Reddy and Reddy, 2015; Singh et al., 2023).

According to Gottlieb (1950) and Grau et al. (2004), the range of pathogens can shift, thereby facilitating the spread of diseases into new areas. For instance, plant pathogenic fungi are responsible for causing more than 80% of diseases, and their spore germination and infection are favored during periods of high humidity. In recent years, plant diseases have caused great negative impacts on human life; a good example is the Great Irish Famine, which lasted from 1845–1855 and was caused by late blight disease in Irish potatoes. This epidemic tragedy caused the county's population to fall to about 20–25% (in some towns, populations fell as much as 67%) between 1841 and 1871, when 1 million people died, and more than a million fled the country. In the entire period between 1845 and 1855, more than 2.1 million people had left Ireland (Jones, 2021; O'Neill, 2009). Dutch elm disease, a disease affecting elm trees, is another example of a disease that has created significant loss in the history of America, Europe, and Asia's forests. For instance, in Toronto, the disease caused about 80% loss of elm trees, and many more were infected in Ottawa, Montreal, and other cities during the 1970s and 1980s (Caron, 2018; Thompson, 2000).

6.2 PLANT DISEASES

6.2.1 CONCEPT OF PLANT DISEASES

An agent that keeps a plant from reaching its full potential is called a plant disease (Agrios, 2008). Steele and Gibbs (2018) define diseases of plants broadly, encompassing both biotic and abiotic conditions. Because it is caused by living things, this infection falls under the category of biological diseases (pathogens). Plants can be infected through these infectious agents, which can infect every type of vegetation tissue, including vascular tissues, leaves, shoots, stems, crowns, roots, tubers, fruit, and seeds (Scheffer, 1997). These infections generally transmit from plant to plant through different types of vectors or spores produced by causal organisms (Agrios, 2008; Baker et al., 1966). On the other hand, abiotic factors and diseases can be caused by external factors rather than by living things (Seleiman et al., 2021). Though these diseases are quite widespread, they are unable to spread from one plant to another and should be taken into account when evaluating the health of any given plant. Plant-borne pathogens comprise fungi, fungi-like organisms, bacteria, phytoplasmas, viruses, viroids, nematodes, and parasitic plants (Cohen and Leach, 2019; Dongre et al., 2023).

According to the disease triangle, a virulent pathogen, a susceptible host plant, and a favorable environment should all interact for a disease to appear in a plant system (Scholthof, 2007; Munir, 2018). A combination of all three determines the onset of disease. Another element in this process is time; thus, it is possible to modify any one of the three criteria to lower the probability of a disease developing. A plant's vulnerability to the pathogen, for instance, can be reduced by the introduction of types of crops that are resistant or tolerant to disease (Noman et al., 2016).

Disease transmission may be halted by the application of pesticides, biological controls, and cultural practices (Rahman et al., 2017). It is difficult for humans to change their environments since they are natural (Fuhrer et al., 2016; Tilman et al., 2002). However, by acting promptly, we can control time and manage the disease before favorable environmental conditions for the infection arise. Every pathogen causes infections in the host plant at a different time, and the timing is particularly important from a disease severity point of view (Agrios, 2008). When a specific species serves as the disease's vector, like in the case of begomovirus, for instance, which is spread by whiteflies, the vector populations contribute to the development of the disease in plants in a complementary way (Islam and Wu, 2017; Islam et al., 2017b; Tiwari et al., 2013).

6.2.2 PLANT DISEASE EPIDEMIOLOGY

Disease epidemiology is the investigation of disease in plant communities. Similar to human and other animal diseases, plant diseases are caused by pathogens such as bacteria, viruses, fungi, oomycetes, nematodes, phytoplasmas, protozoa, and parasitic plants (Ikotun, 2011). It exclusively examines changes brought about by epidemics in terms of time and space (Campbell and Madden, 1990). In addition to creating massive losses in the productivity of crops, plant disease epidemics can wipe out entire species, as was the case with Dutch Elm Disease, and could happen with Sudden Oak Death. A potato late-stage blight epidemic caused by *Phytophthora infestans* led to the Great Irish Famine and several fatalities (Munir, 2018). Epidemics can be classified as either polycyclic or monocyclic. The former is caused by pathogens that can cause multiple infection cycles during a growing season, mainly soil-borne diseases like Fusarium wilt of flax. Monocyclic epidemics are characterized by low birth and death rates, implying that they only have one infection cycle per season. Powdery mildew is one example of an airborne plant infection. In bimodal polycyclic epidemics, for instance, the fruits and blooms get infected at separate times, as in the case of brown rot of stone fruits (Van Maanen and Xu, 2003).

Additionally, both monocyclic and polycyclic diseases can create poetic outbreaks. Examples of this are the poetic epidemics induced by monocyclic pathogens, such as Dutch Elm disease, and poetic epidemics caused by polycyclic pathogens, such as apple powdery mildew (Parthasarathy, 2024). Finding the causes of disease occurrence is crucial in the field of plant disease epidemiology. Three elements are necessary for plant infections to establish and spread: a pathogen, a vulnerable host, and a favorable environment (Lucas, 2009). Environmental, physiological, and genetic factors all affect a host plant's vulnerability to disease. Disease-causing organisms can infect plants in a multitude of mechanisms and differ in their biological composition. Crops can be infected by seeds, colonize the plant's surface, or penetrate through wounds or other natural openings. In addition to affecting the pathogen's and the host plant's development and proliferation, temperature, moisture, light, and nutrients are environmental factors that contribute to the development of a disease (Colhoun, 1973).

Finding the elements that contribute to the transmission of disease is another important aspect of disease epidemiology. Pathogens are spread by a variety of means, such as wind, water, soil, and insects. Certain infections can also endure for a long time in soil or plant detritus while they wait for the right circumstances to infect new hosts. The effectiveness of the pathogen's dispersal mechanisms, the density of susceptible host plants, and the environmental factors that support pathogen development and reproduction all affect the rate and extent of disease transmission (Fitt et al., 2006). Creating efficient disease management plans that fall into the categories of cultural, chemical, biological, and genetic is also crucial. Crop rotation, planting resistant cultivars, and keeping soils in good condition are examples of cultural methods in action. Pesticides and fungicides are used in chemical techniques to manage disease populations. Using natural enemies of the virus, such as helpful bacteria or predatory insects, to stop its spread is known as a biological strategy. Breeding plants resistant to the pathogen or utilizing genetic engineering to create plants with increased resistance are examples of genetic techniques (Ab Rahman et al., 2018; Zadoks and Schein, 1979).

6.2.3 EFFECT OF CLIMATE CHANGE ON PLANT DISEASES

Sea level rise-related floods, which may result in large losses in crops, including the displacement of low-lying regions, are an excellent illustration of the devastating repercussions of climate change (Marks, 2011). A further example is dryness, which results in low soil water levels that deprive plants of their bio capacities and even make them more vulnerable to pests and illnesses (Al-Kaisi, 2013). Scholthof (2007) stated that the presence of an infectious agent, a susceptible host, and suitable surroundings are necessary for the occurrence of an infection (Figure 6.1). These variables may change due to changes in the environment, favoring the pathogen or the host (Maitra et al., 2023). Some instances of environmental factors include temperature, dew, rain, sunshine intensity, aeration (wind), and humidity levels.

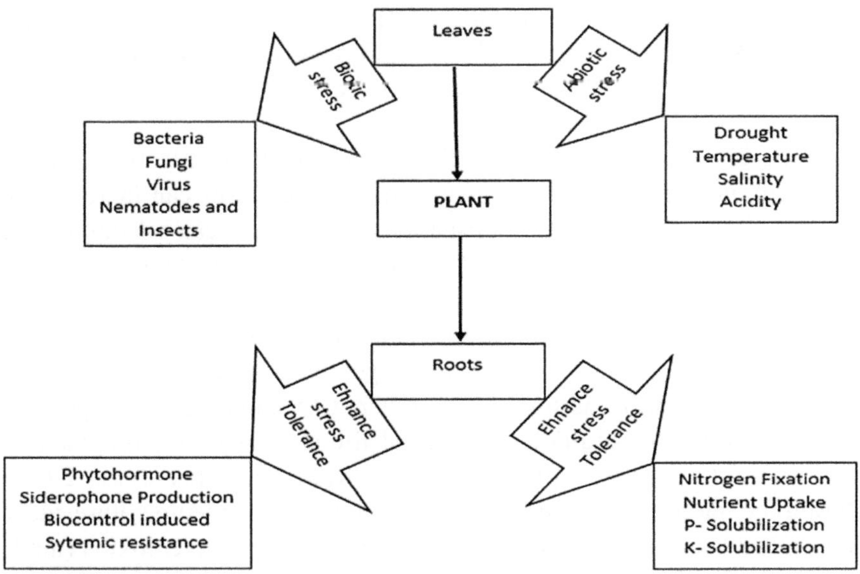

FIGURE 6.1 Effect of climate change on plant diseases.

Depending on the optimal temperature for the crop's growth and reproduction, high temperatures can have a variety of consequences on crops. Warmer weather in certain regions can benefit the crops that are normally cultivated there or enable producers to plant crops that thrive in higher temperatures. This isn't always the case, however; if the outside temperature advances above what a crop can tolerate, the output might drop, or even worse, diseases can spread and plague the crop as a result (Krishnan et al., 2011). According to Amthor (2001), elevated CO_2 levels have the potential to impact crop productivity. A few studies in the lab suggested that elevated CO_2 could have a beneficial effect on growth. Some factors, meanwhile, like fluctuating temperatures, water, ozone, and low nutrition levels can be in opposition to those potential output gains.

For example, an agricultural crop may not yield more if the temperature is higher than its ideal temperature or if it receives inadequate amounts of water and nutrients. Lower nitrogen and protein contents in alfalfa and soybean plants are associated with higher CO_2 concentrations; this leads to a significant drop in quality (Morison and Lawlor, 1999). Furthermore, it has been stated that dryness has become an important concern in areas where temperatures during the summer are higher since this results in soil that is dehydrated. While bigger irrigation systems would be feasible in some areas, other places might see reduced water supplies, which would result in less water available for irrigation when more is required. Warmer temperatures, wet climates, and higher CO_2 levels favor the development of numerous weeds, pests, and fungi (Kneebone et al., 1992; Majumdar, 2001).

Some climate change characteristics can affect the infection-causing pathogen and the host. For instance, some temperatures can increase the host's resistance to pathogenic diseases, while others can encourage the growth of pathogens (Dobson, 2009; Eastburn et al., 2011). Using wheat and oats as an example, these occurrences can be observed in the way that certain species of forage become hardier and that they become more susceptible to rust infections when temperatures increase. In addition, temperature variations as small as swings in CO_2 can cause some pests to undergo one to five additional phases of life each season, strengthening their ability to overcome plant resistance (Martín Sánchez et al., 2017). During the harvesting process, excessive humidity and temperature cause concentrations of some mycotoxins, such as fusarium mycotoxins (made by *Fusarium* spp.), to increase. According to Alhammad et al. (2023) and

Ferrigo et al. (2016), humidity contributes to the growth of weeds, and the biomass of weeds rises with rising temperatures.

In contrast, Paterson et al. (1997) and Zayan (2019) indicated that some factors can have varied effects based on the physiology of the plant. Lower vegetation disintegration rates, for example, may be caused by elevated CO_2 levels, and these may increase fungal inoculum concentrations and promote the formation of more fungal spores. Conversely, elevated CO_2 levels can change a plant's physiological responses and make it more resistant to some diseases. Two other extreme conditions that might lower soil fertility and, thus, plant health are low water levels and soil erosion (Lal, 1998).

6.3 EFFECT OF CLIMATE CHANGE ON PLANT DISEASES AND MANAGEMENT

Many pathogenic organisms, such as bacteria, fungi, oomycetes, viruses, and nematodes, are capable of causing diseases in plants. These pathogens have different lifestyles: biotrophs obtain nutrients from living cells, whereas necrotrophs obtain nutrients from dead cells and can use either intracellular or extracellular infection strategies (Boddy, 2016). Climate change is thought to increase plant infection in several ways, including changing host-pathogen relationships, host-pathogen evolution, and vector physiology. This is also thought to contribute to the emergence of new pathogen strains that can reduce host-plant resistance (Velásquez et al., 2018). Consequently, changes in the pathogen and host lifestyle can exacerbate the disease and cause it to spread to new areas (Laine, 2023). For example, when environmental conditions favor pathogen replication and vulnerable hosts are available, many soil opportunistic pathogens will increase disease outbreaks (Cheng et al., 2019; Trivedi et al., 2020). Plant–pathogen interactions can also be directly altered by climate change; this is through changes in the biochemical, physiological, ecological, and evolutionary processes of the plant host and/or pathogen (Cheng et al., 2019; Desaint et al., 2021; Guan, Elad, and Pertot, 2014). When there is prolonged drought, it causes an increase in the chances of an outbreak of dieback disease, as drought causes water stress in forest trees (Desprez-Loustau et al., 2006; Seleiman et al., 2021; Trivedi et al., 2022). Overall, the direct impacts of climate change are likely to vary depending on the pathogen, host identity, and properties of biomes. For instance, many soil-dwelling opportunistic pathogens will intensify disease outbreaks when environmental factors support pathogen replication and susceptible hosts are accessible (Cheng et al., 2019; Trivedi et al., 2020). Through modifications to the biological, physiological, ecological, and evolutionary processes of the plant host and/or pathogen, climate change can also directly affect plant-pathogen interactions (Cheng et al., 2019; Desaint et al., 2021). Prolonged drought increases the likelihood of a dieback disease epidemic because it impairs forest plants' ability to absorb water (Desprez-Loustau et al., 2006; Trivedi et al., 2022). In general, the pathogen, host, and biome characteristics are anticipated to influence the direct effects of climate change differently.

6.3.1 ELEVATED TEMPERATURE

Climate change can have significant effects on the number of generations of polycyclic species, the rate of population expansion, overwintering, and longevity rates, among other aspects of pathogen population structure. Lower diurnal temperatures, for illustration, promote rust epidemics in Central America by reducing the latency period of the pathogen *Hemileia vastatrix* (Toniutti et al., 2017). The pathogen's incubation period shortens in response to warming temperatures, which increases the pathogen's prevalence during the growing season. Higher temperatures and greater levels of humidity are linked with higher severity of the oilseed rape phoma stem canker infections and the potato blight pathogen (*Phytophthora infestans*). The global chestnut fungus *Phytophthora* spp. infection rates have been made worse by elevated average winter conditions, which caused serious tree mortality episodes across North America (Castro Verde, and Dina, 2021; Gustafson et al., 2022; Sánchez-Martín et al., 2017).

The frequency of pathogens in natural and agricultural settings is greatly impacted by variations in world temperature, which increases the danger of coming into contact with new pests and illnesses. Global warming is expected to increase the prevalence of many fungal soil-borne plant diseases, which would significantly affect primary production (Delgado-Baquerizo, 2020). Newly aggressive and better-adapted disease strains may arise as a result of rising temperatures (Cohen and Leach, 2020; Velásquez et al., 2018). According to Vaughan et al. (2016), it is anticipated that *Fusarium graminearum*, a more aggressive species that favors warm, humid environments, will displace the milder *Fusarium culmorum*, which prefers cold, wet conditions, in wheat, increasing the severity of Fusarium head blight (Parikka et al., 2012).

Although there hasn't been enough research done to understand the molecular basis of why plants are more vulnerable to infections under high temperatures, increasing temperatures can depress plant immunity, which increases the risk of pathogen infection (Cheng et al., 2019; Goher et al., 2023). Heat-induced suppression of salicylic acid production in Arabidopsis is attributed to compromised activation of master immune transcription factors, including CBP60g (Castroverde and Dina, 2021; Kim et al., 2022). Salicylic acid is a hormone essential to plant defense (Loake and Grant, 2007). Warm temperatures in rice cause the formation of abscisic acid and the expression of response genes, which are associated with a higher risk of bacterial blight illness; interestingly, resistance to high temperatures was linked to regulation of the abscisic acid pathways (Cohen et al., 2017; Khan et al., 2020).

6.3.2 ELEVATED CO_2

A shifting disease incidence is linked to an elevated concentration of carbon dioxide (CO_2), indicating that the pathogen and host react differently to CO_2. According to Khan et al. (2020), elevated CO_2 levels aggravated powdery mildew on cucurbits (*Sphaerotheca fuliginea*) and head blight and blotch on wheat (*Fusarium* spp. and *Septoria tritici*, respectively). *Peronospora manshurica*, the fungus that causes downy mildew, was less likely to affect soybeans, on the other hand. Changes in the characteristics of the surface of the leaf caused by higher CO_2 treatment further exacerbated the condition known as aspen tree rust disease (Eastburn et al., 2011). Plant immune responses and hormone levels are impacted by atmospheric CO_2, which can affect the interactions between plants and pathogens. For instance, in conditions of elevated CO_2, greater basal expression of genes responding to jasmonic acid strengthened resistance against the necrotrophic leaf pathogen *Botrytis cinerea* but decreased resistance against the hemi-biotrophic leaf pathogen *Pseudomonas syringae* pv. tomato (Gentine et al., 2019; Zhou et al., 2019).

Wheat becomes more vulnerable to two major diseases, *Zymoseptoria tritici* and *Fusarium graminearum*, under high CO_2 conditions, where plant defense pathways are less effective. These pathogens cause *Septoria tritici* blotch and *Fusarium* head blight, respectively (Andresen, 2001; Vary et al., 2015). Increased CO_2 affects the tripartite biotic relationships that exist between the wheat, barley yellow dwarf virus (BYDV), and *Rhopalosiphum padi*, the virus's aphid vector. In comparison to non-infected plants, BYDV contamination increased the aboveground nitrogen concentration of wheat growing under elevated CO_2, which decreased vector performance and phloem consumption. Increased CO_2 affects how plant–pathogen interactions turn out, but as of right now, no comprehensive framework is available to comprehend and forecast its impacts (Trębicki et al., 2016, 2017).

6.3.3 CLIMATE CHANGE-INDUCED VARIABILITY IN WATER AVAILABILITY

Climate-induced changes in humidity will probably have an impact on future plant disease breakouts since variations in relative humidity and soil moisture are two of the primary factors influencing the quantity and infectivity of plant diseases. High humidity is necessary for the spores of several fungal diseases to germinate and infect their host plants (Romero et al., 2022). In general, pathogens invading aerial plant tissues become more virulent when there is high humidity in the air. Higher humidity levels are associated with higher infection rates of the stem rot pathogen *Phytophthora sojae* and *Sclerotinia sclerotiorum* in lettuce (Mamo et al., 2021). Plant immunity responses are modified by bacterial effectors expressed in response to humidity, which facilitates *P. syringae's* establishment in the apoplast, or aqueous intercellular space, of Arabidopsis leaves (Xin et al., 2016).

However, the impact of dryness on pathogen infection rates and sickness severity can vary greatly. According to Wakelin et al. (2018), several plant diseases become more severe as drought conditions persist and intensify. These include grapevine black foot (*Ilyonectria/ Dactylonectria* spp.), brassica blackleg (*Leptosphaeria maculans*), pea root rot (*Aphanomyces euteiches*), onion white rot (*Sclerotium cepivorum*), wheat take-all (*Gaeumannomyces graminis* var. *tritici*), wheat crown rot

(*Fusarium* spp.), and wheat take-all (*Gaeumannomyces graminis* var. *tritici*). Drought-induced impairment of plant basal immune responses led to an increase in signs of yellow vein disease and infection with the potato yellow vein virus (Vasquez et al., 2022). The interactions between the virus, host, and vector of the greenhouse whitefly are also impacted by these modifications, which enhance the virus's capacity for horizontal transmission.

6.3.4 EFFECTS ON PLANT BIOME

Climate change-related modifications to the plant microbiome may also have an indirect effect on the occurrence of disease. Plant microbiomes are predicted to be impacted by climate change as a result of the host responses, including altered morphology, immunological responses, exudation patterns, and physiology (Trivedi et al., 2022; Wang et al., 2017), as well as by alterations to the original inoculum from bulk soil or the rhizosphere. This is noteworthy because, according to Desprez-Loustau et al. (2006) and Trivedi et al. (2022), plant microbiomes have a substantial influence on the physiology of their hosts and aid in regulating their metabolism, immune response, and fitness in strange situations. Plant microbiomes employ several strategies to manage plant infections, including altering plant immune responses and engaging in competition with pathogens (Figure 6.2).

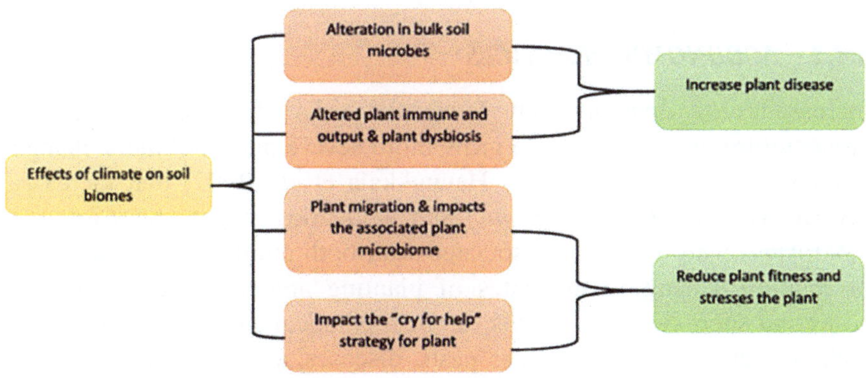

FIGURE 6.2 Effect of climate change on soil biomes.

6.4 PLANT DISEASE MANAGEMENT UNDER CHANGING CLIMATE

Climate change is anticipated to cause changes in plant disease incidence, prevalence, and severity. The fight against diseases will also be impacted by the appropriateness, preference, and efficacy of physical, chemical, and biological control measures, in addition to how integrated pest management (IPM) approaches use them. Extension services, agroindustry, and pragmatic farmers place a high priority on future needs for disease management (Juroszek and Von Tiedeman, 2011). Population growth will have doubled the world's demand for grains by 2050. Competition for land usage, particularly in places with rapid population expansion, will make land for agriculture scarce (Nungula et al., 2023; Tilman et al., 2002). Consequently, it is necessary to raise crop productivity. However, if losses in yield due to pests and diseases stay at current levels or even rise further, it is still conceivable to meet the food demand by not farming more land or planting more productive crops (Chappa, 2023; Oerke and Dehne, 2004). Therefore, as stated in Strand (2000), improvements in crop disease and insect control are necessary to lower pre- and postharvest output losses. Customized treatments are required for pest and disease problems, depending on the crop, the environment, and the type of pathogen involved. It is also necessary to consider the standards that are now in place to reduce contaminants from agriculture, including heavy metals, mycotoxins, plant protection products (PPPs), and hazardous microorganisms from entering areas related to humans and/or agroecosystems. Several climate change-driven management strategies are explained as described in subsections.

6.4.1 AGRONOMIC PRACTICES

Increased crop diversity may lower the risks of diseases that, in a monoculture, might otherwise grow more severe due to climate change. For instance, it was shown by Hannukkala et al. (2007) that increased severity (*Phytophthora infestans*) and epidemics in potatoes (*Solanum tuberosum*) were probably associated with both climate change and lack of rotation. Changing the dates of planting and/or harvesting annual crops and short-lived perennials can be a practical and affordable way to make the crop more resistant to pests, diseases, and unfavorable abiotic conditions (Srivastava et al., 2010). Nevertheless, location, cultivar, degree of temperature change, and market conditions all affect how feasible it is to alter planting and/or harvesting dates (Mirriam et al., 2023; Wolfe et al., 2008). In the highlands of Africa, for instance, potato farmers have to decide whether to raise their risk of drought stress or forego the rainy season to lower the danger of late blight (Hijmans et al., 2000). Agronomic practices such as crop rotation, liming, fertilization, soil tillage, irrigation, and others are crucial for managing and reducing disease risk. Such agronomic practices are expected to have a bigger short-term impact than long-term climate changes, given their well-established significant impact on diseases. Therefore, they are useful tactics to counteract potential rises in the risks of disease (Heitefuss and Welch, 1989).

Producers typically have to decide which area is ideal for their crop species first. Additionally, they might bring in new crop species that would benefit from rising CO_2 levels in the atmosphere and climate change (Wolfe et al., 2008). Apart from the possible advantages of cultivating novel crop species, there's a danger that newly developed crops will carry new illnesses (Boland et al., 2004). Warmer temperatures, thereby a longer growing season, and a greater likelihood of summer drought could lead to increased water demand in agriculture (Wolfe et al., 2008). More growth and development in plants brought about by the CO_2 fertilization effect may also be supported by proper watering and fertilizing. Enhanced vigor of plants may also promote disease aggressiveness (Lamichhane et al., 2017). Managing overwintering pathogens in crop residues left on the field is an additional essential component of agronomic management. Many residue-borne diseases of plants caused by necrotrophic pathogenic organisms can be treated using rotations of crops and other agronomic practices meant to reduce the inoculum levels (Melloy et al., 2010). Farmers who plow the ground to bury toxic leftovers and turn the soil under extreme conditions forfeit the benefits of conservation tillage, which include enhanced soil structure and water retention. Consequently, there may be a trade-off between reducing climate change and possible health effects. In light of climate change, conservation agriculture makes sense because it boosts crop resilience under trying circumstances (Ortiz et al., 2008).

6.4.2 HOST-PLANT RESISTANCE AND PATHOGEN AGGRESSIVENESS

Legrève and Duveiller (2010) state that genotypes that are resilient to or capable of withstanding pathogen attack are what lead to host-plant resistance. Using tolerant or resistant cultivars is simple, affordable, ecologically friendly, and effective—that is until viruses can overcome the resistance. Climate and cultural practices, which make up the outside environmental effect, interact with genetic factors in the pathogen and the host plant to determine the manifestation of plant resistance (Dodds and Rathjen, 2010). Loss of resistance may result from exposure to inappropriate environmental conditions (Strand, 2000). For instance, disease resistance may be influenced by temperature (Hartleb and Heitefuss, 1997). In the case of oilseed rape's phoma stem canker (*Leptosphaeria maculans*), resistance appears at 15°C yet does not develop at 25°C (Huang et al., 2006). According to Chakraborty and Pangga (2004), crop plants may experience a sharp decrease in disease resistance due to the introduction of new, highly pathogenic strains of the pathogen as a result of climate and changes in the atmosphere. According to Legrève and Duveiller (2010), developing a tolerance for environmental factors like heat and drought should be done in tandem with developing resistance to many infections. Nonetheless, an arsenal of additional plant protection techniques, such as crop rotation and the application of spatial and temporal diversity of crops, may be necessary when numerous plant diseases arise concurrently. Numerous resistances from the host plant may not be sufficient in such cases (Tilman et al., 2002).

6.4.3 FUNGICIDES

Oomycetes and pathogenic fungi are managed with fungicides. Furthermore, according to Wu and von Tiedemann (2001), they might exhibit advantageous adverse effects on the physiological functioning of the crop that raise production and overall stress tolerance. Usually, fungicide applications are applied externally or utilized as seed dressing. Research on the potential effects of climate change on fungicide applications is scarce (Ghini et al., 2008). Patterns of precipitation have been the focus of most writers, with very few discussing the temperature dependence of fungicide efficacy. Extreme temperatures can change the efficacy of PPPs, like fungicides, or lessen their plant toxicity (Strand, 2000). Changes in soil/air moisture content, precipitation, temperature, direction of the wind, and light levels can all affect how effective PPP interventions are (Bedos et al., 2002; Ziska and Runion, 2007). As noted by Bedos et al. (2002) and Monkiedje et al. (2007), several environmental factors may alter the dynamics of fungicides in the soil and on/in the leaves, including uptake, degradation, and volatilization. Precipitation is especially significant during the post-application phase. Chakraborty et al. (2000) suggest that the fungicide's efficiency may be altered by either a decrease in the fungicide layer on the leaves or an increase in its spread.

6.4.4 BIOLOGICAL CONTROL

Introducing biological control agents (BCAs) or promoting their natural occurrence can both be advantageous methods of introducing them. For them to be effective, conditions in the environment are typically required. A lack of ongoing access to appropriate moisture and temperature may be the reason for BCA communities' incapacity to reduce the frequency and severity of disease, as well as their slower pace of recovery when good conditions return. The assumption made by Garrett et al. (2006) was that as climate change occurs, applied antagonists will have a tougher time surviving and functioning due to an increase in climate unpredictability. They claimed, nevertheless, that notwithstanding potential problems with BCA use, research needs to be supported to develop biocontrol methods that are more adaptable to shifting conditions. However, this has consistently been a prerequisite for efficient biocontrol since the first BCAs were found, so it's not a novel problem (Ghini et al., 2008).

6.4.5 INTEGRATED DISEASES MANAGEMENT

Plant pathology research and extension activities have emphasized the combination of different methods for control under an IPM framework; entomologists laid the framework's preliminary foundations (Jacobsen, 1997). Integrated pest management, or IPM, is an ecosystem-based strategy that uses a range of tactics to prevent pests and the long-term damage they cause. These techniques include biological control, the use of resistant cultivars, habitat management, and cultural practices (Strand, 2000). PPPs are implemented after careful evaluation and after establishing monetary impact thresholds. Integrated crop management (ICM) integrates components of IPM, such as fertilizer supply, water management, and soil tillage, to establish environmentally friendly farming methods (Juroszek and Von Tiedeman, 2011).

6.4.6 QUARANTINE

As per Garrett et al. (2010), various diseases may pose a greater risk in a new setting. According to Wolfe et al. (2008), quarantines may not be as effective in preventing the spread of new or exotic diseases since these infections can be transported by imported plant products or by airborne drift. There will be fresh possibilities for crops and cultivars to be developed in previously unexplored regions. However, effective systems, which include diagnostic tools, are required to detect and monitor introduced pathogens and notice their behavior in such altered conditions (Mithöfer et al., 2004). Considering both present and potential climate fluctuations, quarantine agencies and workers may find it simpler to evaluate the risk posed by specific pathogens in various areas.

6.4.7 FORECASTING AND EARLY WARNING SYSTEM

Some of the disease outbreaks could have been overcome or minimized by forecasting and developing early warning signs. Forecasting can be done using the existing weather data from meteorological stations, which can help to predict the weather in the future and the probability of disease outbreaks (Andersson et al., 2020; Southall et al., 2021). The idea to start utilizing the forecasting and early warning sign system started as early as 1926 when computer technology and software were available. The main concept behind this is to use the automated weather station to monitor the environmental weather changes, and then data is automatically entered into a computer. The computer software interprets the data, and then the decision is synthesized regarding the plant health to be maintained or protected. An example of an early warning sign that has been used for the management of beans' chocolate spot was developed by Reddy (2015).

6.5 CHALLENGES AND OPPORTUNITIES IN DISEASE MANAGEMENT UNDER CLIMATE CHANGE

6.5.1 ADEQUATE RESOURCES FOR DISEASE FORECASTERS

The lack of thorough, current multi-factor and multi-species datasets (Ziska and Runion, 2007) and the diversity and adaptability of pathogen populations (Garrett, 2006) will restrict the accuracy of pathogen responses to climate change (Dukes et al., 2009). Long-term forecasts will continue to be especially difficult because of the numerous linkages between processes in ecosystems, human effects, environmental factors, disease populations, and their potential for adaptability (Fuhrer, 2003; Martinelli et al., 2015). This also holds for forecasts of upcoming modifications regarding plant disease management techniques. Utilizing weather data, disease-forecasting models are useful in determining the meteorological variables (as well as the period) that have a strong correlation with disease (Coakley et al., 1988). Though the majority of general circulatory models work at higher resolutions, these kinds of disease forecasting models can be used in conjunction with each other to predict potential disease outbreak scenarios. One way to close this gap is via scaling down climate simulations (Soussana et al., 2010). However, taking into account the diversity in illness epidemiology continues to be difficult (Legrève and Duveiller, 2010). Disease management strategies can be recommended or enhanced in light of potential future epidemic scenarios (Salinari et al., 2004).

The vast majority of assessments that currently exist on the subject of plant diseases and climate change concur that additional empirical evidence is necessary (Chakraborty et al., 2000). The prediction's reliability and precision will rise with more and better data (VanRaden et al., 2009). The forecasts will support the proven effects of environmental change on plant pathosystems and the ongoing attempts to mitigate them (Roos et al., 2010). For instance, models show that both the north and the south of the United Kingdom will see distinct impacts from climate change on phoma stem canker on oilseed rape (Butterworth et al., 2010). While extrapolating from individual studies to regional epidemics remains difficult (Burdon et al., 2006; Garrett et al., 2006), the phoma study demonstrates that predictions regarding regional trends in the development of diseases under climate change are feasible. However, there is a practical test that will require to be awaited. There is a growing recognition that controlling plant diseases throughout entire landscapes requires a knowledge of the relationships between weather and the geographical distribution of susceptible/resistant host patches (Seem, 2004). According to Holdenrieder et al. (2004) and Pinkard et al. (2010), one of the major issues facing landscape epidemiology is the integration of multi-scale epidemic simulations with scenarios related to climate change.

6.5.2 DEVOTION OF RESOURCES TO BREEDERS AND RESEARCHERS

The need to develop new varieties with greater resistance in less time may make biotechnological methods crucial to crop adaptation to climate change. To develop long-lasting resistance in a variety of settings, conventional breeding programs will continue to be equally important in accumulating minor genes and racialized resistance to infections (Raza et al., 2019). Increased cropping system heterogeneity that lowers disease risks, the use of more effective cultivars resistant to biotic and abiotic stress, and trustworthy methods for predicting infectious agents' occurrence to quickly respond to emerging pathogens are a few examples of these situations (Juroszek and Von Tiedemann, 2011).

6.5.3 STRENGTHENING QUARANTINE SYSTEMS

As per Garrett et al. (2010), certain diseases might pose a greater threat in an unfamiliar environment. Due to the increased likelihood of novel or invasive infections spreading by air-stream drift or imported plant products (Wolfe et al., 2008), authorities may find it more challenging to prevent disease transmission through quarantine. Growing crops and cultivars in previously uncultivated areas and locations will present novel possibilities; however, efficient systems, including diagnostic tools, are required to identify and track pathogenic organisms and their behavior in these altered environments (Boland et al., 2004). Quarantine authorities may find it easier to assess the threat posed by a particular pathogen in several areas under uncertainty if they use methods for matching climates and geographic information systems under present and projected climatic changes in various areas caused by a specific pathogen (Coakley et al., 1999; Legrève and Duveiller, 2010).

6.6 INNOVATIONS AND FUTURE DIRECTIONS

6.6.1 ADVANCED DISEASE MODELING AND SURVEILLANCE SYSTEMS

Considering all of the variables that influence plant health in a changing environment, as well as how they interact, direct effects, and indirect effects, it is uncertain if a predictive understanding of these complex systems can be achieved (Garrett et al., 2006). Predictability is a critical need to develop remedies for the multitude of new plant health problems that are either already becoming more significant or are sure to arise. Several research studies have demonstrated that to comprehend plant disease and climate change, better and more accurate information is required (Desprez-Loustau et al., 2007). The best approach is to employ improved empirical data to deliver accurate and/or trustworthy information. What is currently being done, the projections will assist in documenting the effects of climate change (Desprez-Loustau, 2007). In order to manage plant diseases across large landscapes, it is becoming increasingly vital to understand how climate variables relate to the geographic distribution of susceptible/resistant host areas (Seem, 2004). Among the primary issues in landscape epidemiology, according to Holdenrieder et al. (2004) and Pinkard et al. (2010), is combining scale disease models with changing climate models.

Although there has been progress in understanding the relationship between the pathosystems and climate change, it is important to recognize that plant disease development models under climate change are inherently uncertain. Increased climatic unpredictability and the effects of climate change on plant health are predicted by studies on climate. The interplay of multiple causes, including pollution, urbanization, trade expansion, and land use change, leads to climate change (Sahoo et al., 2024). Apart from the indirect effects on host-pathogen connections, disease dispersal, and host migration, there are also indirect impacts on diversity in genes and phenology. According to Moricca and Ragazzi (2009) and Pautasso et al. (2012), increased monitoring and modeling are necessary due to the likelihood that invasive plant diseases may spread in areas with naïve host populations—that is, populations without a history of co-evolution. Though they acknowledge that forest managers have not always taken climate change into account in their strategic and operational planning, Ogden and Innes (2007) lament the lack of action to date in their global overview.

6.7 CONCLUSION

Climate change directly and/or indirectly influences plant diseases. It consequently leads to the increased prevalence, incidence, severity, and epidemics of infections and diseases, which poses more difficulty for farmers to manage them. To ensure food security, this condition necessitates that farmers re-plan the management of such diseases as it may require more resources, time, and knowledge. This chapter has revealed the climate change-driven disease management practices to be adopted. However, for these management practices to be successful, there is a need for policymakers, researchers, breeders, forecasters, disease modelers, and farmers to work together to achieve the common goal of minimizing crop losses due to disease in a changing climate.

KEYWORDS

- **anthropogenic activity**
- **carbon dioxide levels**
- **climate change**
- **cryosphere**
- **fungal pathogens**
- **global warming**
- **greenhouse gases**
- **hydrosphere**
- **integrated pest management**
- **pathogen evolution**
- **pathogenicity**
- **spore germination**

REFERENCES

Ab Rahman, S. F. S., Singh, E., Pieterse, C. M., & Schenk, P. M. (2018). Emerging microbial biocontrol strategies for plant pathogens. *Plant Science, 267*, 102–111.

Agrios, G. N. (2008). Transmission of plant diseases by insects. *Encyclopedia of Entomology*, 3853–3885.

Alhammad, B. A., Roy, D. K., Ranjan, S., Padhan, S. R., Sow, S., Nath, D., Seleiman, M. F., & Gitari, H. (2023). Conservation tillage and weed management influencing weed dynamics, crop performance, soil properties, and profitability in rice-wheat-green gram system in Eastern Indo-Gangetic Plains. *Agronomy, 13*(7), 1953.

Al-Kaisi, M. M., Elmore, R. W., Guzman, J. G., Hanna, H. M., Hart, C. E., Helmers, M. J., & Sawyer, J. E. (2013). Drought impact on crop production and the soil environment: 2012 experiences from Iowa. *Journal of Soil and Water Conservation, 68*(1), 19A–24A.

Andersson, L., Wilk, J., Graham, L. P., Wikner, J., Mokwatlo, S., & Petja, B. (2020). Local early warning systems for drought–Could they add value to nationally disseminated seasonal climate forecasts? *Weather and Climate Extremes, 28*, 100241.

Andresen, J. S. (2001). Effects of atmospheric CO_2 concentration on wheat yield: Review of results from experiments using various approaches to control CO_2 concentration. *Field Crops Research, 73*(1), 1–34.

Andresen, J., Hilberg, S., Kunkel, K., & Center, M. R. C. (2012). Historical climate and climate trends in the Midwestern USA. *US National Climate Assessment Midwest Technical Input Report*, 1–18.

Baker, K. F., & Smith, S. H. (1966). Dynamics of seed transmission of plant pathogens. *Annual Review of Phytopathology, 4*(1), 311–332.

Bedos, C., Cellier, P., Calvet, R., Barriuso, E., & Gabrielle, B. (2002). Mass transfer of pesticides into the atmosphere by volatilization from soils and plants: Overview. *Agronomie, 22*(1), 21–33.

Boddy, L. (2016). Pathogens of autotrophs. In *The Fungi* (pp. 245–292). Academic Press.

Boland, G. J., Melzer, M. S., Hopkin, A., Higgins, V., & Nassuth, A. (2004). Climate change and plant diseases in Ontario. *Canadian Journal of Plant Pathology, 26*(3), 335–350.

Bootsma, A. (1994). Long-term (100 yr.) climatic trends for agriculture at selected locations in Canada. *Climatic Change, 26*(1), 65–88.

Burdon, J. J., Thrall, P. H., & Ericson, A. L. (2006). The current and future dynamics of disease in plant communities. *Annual Review of Phytopathology, 44*, 19–39.

Butterworth, M. H., Semenov, M. A., Barnes, A., Moran, D., West, J. S., & Fitt, B. D. (2010). North–South divide: Contrasting impacts of climate change on crop yields in Scotland and England. *Journal of the Royal Society Interface, 7*(42), 123–130.

Caron, M. (2018). Taming the Jungle in the City: Uprooting Trees, Bushes, and Disorder from *Mount Royal Park. Urban History Review, 47*(1), 39–53.

Castro Verde, C. D. M., & Dina, D. (2021). Temperature regulation of plant hormone signaling during stress and development. *Journal of Experimental Botany, 72*(21), 7436–7458.

Chakraborty, S., & Pangga, I. B. (2004). Plant disease and climate change. *Plant Microbiology*, 175–193.

Chakraborty, S., Tiedemann, A. V., & Teng, P. S. (2000). Climate change: Potential impact on plant diseases. *Environmental Pollution, 108*(3), 317–326.

Chappa, L. R. (2023). Soil chemical properties and sunflower (*Helianthus annuus*) productivity responses to sunn Hemp intercropping and mineral fertilizer application in Dodoma, Tanzania. *MSc Thesis, Kenyatta University, Nairobi–Kenya.*

Chen, G. Q., & Zhang, B. (2010). Greenhouse gas emissions in China 2007: Inventory and input–output analysis. *Energy Policy, 38*(10), 6180–6193.

Cheng, Y. T., Zhang, L., & He, S. Y. (2019). Plant-microbe interactions facing environmental challenge. *Cell Host & Microbe, 26*(2), 183–192.

Coakley, S. M., Scherm, H., & Chakraborty, S. (1999). Climate change and plant disease management. *Annual Review of Phytopathology, 37*(1), 399–426.

Cohen, S. P., & Leach, J. E. (2019). Abiotic and biotic stresses induce a core transcriptome response in rice. *Scientific Reports, 9*(1), 6273.

Cohen, S. P., & Leach, J. E. (2020). High temperature-induced plant disease susceptibility: More than the sum of its parts. *Current Opinion in Plant Biology, 56*, 235–241.

Cohen, S. P., Liu, H., Argueso, C. T., Pereira, A., Vera Cruz, C., Verdier, V., & Leach, J. E. (2017). RNA-Seq analysis reveals insight into enhanced rice *Xa7*-mediated bacterial blight resistance at high temperatures. *PloS One, 12*(11), e0187625.

Colhoun, J. (1973). Effects of environmental factors on plant disease. *Annual Review of Phytopathology, 11*(1), 343–364.

Collett, S. R., Smith, J. A., Boulianne, M., Owen, R. L., Gingerich, E., Singer, R. S., & Stewart-Brown, B. (2020). Principles of disease prevention, diagnosis, and control. *Diseases of Poultry*, 1–78.

Crickenberger, S., & Wethey, D. S. (2018). Annual temperature variation as a time machine to understand the effects of long-term climate change on a poleward range shift. *Global Change Biology, 24*(8), 3804–3819.

Delgado-Baquerizo, M. (2020). The proportion of soil-borne pathogens increases with warming at the global scale. Dataset. *Nature Climate Change, 10*, 550–554. https://doi.org/10.1038/s41558-020-0759-3.

Desaint, H., Aoun, N., Deslandes, L., Vailleau, F., Roux, F., & Berthomé, R. (2021). Fight hard or die trying: When plants face pathogens under heat stress. *New Phytologist, 229*(2), 712–734.

Desprez-Loustau, M. L., Marçais, B., Nageleisen, L. M., Piou, D., & Vannini, A. (2006). Interactive effects of drought and pathogens in forest trees. *Annals of Forest Science, 63*(6), 597–612.

Desprez-Loustau, M. L., Robin, C., Buee, M., Courtecuisse, R., Garbaye, J., Suffert, F., & Rizzo, D. M. (2007). The fungal dimension of biological invasions. *Trends in Ecology & Evolution, 22*(9), 472–480.

Dobson, A. (2009). Climate variability, global change, immunity, and the dynamics of infectious diseases. *Ecology, 90*(4), 920–927.

Dodds, P. N., & Rathjen, J. P. (2010). Plant immunity: Towards an integrated view of plant–pathogen interactions. *Nature Reviews Genetics, 11*(8), 539–548.

Dongre, M. A., Ashraf, M. S., Koshariya, A. K., & Reddypriya, P. (2023). *Plant Pathology and Plant Diseases*. AG PUBLISHING HOUSE (AGPH Books).

Dukes, J. S., Pontius, J., Orwig, D., Garnas, J. R., Rodgers, V. L., Brazee, N., & Ayres, M. (2009). Responses of insect pests, pathogens, and invasive plant species to climate change in the forests of northeastern North America: What can we predict? *Canadian Journal of Forest Research, 39*(2), 231–248.

Eastburn, D. M., McElrone, A. J., & Bilgin, D. D. (2011). Influence of atmospheric and climatic change on plant–pathogen interactions. *Plant Pathology, 60*(1), 54–69.

Elad, Y., & Pertot, I. (2014). Climate change impacts on plant pathogens and plant diseases. *Journal of Crop Improvement, 28*(1), 99–139.

Ferrigo, D., Raiola, A., & Causin, R. (2016). Fusarium toxins in cereals: Occurrence, legislation, factors promoting the appearance and their management. *Molecules, 21*(5), 627.

Fitt, B. D., McCartney, H. A., & West, J. S. (2006). Dispersal of foliar plant pathogens: Mechanisms, gradients, and spatial patterns. *The Epidemiology of Plant Diseases*, 159–192.

Fuhrer, J. (2003). Agroecosystem responses to combinations of elevated CO_2, ozone, and global climate change. *Agriculture, Ecosystems & Environment, 97*(1–3), 1–20.

Fuhrer, J., Val Martin, M., Mills, G., Heald, C. L., Harmens, H., Hayes, F., & Ashmore, M. R. (2016). Current and future ozone risks to global terrestrial biodiversity and ecosystem processes. *Ecology and Evolution, 6*(24), 8785–8799.

Garrett, K. A., Dendy, S. P., Frank, E. E., Rouse, M. N., & Travers, S. E. (2006). Climate change effects on plant disease: Genomes to ecosystems. *Annual Review of Phytopathology, 44*, 489–509.

Garrett, K. A., Jumpponen, A., & Montano, L. G. (2010). Emerging plant diseases: What are our best strategies for management. *Controversies in Science & Technology, 3*, 152–160. http://hdl.handle.net/10919/68898 (accessed on 16 January 2025).

Gentine, P., Green, J. K., Guérin, M., Humphrey, V., Seneviratne, S. I., Zhang, Y., & Zhou, S. (2019). Coupling between the terrestrial carbon and water cycles—a review. *Environmental Research Letters, 14*(8), 083003.

Ghini, R., Hamada, E., & Bettiol, W. (2008). Climate change and plant diseases. *Scientia Agricola, 65*, 98–107.

Goher, R., Alkharabsheh, H. M., Seleiman, M. F., Diatta, A. A., Gitari, H., Wasonga, D. O., Khan, G. R., & Akmal, M. (2023). Impacts of heat shock on productivity and quality of *Triticum aestivum* L. at different growth stages. *Notulae Botanicae Horti Agrobotanici Cluj-Napoca, 51*(1), 13090.

Gottlieb, D. (1950). The physiology of spore germination in fungi. *The Botanical Review, 16*, 229–257.

Gowda, P., McCartney, H. A., & West, J. S. (2018). Ch. 10: Agriculture and rural communities. In: *Impacts, Risks, and Adaptation in the United States: Fourth National Climate Assessment, Volume II*. U.S. Global Change Research Program, Washington, DC, p. 401.

Grau, C. R., Dorrance, A. E., Bond, J., & Russin, J. S. (2004). Fungal diseases. In *Soybeans: Improvement, Production, and Uses* (pp. 679–763).

Guan, J., & Nutter Jr, F. W. (2001). Factors that affect the quality and quantity of sunlight reflected from alfalfa canopies. *Plant Disease, 85*(8), 865–874.

Gustafson, E. J., Miranda, B. R., Dreaden, T. J., Pinchot, C. C., & Jacobs, D. F. (2022). Beyond blight: *Phytophthora* root rot under climate change limits populations of reintroduced American chestnut. *Ecosphere, 13*(2), e3917.

Hannukkala, A. O., Kaukoranta, T., Lehtinen, A., & Rahkonen, A. (2007). Late-blight epidemics on potatoes in Finland, 1933–2002; increased and earlier occurrence of epidemics associated with climate change and lack of rotation. *Plant Pathology, 56*(1), 167–176.

Hartleb, H., & Heitefuss, R. (1997). Abiotic and biotic influences on resistance of crop plants against fungal pathogens. In *Resistance of Crop Plants Against Fungi*.

Heitefuss, R., & Welch, J. (1989). *Crop and Plant Protection: The Practical Foundations*. Ellis Horwood Limited.

Hijmans, R. J., Forbes, G. A., & Walker, T. S. (2000). Estimating the global severity of potato late blight with GIS-linked disease forecast models. *Plant Pathology, 49*(6), 697–705.

Holdenrieder, O., Pautasso, M., Weisberg, P. J., & Lonsdale, D. (2004). Tree diseases and landscape processes: The challenge of landscape pathology. *Trends in Ecology & Evolution, 19*(8), 446–452.

Huang, Y. J., Evans, N., Fitt, B. D. L., Li, Z. Q., Rouxel, T., & Balesdent, M. H. (2006). Fitness cost associated with loss of the *AvrLm4* avirulence function in *Leptosphaeria maculans* (phoma stem canker of oilseed rape). In *Sustainable Strategies for Managing Brassica napus* (oilseed rape) Resistance to *Leptosphaeria maculans* (phoma stem canker) Under the aegis of a European-Australian workshop held at INRA, Versailles, France, in association with the EU SECURE project (QLK5-CT-2002-01813) (pp. 77–89). Springer Netherlands.

Ikotun, B. A. (2011). *Plant Pathology: Plant Afflictions and Man's Interventions*. Ibadan University Press.

Islam, W., & Wu, Z. (2017). Genetic defense approaches against begomoviruses. *Journal of Applied Virology*. https://doi.org/10.21092/jav.v6i3.81.

Islam, W., Zaynab, M., Qasim, M., & Wu, Z. (2017b). Plant-virus interactions: Disease resistance in focus. *Hosts and Viruses, 4*(1), 1–16.

Jacobsen, B. J. (1997). Role of plant pathology in integrated pest management. *Annual Review of Phytopathology, 35*(1), 373–391.

Jansson, J. K., & Hofmockel, K. S. (2020). Soil microbiomes and climate change. *Nature Reviews Microbiology, 18*(1), 35–46.

Jones, G. (2021). Doctors for export: Medical migration from Ireland c. 1860 to 1960. In *'Doctors for Export'* (pp. 1–15). Brill.

Juroszek, P., & Von Tiedemann, A. (2011). Potential strategies and future requirements for plant disease management under a changing climate. *Plant Pathology, 60*(1), 100–112.

Khan, A., Ahmad, M., Ahmed, M., & Iftikhar Hussain, M. (2020). Rising atmospheric temperature impact on wheat and thermotolerance strategies. *Plants, 10*(1), 43. https://doi.org/10.3390/plants10010043.

Khan, M. R., & Rizvi, T. F. (2020). Effect of elevated levels of CO2 on powdery mildew development in five cucurbit species. *Scientific Reports, 10*(1), 4986. https://doi.org/10.1038/s41598-020-61716-9.

Khoury, W. E., & Makkouk, K. (2010). Integrated plant disease management in developing countries. *Journal of Plant Pathology, 92*(S35), S35–S42.

Kim, Y. W., Youn, J. H., Roh, J., Kim, J. M., Kim, S. K., & Kim, T. W. (2022). Brassinosteroids enhance salicylic acid-mediated immune responses by inhibiting BIN2 phosphorylation of clade I TGA transcription factors in *Arabidopsis*. *Molecular Plant, 15*(6), 991–1007.

Kneebone, W. R., Kopec, D. M., & Mancino, C. F. (1992). Water requirements and irrigation. *Turfgrass, 32*, 441–472.

Krishnan, P., Ramakrishnan, B., Reddy, K. R., & Reddy, V. R. (2011). High-temperature effects on rice growth, yield, and grain quality. *Advances in Agronomy, 111*, 87–206.

Laine, A. L. (2023). Plant disease risk is modified by multiple global change drivers. *Current Biology, 33*(11), 574–583.

Lal, R. (1998). Soil erosion impact agronomic productivity and environmental quality. *Critical Reviews in Plant Sciences, 17*(4), 319–464.

Lamichhane, J. R., Dürr, C., Schwanck, A. A., Robin, M. H., Sarthou, J. P., Cellier, V., & Aubertot, J. N. (2017). Integrated management of damping-off diseases: A review. *Agronomy for Sustainable Development, 37*(1), 1–25.

Legrève, A., & Duveiller, E. (2010). Preventing potential diseases and pest epidemics under a changing climate. In *Climate Change and Crop Production* (pp. 50–70). CABI Publishing.

Ljuša, M., & Sitaula, B. K. (2016). Adaptation to climate change in agriculture. *Agroecology and Sustainable Food Systems, 40*(3), 228–249.

Loake, G., & Grant, M. (2007). Salicylic acid in plant defence—The players and protagonists. *Current Opinion in Plant Biology, 10*(5), 466–472.

Lucas, J. A. (2009). *Plant Pathology and Plant Pathogens*. John Wiley & Sons.

Maitra, S., Sahoo, U., Sairam, M., Gitari, H., Rezaei-Chiyaneh, E., Battaglia, L., & Hossain, A. (2023). Cultivating sustainability: A comprehensive review on intercropping in a changing climate. *Research on Crops, 24*(4), 702–715.

Majumdar, D. K. (2001). *Irrigation Water Management: Principles and Practice*. PHI Learning Pvt. Ltd.

Mamo, B. E., Eriksen, R. L., Adhikari, N. D., Hayes, R. J., Mou, B., & Simko, I. (2021). Epidemiological characterization of lettuce drops (*Sclerotinia* spp.) and biophysical features of the host identify soft stem as a susceptibility factor. *PhytoFrontiers™, 1*(3), 182–204.

Maponya, P., & Mpandeli, S. (2012). Climate change and agricultural production in South Africa: Impacts and adaptation options. *Journal of Agricultural Science, 4*(10), 48.

Marks, D. (2011). Climate change and Thailand: Impact and response. *Contemporary Southeast Asia*, 229–258.

Martinelli, F., Scalenghe, R., Davino, S., Panno, S., Scuderi, G., Ruisi, P., & Dandekar, A. M. (2015). Advanced methods of plant disease detection: A review. *Agronomy for Sustainable Development, 35*, 1–25.

Matsaba, E. O. (2021). Modeling of urban heat island and quantification of amelioration effects of plant species on microclimate and human thermal comfort in Nairobi City, Kenya (Doctoral dissertation, JKUAT-COETEC).

Melillo, J. M., Richmond, T. T., & Yohe, G. (2014). Climate change impacts in the United States. In *Third National Climate Assessment* (pp. 150–174). U.S. Global Change Research Program.

Melloy, P., Hollaway, G., Luck, J. O., Norton, R. O. B., Aitken, E., & Chakraborty, S. (2010). Production and fitness of *Fusarium pseudograminearum* inoculum at elevated carbon dioxide in FACE. *Global Change Biology, 16*(12), 3363–3373.

Mirriam, A. (2023). Optimizing soybean yield, phosphorus uses efficiency and economic returns using phosphatic fertilizer and inoculation with *Bradyrhizobium japonicum* in Lira City, Uganda (Master's thesis, Kenyatta University, Nairobi, Kenya).

Mithöfer, A., Schulze, B., & Boland, W. (2004). Biotic and heavy metal stress response in plants: Evidence for common signals. *FEBS Letters, 566*(1–3), 1–5.

Mohajan, H. (2013). *Global Greenhouse Gas Emissions and Climate Change*. LAP LAMBERT Academic Publishing.

Monkiedje, A., Spiteller, M., Maniepi, S. J. N., & Sukul, P. (2007). Influence of metalaxyl-and mefenoxam-based fungicides on chemical and biochemical attributes of soil quality under field conditions in a southern humid forest zone of Cameroon. *Soil Biology and Biochemistry, 39*(4), 836–842.

Moricca, S., & Ragazzi, A. (2009). Lusus naturae: Climate and invasions of plant pathogens modify agricultural and forest lands. *Italian Journal of Agronomy, 4*(s3), 13–18.

Morison, J. I. L., & Lawlor, D. W. (1999). Interactions between increasing CO_2 concentration and temperature on plant growth. *Plant, Cell & Environment, 22*(6), 659–682.

Mote, P. W., Abatzoglou, J. T., & Kunkel, K. E. (2013). Climate: Variability and change in the past and the future. In *Climate Change in the Northwest: Implications for our Landscapes, Waters, and Communities* (pp. 25–40).

Munir, M. (2018). Plant disease epidemiology: Disease triangle and forecasting mechanisms in highlights. *Hosts Virus, 5*(1), 7–11.

Noman, A., Bashir, R., Aqeel, M., Anwer, S., Iftikhar, W., Zainab, M., & Adnan, M. (2016). Success of transgenic cotton (*Gossypium hirsutum* L.): Fiction or reality? *Cogent Food & Agriculture, 2*(1), 1207844.

Nungula, E. Z., Mugwe, J., Nasar, J., Massawe, B. H. J., Karuma, A. N., Maitra, S., Seleiman, M. F., Dindaroglu, T., Khan, N., & Gitari, H. I. (2023). Land degradation unmasked as the key constraint in sunflower (*Helianthus annus*) production: Role of GIS in revitalizing this vital sector. *Cogent Food & Agriculture, 9*(2), 2267863.

Oerke, E. C., & Dehne, H. W. (2004). Safeguarding production—Losses in major crops and the role of crop protection. *Crop Protection, 23*(4), 275–285.

Ogden, A. E., & Innes, J. (2007). Incorporating climate change adaptation considerations into forest management planning in the boreal forest. *International Forestry Review, 9*(3), 713–733.

O'Neill, J. R. (2009). *Irish Potato Famine*. ABDO.

Ortiz, R., Sayre, K. D., Govaerts, B., Gupta, R., Subbarao, G. V., Ban, T., & Reynolds, M. (2008). Climate change: Can wheat beat the heat? *Agriculture, Ecosystems & Environment, 126*(1, 2), 46–58.

Parikka, P., Hakala, K., & Tiilikkala, K. (2012). Expected shifts in *Fusarium* species' composition on cereal grain in Northern Europe due to climatic change. *Food Additives & Contaminants: Part A, 29*(10), 1543–1555.

Parry, M., Rosenzweig, C., Iglesias, A., Fischer, G., & Livermore, M. (1999). Climate change and world food security: A new assessment. *Global Environmental Change, 9*, 51–67.

Parthasarathy, S. (2024). *Fundamentals of Plant Pathology*. CRC Press.

Paterson, E., Hall, J. M., Rattray, E. A. S., Griffiths, B. S., Ritz, K., & Killham, K. (1997). Effect of elevated CO_2 on rhizosphere carbon flow and soil microbial processes. *Global Change Biology, 3*(4), 363–377.

Pautasso, M., Döring, T. F., Garbelotto, M., Pellis, L., & Jeger, M. J. (2012). Impacts of climate change on plant diseases: Opinions and trends. *European Journal of Plant Pathology, 133*, 295–313.

Peixóto, J. P., & Oort, A. H. (1984). Physics of climate. *Reviews of Modern Physics, 56*(3), 365.

Pinkard, E. A., Kriticos, D. J., Wardlaw, T. J., Carnegie, A. J., & Leriche, A. (2010). Estimating the spatio-temporal risk of disease epidemics using a bioclimatic niche model. *Ecological Modelling, 221*(23), 2828–2838.

Rahman, M. U., Khan, A. Q., Rahmat, Z., Iqbal, M. A., & Zafar, Y. (2017). Genetics and genomics of cotton leaf curl disease, its viral causal agents and whitefly vector: A way forward to sustain cotton fiber security. *Frontiers in Plant Science, 8*, 1157.

Rasool, F., Singh, P., Anwar, A., Mughal, N., & Kumar, S. (2022). Disease management approaches-strategies and concepts. *The Pharma Innovation Journal, SP-11*(7), 413–417.

Razdan, V. K., & Sabitha, M. (2009). Integrated disease management: Concepts and practices. In *Integrated Pest Management: Innovation-Development Process* (pp. 369–389).

Reddy, P. P. (2015). *Climate Resilient Agriculture for Ensuring Food Security* (Vol. 373). Springer India.

Reddy, P. P., & Reddy, P. P. (2015). Impacts on plant pathogens. In *Climate Resilient Agriculture for Ensuring Food Security* (pp. 151–177).

Romero, F., Cazzato, S., Walder, F., Vogelgsang, S., Bender, S. F., & van der Heijden, M. G. (2022). Humidity and high temperatures are important for predicting fungal disease outbreaks worldwide. *New Phytologist, 234*(5), 1553–1556.

Sahoo, U., Maitra, S., Hossain, A., Santosh, D. T., Nath, S., Sairam, M., Sagar, L., Jena, J., Pattanayak, S., Gitari, H. I., & Rezaei-Chiyaneh, E. (2024). Biochar for soil health improvement in the present context of climate change: A reality or fantasy. In S. Fahad, M. Adnan, R. Zhou, T. Nawaz, & S. Saud (Eds.), *Biochar-Assisted Remediation of Contaminated Soils Under Changing Climate* (pp. 1–17). Elsevier. https://doi.org/10.1016/B978-0-443-21562-9.00001-3.

Salinari, F., Rettori, A., Gullino, M. L., Giosuè, S., Rossi, V., & Spanna, F. (2004). Effect of climate on epidemiology of *Plasmopara viticola* (Berk et Curt.) Berl. et de Toni in Piedmont (*Vitis vinifera* L). *Informatore Fitopatologico (Italy), 54*(4).

Sánchez-Martín, J., Rispail, N., Flores, F., Emeran, A. A., Sillero, J. C., Rubiales, D., & Prats, E. (2017). Higher rust resistance and similar yield of oat landraces versus cultivars under high temperature and drought. *Agronomy for Sustainable Development, 37*, 1–14.

Scheffer, R. P. (1997). *The Nature of Disease in Plants*. Cambridge University Press.

Scholthof, K. B. G. (2007). The disease triangle: Pathogens, the environment, and society. *Nature Reviews Microbiology, 5*(2), 152–156.

SDA, ERS. (2022). Agriculture and food statistics: Charting the essentials.

Seem, R. C. (2004). Forecasting plant disease in a changing climate: A question of scale. *Canadian Journal of Plant Pathology, 26*(3), 274–283.

Seleiman, M. F., Aslam, M. T., Alhammad, B. A., Hassan, M. U., Maqbool, R., Chattha, M. U., Khan, I., Gitari, H. I., Uslu, O. S., Roy, R., & Battaglia, M. L. (2021). Salinity stress in wheat: Effects, mechanisms and management strategies. *Phyton-International Journal of Experimental Botany, 91*(4), 667–694.

Singh, B. K., Delgado-Baquerizo, M., Egidi, E., Guirado, E., Leach, J. E., Liu, H., & Trivedi, P. (2023). Climate change impacts on plant pathogens, food security, and paths forward. *Nature Reviews Microbiology*, 1–17.

Southall, E., Brett, T. S., Tildesley, M. J., & Dyson, L. (2021). Early warning signals of infectious disease transitions: A review. *Journal of the Royal Society Interface, 18*(182), 20210555.

Srivastava, A., Kumar, S. N., & Aggarwal, P. K. (2010). Assessment on vulnerability of sorghum to climate change in India. *Agriculture, Ecosystems & Environment, 138*(3, 4), 160–169.

Steele, P., & Gibbs, M. (2018). *Crop Diseases and Their Control*. Scientific e-Resources.

Strand, J. F. (2000). Some agrometeorological aspects of pest and disease management for the 21st century. *Agricultural and Forest Meteorology, 103*(1, 2), 73–82.

Thomas, M. B. (1999). Ecological approaches and the development of "truly integrated" pest management. *Proceedings of the National Academy of Sciences, 96*(11), 5944–5951.

Thompson, I. D. (2000). Forest vegetation of Ontario: Factors influencing landscape change. In *Ecology of a Managed Terrestrial Landscape: Patterns and Processes of Forest Landscapes in Ontario* (pp. 30–53).

Tilman, D., Cassman, K. G., Matson, P. A., Naylor, R., & Polasky, S. (2002). Agricultural sustainability and intensive production practices. *Nature, 418*(6898), 671–677.

Tiwari, S. P., Nema, S., & Khare, M. N. (2013). Whitefly–A strong transmitter of plant viruses. *International Journal of Phytopathology, 2*(2), 102–120.

Toniutti, L., Breitler, J. C., Etienne, H., Campa, C., Doulbeau, S., Urban, L., & Bertrand, B. (2017). Influence of environmental conditions and genetic background of arabica coffee (*C. arabica* L) on leaf rust (*Hemileia vastatrix*) pathogenesis. *Frontiers in Plant Science, 8*, 2025.

Tran, H. (2013). *The Ecology and Integrated Management of Broom (Cytisus scoparius) in New Zealand Plantation Forests: A Thesis Presented in Partial Fulfillment of the Requirements for the Degree of Doctor of Philosophy in Ecology at Massey University, Manawatu, New Zealand* (Doctoral dissertation, Massey University).

Trębicki, P., Dáder, B., Vassiliadis, S., & Fereres, A. (2017). Insect-plant–pathogen interactions as shaped by future climate: Effects on biology, distribution, and implications for agriculture. *Insect Science, 24*(6), 975–989.

Trębicki, P., Vandegeer, R. K., Bosque-Pérez, N. A., Powell, K. S., Dader, B., Freeman, A. J., & Luck, J. E. (2016). Virus infection mediates the effects of elevated CO_2 on plants and vectors. *Scientific Reports, 6*(1), 22785.

Trivedi, P., Batista, B. D., Bazany, K. E., & Singh, B. K. (2022). Plant–microbiome interactions under a changing world: Responses, consequences, and perspectives. *New Phytologist, 234*(6), 1951–1959.

Trivedi, P., Leach, J. E., Tringe, S. G., Sa, T., & Singh, B. K. (2020). Plant–Microbiome interactions: From community assembly to plant health. *Nature Reviews Microbiology, 18*(11), 607–621.

Van Huis, A. (2009). Challenges of integrated pest management in sub-Saharan Africa. In *Integrated Pest Management: Dissemination and Impact: Volume 2* (pp. 395–417).

Van Maanen, A., & Xu, X. M. (2003). Modeling plant disease epidemics. *European Journal of Plant Pathology, 109*, 669–682.

VanRaden, P. M., Van Tassell, C. P., Wiggans, G. R., Sonstegard, T. S., Schnabel, R. D., Taylor, J. F., & Schenkel, F. S. (2009). Invited review: Reliability of genomic predictions for North American Holstein bulls. *Journal of Dairy Science, 92*(1), 16–24.

Vary, Z., Mullins, E., McElwain, J. C., & Doohan, F. M. (2015). The severity of wheat diseases increases when plants and pathogens are acclimatized to elevated carbon dioxide. *Global Change Biology, 21*(7), 2661–2669.

Vasquez, D. F., Hernandez, A., Torres, D., Borrero-Echeverry, F., Zuluaga, P., & Rincon, D. F. (2022). Drought as a modulator of plant–virus–vector interactions: Effects on symptom expression, plant immunity, and vector behaviour. *Plant Pathology, 71*(6), 1282–1292.

Vaughan, M., Backhouse, D., & Ponte, E. D. (2016). Climate change impacts on the ecology of *Fusarium graminearum* species complex and susceptibility of wheat to Fusarium head blight: A review. *World Mycotoxin Journal, 9*(5), 685–700.

Velásquez, A. C., Castroverde, C. D. M., & He, S. Y. (2018). Plant–pathogen warfare under changing climate conditions. *Current Biology, 28*(10), R619–R634.

Wakelin, S. A., Gomez-Gallego, M., Jones, E., Smaill, S., Lear, G., & Lambie, S. (2018). Climate change induced drought impacts on plant diseases in New Zealand. *Australasian Plant Pathology, 47*, 101–114.

Walsh, M. K., McCartney, H. A., & West, J. S. (2020). Climate indicators for agriculture. USDA Technical Bulletin 1953. Washington, DC, p. 40.

Wang, P., Marsh, E. L., Ainsworth, E. A., Leakey, A. D., Sheflin, A. M., & Schachtman, D. P. (2017). Shifts in microbial communities in soil, rhizosphere, and roots of two major crop systems under elevated CO_2 and O_3. *Scientific Reports, 7*(1), 15019.

Wolfe, B. B., Hall, R. I., Edwards, T. W., Jarvis, S. R., Sinnatamby, R. N., Yi, Y., & Johnston, J. W. (2008). Climate-driven shifts in quantity and seasonality of river discharge over the past 1000 years from the hydrographic apex of North America. *Geophysical Research Letters, 35*(24).

Wu, Y. X., & von Tiedemann, A. (2001). Physiological effects of azoxystrobin and epoxiconazole on senescence and the oxidative status of wheat. *Pesticide Biochemistry and Physiology, 71*(1), 1–10.

Xin, X. F., Nomura, K., Aung, K., Velásquez, A. C., Yao, J., Boutrot, F., & He, S. Y. (2016). Bacteria establish an aqueous living space in plants crucial for virulence. *Nature, 539*(7630), 524–529.

Zadoks, J. C., & Schein, R. D. (1979). *Epidemiology and Plant Disease Management* Epidemiology and Plant Disease Management.

Zayan, S. A. (2019). Impact of climate change on plant diseases and IPM strategies. In *Plant Diseases–Current Threats and Management Trends*. IntechOpen.

Zhou, Y., Van Leeuwen, S. K., Pieterse, C. M., Bakker, P. A., & Van Wees, S. C. (2019). Effect of atmospheric CO2 on plant defense against leaf and root pathogens of Arabidopsis. *European Journal of Plant Pathology, 154*, 31–42.

Ziska, L. H., & Runion, G. B. (2007). Future weed, pest, and disease problems for plants. In *Agroecosystems in a Changing Climate* (pp. 261–287).

Ziska, L., Crimmins, A., Auclair, A., DeGrasse, S., Garofalo, J. F., Khan, A. S., & Walls, I. (2016). The impacts of climate change on human health in the United States: A scientific assessment. Washington, DC: US Global Change Research Program, 189–216.

CHAPTER 7

Accelerating Role of Artificial Intelligence and Robotics in Crop Improvement

M. VENNELA,[1,*] D. DINESH VARMA,[1] SOUVIK SADHU,[2] SMRUTI RANJAN PADHAN,[3] DHIRENDRA KUMAR ROY,[4] and HARUN GITARI[5]

[1]*Department of Genetics and Plant Breeding, Dr. Rajendra Prasad Central Agricultural University, Pusa, Bihar, India*

[2]*Department of Soil Science and Agricultural Chemistry, Bihar Agricultural University, Sabour, Bihar, India*

[3]*Division of Agronomy, ICAR–Indian Agricultural Research Institute, New Delhi, India*

[4]*Department of Agronomy, Dr. Rajendra Prasad Central Agricultural University, Pusa, Samistapur, Bihar, India*

[5]*Department of Agricultural Sciences and Technology, School of Agriculture and Environmental Sciences, Kenyatta University, Nairobi, Kenya*

*Corresponding author

ABSTRACT

Artificial intelligence (AI) is a promising technology with the potential to transform conventional farming into smart agriculture by optimizing resource use, reducing human labor, and improving supply chain management, crop improvement, weed control, and disease management.

Climate-Resilient Agriculture for Sustainable Crop Production: New Adaptation and Mitigation Strategies. Sanjay Kumar, Shivani Ranjan, Sumit Sow, Smruti Ranjan Padhan, and Dhirendra Kumar Roy (Eds.)

DOI: 10.1201/9781998511150-7

Rapid technological advancement, particularly in automation and robotics, offers attractive solutions to meet rising food demands. Plant breeders play a central role in leveraging AI to enhance the sustainability and efficiency of breeding systems through advanced phenotyping and precision farming techniques. The integration of robots, smart sensors, and AI enables non-invasive measurement of physiological plant data, evaluation of complex traits, and prediction of plant performance and yield across various management scales. Machine learning (ML) algorithms can surpass the forecasting accuracy of traditional genotype-to-phenotype models by autonomously identifying data characteristics and relationships at multiple levels of abstraction. Deep learning, a subset of ML, employs artificial neural networks for model training using densely connected networks. AI also has applications in comparative genomics and knowledge transfer from model plants to target crops. To capitalize on the digital revolution, plant breeders must remain informed and evaluate AI-driven insights in alignment with farmers' needs. AI holds significant potential to revolutionize future breeding studies, including omics integration and management through advanced technologies.

7.1 INTRODUCTION

One of the evolving areas of agricultural science is plant breeding. Selection is the key to obtaining plants that have exceptional qualities in any breeding program. Following the discoveries of Gregor Mendel and Sir R. A. Fisher, statistics and genetics were eventually incorporated into traditional plant breeding (Ali et al., 2018). While traditional breeding has certain advantages, its combination with modern biotechnological techniques has proven to be exceptional. The primary techniques used in traditional plant breeding were yield components analysis, yield stability analysis, improved adaptability to both abiotic and biotic stresses, as well as hybrid breeding initiatives. The principal techniques used in *in vitro*-based biotechnological breeding were Agrobacterium-mediated gene transformation, doubled haploid production, artificial polyploidy induction, and *in vitro* micropropagation. Traditional and *in vitro* plant breeding research has both analyzed and interpreted data using conventional statistical techniques (Wijerathna-Yapa et al., 2022). The above-mentioned techniques rely on variability and linear regression approaches in order to find out the connection between the variables and also to know the effect of unrelated variables on the dependent ones (Wright, 1976). Though these approaches are beneficial for understanding linear relationships, they do not provide insights into the nonlinear interactions between the dependent and independent variables, as plant biological systems are intrinsically linked to nonlinear and unpredictable properties (De Souza et al., 2020). Manual measurement-based plant phenotyping is tedious, expensive, error-prone, and time-consuming. Most plant characteristics can be assessed through the use of high-throughput technology that includes robots for collecting the data required in both field and laboratory settings. Thus, plant breeding studies utilize robust data mining tools to anticipate and elucidate intricate data.

According to Tripodi et al. (2018), plant phenotyping is the estimation of both quantitative and qualitative characteristics of plants or their varieties in a specific environment. These characteristics include the plant's morphology, structure, biochemistry, physiology, and performance at different organizational levels. In addition to non-additive interactions between genetic and environmental factors, environmental factors also determine plant traits. The experimental studies conducted proved that there is a mutual relationship between the phenotypic traits, as they influence each other positively (Chawade et al., 2018). Breeders must be able to evaluate a large number of individual plants phenotypically within a variety in different locations to obtain ideal genotypes and observe genetic variation that offers optimal efficiency (Walter et al., 2015). High-throughput operations and systems have been created recently to quickly track and analyze massive numbers of plants in field and greenhouse locations. These devices enable fast assessment and wide area coverage by combining the latest sensor deployment methods with imaging and sensing modalities (Hassan et al., 2019).

Even though there has been disagreement regarding what constitutes intelligence for a long time, intelligence can be broadly defined as the capacity to draw lessons from the past and apply those lessons to relatively novel circumstances. The creation of software with the ability to solve problems and make decisions to duplicate the procedure of "sense, decide, and act" is known as artificial intelligence. It is designed for machine and computer use (Alen, 2016). The discipline of artificial intelligence covers many important areas, including probability, computation, and logical reasoning. Through the use of machine learning algorithms, artificial intelligence generates flexible learning strategies to achieve various environmental objectives. The measurement of plant genotypes and phenotypes has been revolutionized by recent technological advancements, resulting in the regular production of huge quantities of intricate information. Utilizing this genetic, genomic, and phenomics data to forecast or even demonstrate phenotypes derived from underlying genotypes in various environments presents a challenge for both basic and applied plant sciences. It is widely recognized that artificial intelligence and machine learning techniques have the capacity to extract pertinent information from unprocessed phenotype measurement data (Van Dijk et al., 2022). Deep neural networks (DNNs) are popular models in machine learning that have numerous sequentially stacked layers that enable the machine learning algorithm to learn capabilities that encode data gathered from images. The characteristics are generated automatically from input data using a DNN, eliminating the need for manual feature creation. Image classification, object detection, and object counting are among the agricultural tasks that can be automated with increased computing power (Tang and Rollin, 2021).

7.2 CLIMATE CHANGE IN AGRICULTURE

There is a significant influence of climate change on the agricultural sector, as many natural resources like water, land, and soil are primarily impacted by climate. Since most crops depend on the climate, its impact can be observed at different growing stages of crops as well as during different growing seasons, complicating agricultural practices. Its impact will be particularly pronounced in certain areas of the world (Walsh et al., 2020). The effect of climate change on agricultural productivity is increasing the risk of insufficient food supply. Various climate-related extremes, such as heat waves, droughts, heavy rainfall, floods, and other biotic factors, are negatively influencing the livelihoods of many farmers. Future predictions regarding climate change patterns indicate an extreme increase in temperature and irregular rainfall. However, climatic patterns are becoming variable, making it difficult to predict the extremity (Gouldson et al., 2016).

Many experiments have been conducted on different crops, especially cereals like wheat, rice, maize, and soybean, regarding yield and production in relation to the effects of climate change (Wang et al., 2018).

In rice crops, it was found that a minimum increase of 1°C in temperature above the normal growing temperature (32°C) resulted in a 10% reduction in production. The aforementioned climatic anomalies can be mitigated by adopting different strategies, such as identifying the traits that promote growth and development in plants when exposed to various stress conditions and pyramiding the genes controlling these traits to impart tolerance to significant climatic stresses. This, in turn, helps plants increase their survival and development abilities to their full potential. The creation of high-yielding, climate-resilient varieties through climate-smart mitigation technologies and the use of these varieties in farmers' fields will promote the importance of achieving climate-smart solutions.

Addressing the consequences of climate change impacts and assisting farmers in utilizing germplasm, wild relatives, and important varieties for the normal growth and development of plants is the primary duty of plant breeders. Conventional plant breeding strategies are being replaced by advanced technologies like marker-assisted selection, transgenic technology, genome editing, and genomic selection, which are of utmost importance for developing varieties resistant or tolerant to different climate conditions (Antoine et al., 2019). Pre-breeding, which involves the use of wild crop relatives, is also one of the novel techniques that expand the range of crop plants to meet the needs and demands of a growing population. The most important part of a breeding program, that is, selection, depends on phenotypic observations that can be strengthened by the development of fully effective and precise methods. This can ultimately be achieved by using techniques like omics studies and implementing artificial intelligence, the Internet of Things, and big data analysis in plant breeding (Srinivasarao et al., 2020).

7.3 ARTIFICIAL INTELLIGENCE AND ITS TYPES

The fundamental principle of artificial intelligence is the notion that human intelligence can be simplified for a machine to replicate and carry out tasks, from the most basic to the most complicated. Artificial intelligence aims to accomplish the following: perception, reasoning, and learning. Even though there are numerous approaches to categorize artificial intelligence, there are primarily two: capabilities-based and functional-based.

7.3.1 TYPE 1 AI: CAPABILITIES-BASED

1. **Weak AI:** It as the name suggests, possesses intelligence that is sufficient to carry out a single task; hence, it can also be named narrow AI. Among the other AI methods, it is observed to be the most widespread type that is currently accessible. As weak AI is used only to perform a particular task, it is limited to its field of expertise and cannot perform beyond it. Consequently, it is known as weak AI. If narrow AI is allowed to operate for too long, it may fail in unexpected ways.
2. **General AI:** It is the advanced approach that is capable enough to perform all tasks as efficiently as humankind. The goal of general artificial intelligence is to create a system that is capable of thinking like a human on its own. There is not currently a system that fits under general artificial intelligence or that can perform any task with the accuracy of a human. The goal of current global research is to create machines with general artificial intelligence. Because general artificial intelligence (AI) systems are still being researched, developing them will require a lot of time and work.
3. **Super Artificial Intelligence:** It pertains to a system intelligence level where machines are able to outperform humans in any task and have cognitive qualities. It is a result of AI in general. Strong artificial intelligence is characterized by several essential traits, such as autonomous thinking, reasoning, problem-solving, judgment, planning, learning, and communication. Super artificial intelligence is still just a theoretical idea. The task of developing such systems in reality continues to change the world.

7.3.2 TYPE 2 AI: FUNCTION-BASED

1. Machines that are solely reactive are the most fundamental forms of artificial intelligence, thus called reactive machines. These AI systems do not save information from the past or memories for use in the future. These devices only consider the situations that are occurring right now and respond in the best way possible.

2. For a brief amount of time, limited memory devices can store information or memories from the past. The amount of time that these devices may utilize saved data is limited.
3. Another function-based approach, that is, the theory of mind AI, has the capacity to interact and communicate with human beings. It also has an additional feature to understand human feelings and beliefs. Though it is theoretically possible, its practical application still requires a lot of effort, and currently, researchers are working on this progress.
4. Future Artificial Intelligence is known as Self-awareness AI. These robots will be extremely intelligent and possess sentience, emotions, and self-awareness. The intelligence of these machines will surpass that of humans. It is still a theoretical concept and does not currently exist in reality.

7.4 SUBSETS OF AI

7.4.1 MACHINE LEARNING

There are three primary divisions for machine learning models:

1. **Supervised Machine Learning, or Supervised Learning:** It is characterized by using labeled datasets to create algorithms for precise outcome prediction or data classification. The model modifies its weights as source data is entered until an appropriate fit is achieved. This happens during the cross-validation process, which ensures the model does not overfit or underfit. In a daily example, supervised ML helps us differentiate and protect important emails from spam, which is one of its vital properties. Support vector machines, random forests, logistic regression, neural networks, naïve Bayes, and linear regression are a few techniques used in supervised learning.
2. **Unlike Supervised ML:** The evaluation and differentiation of unlabeled data using different algorithms can be achieved through unsupervised learning. The algorithms used analyze the hidden patterns observed in the provided data without any human intervention. This approach is perfect for customer segmentation, cross-selling tactics, exploratory data analysis, and pattern and

image recognition because it can identify patterns and similarities in data. Two common techniques for this are singular value decomposition and principal component analysis. Neural networks, probabilistic clustering techniques, and k-means clustering are some additional algorithms employed for unsupervised learning.

3. **The Other Learning Method:** It has the features of the above-mentioned methods is known as semi-supervised machine learning. This method uses labeled datasets to classify and extract features based on unlabeled data. To run the algorithm, more unlabeled data is required than labeled data, thus solving the problem of insufficient labeled data requirements in supervised ML. The primary benefit of semi-supervised learning is its lower cost due to its reduced usage of labeled data.

7.4.2 DEEP LEARNING

1. **Artificial Neural Networks:** It is further referred to as deep learning, use a combination of measurements, bias, and data elements to simulate the functioning of the human brain. Together, these components enable precise object identification, classification, and description within the data.
2. **Deep Neural Networks:** They are made up of several layers of nodes that are linked together, each of which builds on the one before it to improve and optimize the classification or prediction. Forward propagation is the term used to describe this computation-based network progression. Deep learning is more like a neural network of the human brain that pertains to both the input layer, where the processing of data takes place, and the output layer, where the ultimate categorization is made.
3. **Backpropagation:** In an additional procedure known as backpropagation, the model is trained by first calculating prediction errors using algorithms such as gradient descent. The function's weights and biases are then modified by going in reverse through the layers. A neural network can make estimations and adjust for errors by utilizing the methods of forward and backpropagation in tandem. The algorithm gradually improves in accuracy over time (Table 7.1).

TABLE 7.1 The Key Characteristics of Various Machine Learning Techniques for Plant Breeding According to Ranking

ML Method	Hyperparameter Tuning	Overfitting Risk	Explainability	Comparative Accuracy	Complexity	Samples Needed	Computation Cost	Implementation Time
Deep neural network	Very high	Low	Low	Very high	Very high	Very high	Very high	Very high
Artificial neural network	High	High	Low	High	High	Medium	High	High
Random forest	High	Medium	High	High	Medium	Low	Medium	Low
K nearest neighbor	Medium	High	Medium	Medium	High	Medium	Low	Low
Linear SVM	High	High	Low	High	Medium	High	Medium	Medium
Decision tree	Medium	High	High	Medium	Medium	Medium	Medium	Low
Gradient boosting	Very high	High	Low	High	Very high	Medium	High	Medium
Logistic regression	Low	High	High	Low	Medium	Medium	Medium	Low
Naive bayes	Medium	Medium	Medium	Medium	High	Medium	High	Medium
Bayesian network	High	Medium	Medium	Medium	High	Medium	High	Medium

Source: Adapted from: Yoosefzadeh et al. (2023).

7.5 OBSTACLES IN TRADITIONAL AGRICULTURE PRACTICES

Prior to comprehending the implications and applications of AI in agriculture, it is important to recognize the difficulties associated with traditional agricultural practices (Singh and Singh, 2017). These difficulties are listed below:

- A variety of weather conditions, including humidity, temperature, and rainfall, are significant in farming. Because of pollution, the climate can fluctuate suddenly, making it challenging for farmers to plan ahead for harvesting, seeding, and soil preparation.
- Fertile soil with the necessary nutrients, including potassium, phosphorus, and nitrogen, is essential for better crop yields. The determination of soil quality through conventional methods is a challenging task as it ultimately affects the quality of crops through insufficient supply of nutrients that are required for normal plant growth.
- Protecting the crops from weeds is of utmost importance as they disturb the normal agricultural cycle and become competitors by absorbing soil nutrients and water, which in turn increases the cost of production. However, crop identification and weed prevention using conventional methods are ineffective.

7.6 THE OUTCOMES OF ARTIFICIAL INTELLIGENCE ON AGRICULTURE

AI-based technologies help to increase productivity across the board and manage the problems that different sectors, including the agricultural sector, encounter, such as crop production, irrigation, soil content sensing, crop surveillance, weeding, and cultivation (Kim et al., 2008). The purpose of agricultural robots is to provide high-value AI applications in the aforementioned industry. The agricultural industry is in crisis due to the world's rapidly growing population, but artificial intelligence may be able to provide much-needed assistance.

The following are several methods by which AI has benefited the agricultural industry:

1. Autonomous UAVs have drawn more attention in recent years due to their potential uses in recognition and surveillance. Drones, or

unmanned aerial vehicles, are gaining popularity as a means to reach enormous heights and perform numerous tasks due to their flexibility and amazing imaging technology, which encompasses everything from delivery to photography (Tomic et al., 2012).

2. In a few decades, farming will be discovered to be a combination of natural and computational abilities due to artificial intelligence. This will not only improve the quality of the product for all farmers but also reduce their expenses and duties.
3. The best crop selection has been made easier by developing technologies, which have also enhanced the choice of hybrid seed options that are most suitable for farmers' requirements. This was successfully implemented through an understanding of how various soil types and weather patterns impact seeds. With this information gathered, the likelihood of plant diseases decreases.
4. Robots are used extensively to carry out complicated tasks across a range of industries, primarily in manufacturing. Currently, many robots are used in the agricultural sector, and to fulfill this criterion, numerous AI startups are being established. These AI robots are designed to be capable of carrying out various farming tasks (Khan et al., 2018).
5. Weeds are easily detected by AI sensors, which can also identify areas where weed growth has occurred. Herbicides can be accurately sprayed in these areas to minimize their use and save time and reduce crop damage. Numerous artificial intelligence companies are developing robots that combine AI and computer vision to accurately spray herbicides.
6. To put it clearly, chatbots are conversational virtual assistants that perform user interactions automatically. Chatbots driven by artificial intelligence and machine learning techniques have made it possible for us to comprehend natural language and communicate with users in a more tailored manner. Agriculture has made use of this facility by helping farmers find answers to their unresolved inquiries while also offering them guidance and a range of suggestions (Tanha et al., 2020).
7. Precision farming is the most accurate method that has many beneficial abilities. It collects precise data by using drone technology in various field aspects and ultimately helps in reducing labor requirements. It also helps in identifying plant stress levels through sensor technology and by capturing high-resolution

pictures. After that, a machine learning model receives the sensor data as input to identify stress (Figure 7.1).

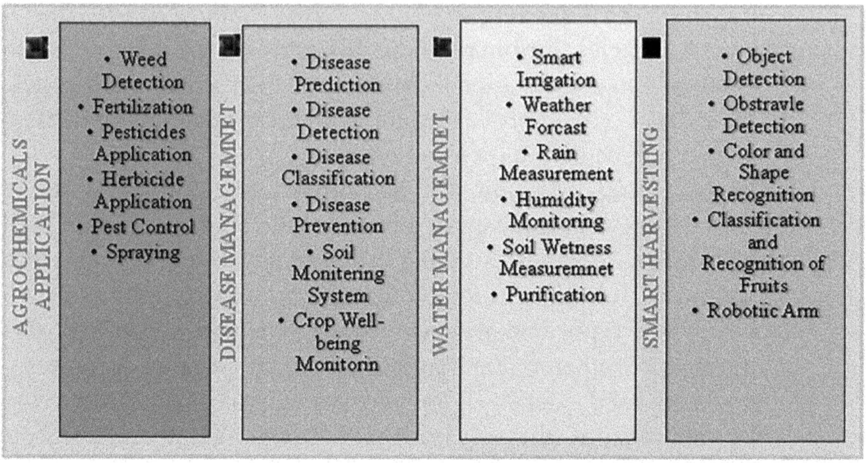

FIGURE 7.1 Applications of AI and IoT for smart agriculture.

Source: Modified from: Katiyar and Farhana (2021). https://creativecommons.org/licenses/by/4.0/

7.7 AGRICULTURAL ROBOTICS AND ITS USE IN BREEDING

In a number of ways, agricultural robots are helping farmers increase their production yields. The advanced equipment is being used in inventive and creative ways for everything from robotic arms to drones and self-driving tractors. Numerous ground-based agricultural robots were recently designed to carry out precision farming tasks and supplement or replace humans in specific roles. There are two primary categories of these robots: (i) mobile robots that can move on their own; and (ii) robotic "smart" tools that are transported by a vehicle.

Type-I robots come in abundant shapes and sizes. With the advent of GPS/GNSS auto-guidance systems, traditional agricultural self-propelled machinery like tractors and sprayers, and in the last 10 years, the development of "robotized" combine harvesters has been witnessed. Type-II robots, also known as "smart" tools, are being created for numerous applications; some of these uses include mechanical weeding, lettuce thinning, and transplanting. Some of these uses are currently readily accessible (Bechar and Vigneault, 2017).

There are many new obstacles in modern breeding for mechanized, engineered, and facility agriculture. AI is necessary to solve these problems, which are brought on by growing data volumes, an assortment of samples, low valid-event rates, structural difficulty, diverse information, and the requirement for data sharing. Expert systems that can predict or classify data based on input are created using AI algorithms. Breeding information systems will be significantly impacted by AI. Later AI-assisted breeding applications can benefit from the abilities as well as understanding of breeders, and the virtualization of breeding will accelerate the shift from direct to AI-driven breeding (Araus and Cairns, 2014). Artificial intelligence-equipped robots have been successful in resolving scientific issues. For instance, the AI-robot system AlphaFold 2 currently estimated the 3D structures of nearly all proteins and greatly surpassed other groups in a protein-folding competition (Yunbi et al., 2022).

Therefore, robotically automated AI technologies could support agricultural cultivation and breeding in numerous ways, including gathering information like recognition and assessment of data related to breeding, data interpretation, that is, the fusion of data and scientific breeding details to create selection models, and breeding choices through the selection of genotypes to begin the subsequent breeding cycles. To produce datasets that may be applied effectively to train a predictive breeding approach, it is imperative to apply an attentive experimental design, use an adequate number of biological replicates, and capture as many details as feasible regarding variation in the environment across and within field sites.

7.7.1 ROBOTS USED TO MEASURE PLANT TRAITS

There are two categories of robots designed for plant phenotyping applications: those intended for regulated indoor environments like greenhouses or laboratories, and those intended for outdoor environments like fields (Shafiekhani et al., 2017). Plants in controlled circumstances are either shifted by a belt system or through other forms of automation to a particular spot where the working process of the robot starts, or the plants are set up in a particular position and the machine shifts within the premises to contact the plants. The robotic system rarely needs to come into contact with the plants.

Time of Flight cameras or 3D laser scanners are used as depth sensors or RGB cameras on the robotic arm in order to capture visible images and point cloud data. Next, using a three-dimensional representation of the plants that has been rebuilt, the morphological features of the plants are determined. An experiment conducted by Lu et al. (2017) resulted in the measurement of stem height and leaf length being possible by using a TOF lens and a robotic arm. A robotic system with a laser scanner was used to calculate the surface area and volume in crops like barley and Arabidopsis (Chaudhury et al., 2017). An additional set of indoor plant phenotyping robots aimed to interact with plants or plant organs, expanding the scope of robotic phenotyping beyond the external morphological characteristics of plants to include intrinsic physiological and biochemical characteristics.

Certain physiological and biochemical characteristics can only be measured when sensors and plants come into direct contact. Therefore, a robot that can hold or touch plant organs is needed to measure these attributes. In turn, capturing plant organs necessitates a skilled robotic arm in addition to being fitted with sensors and algorithms that can recreate the target plant's three-dimensional geometry. To determine the stalk strength of sorghum, a UGV outfitted with a customized camera was developed (Mueller-Sim et al., 2017). It had a three-degree-of-freedom robotic arm and a specially made end-effector. The end-effector was an altered rind penetrometer with a needle and force gauge attached. To take the measurement, they were driven into the stalk once the end-effector was around it.

7.8 CONVENTIONAL TO BREEDING THROUGH AI

The ancient practice of plant breeding carries on the legacy of creating better plant cultivars since the beginning of agriculture. Soon after the first domestication of cereal grains, humans started to recognize the superior performance of a few plants phenotypically in their fields, and they saved seeds that were of better quality to cultivate new crops (Khan et al., 2022). The science of breeding plants has been constantly improved by new methods mediated by various advances in technology since the process's inception, beginning with crop domestication. This has increased the speed, accuracy, and precision of plant breeding. Basic selection strategies were modeled after early plant-breeding methods. Approximately 7000 distinct kinds of food plants were discovered and grown by people from around the globe about 10,000 years ago, which resulted in the first stage of breeding, that is, Breeding 1.0. Following the discovery of inbreeding depression in the late 19th and early 20th centuries, breeding stage 2.0 got underway.

The science of breeding itself saw many advancements in plant breeding during this period, including authorized designs for experiments, regulated crossings, replicated field trials, statistical assessments, hybrid breeding, and pedigree-based predictions of breeding values. Breeding 3.0 started around 30 years ago when genomic and DNA marker contents began to support phenotypic data. With the abundance of omics data and the quick development of informatic technologies, Breeding 4.0 is quickly approaching (Wallace et al., 2018). The discipline of artificial intelligence is still new but has been shown to have tremendous potential to help with the breeding of smart crops resistant to climate change. Crops that are climate-resilient are able to withstand a variety of biotic and abiotic stresses and either maintain or improve crop yields and quality with the changing climate (Harfouche et al., 2019).

7.9 AI BENEFITING CROP BREEDING

Using high-throughput genomics and phenomics for sophisticated breeding, AI has sped up the intensity of creating new plant varieties (Esposito et al., 2020). Marker-assisted selection, genomic prediction, and genomic selection have all seen a rise in the application of ML methodology. Several farming companies have made investments of several billion dollars, including Monsanto and John Deere, to develop methods that can use an abundance of data on weather, seed viability, and soil type to help farmers lessen production costs and ultimately increase productivity (Crossa et al., 2017). Recent studies on AI tools, such as machine learning (ML), deep learning, and predictive analysis, aimed to improve people's capacity for studying, organizing, deductively reasoning, and taking initiative. To help with an improved comprehension of plant behavior under various climatic conditions, plant breeders are creating systems. AI has many potential uses for in-depth genome investigation and is currently being employed thoroughly in plant genomics. For various types of computational biology examination, a variety of machine learning tools and algorithms are readily available (Mahood et al., 2020). The successful advancements in high-throughput sequencing technology have yielded a substantial amount of genomic data; however, the massive volume of data generated poses a major obstacle to the evaluation and preservation of data. The measurement of many genes' simultaneous expressions, or even all the genes in the genome, under a variety of conditions, is made possible by bioinformatics AI technology.

7.9.1 THE PREDICTION OF PHENOTYPE FROM GENOTYPE USING MACHINE LEARNING TECHNIQUES

Machine learning algorithms can independently gather characteristics of data and indicate their associations at various degrees of abstraction; they have the potential to outperform current genotype-to-phenotype prediction tools with regard to prediction accuracy (Danilevicz et al., 2022). Estimating genotype to phenotype has become more accurate with the use of GS. Both Bayesian systems and GBLUP have been widely used in GS (Meuwissen et al., 2001). The substantial dimension of marker data relative to certain individuals and the existence of intricate connections that are challenging to account for can present difficulties for linear methods in GS, despite their successes (Crossa et al., 2017). Nonlinear techniques like ML and DL frameworks were used frequently to forecast plant phenotypes in GS in an effort to enhance linear models (Montesinos-Lopez et al., 2021).

Additionally, multimodal information and information types that are unsuited to straightforward tabular formats can be incorporated in ML and DL architectures. For instance, the machine learning technique random forest may compensate for non-additive effects and identify patterns in data with multiple dimensions to produce precise predictions. According to Spindel et al. (2015), the genomic selection model had better prediction of genetic variance present in an important trait of the crops over other linear models that are included in the study, as it also analyzes the data based on the nonlinear relationship. The application of different ML approaches, such as conventional and deep neural networks, has promised the prediction of accurate results in the genomic selection model. With the right hyperparameter optimization, these models may outperform linear methods (Sandhu et al., 2020).

7.10 APPLICATIONS IN PLANT BREEDING

7.10.1 GENETIC DIVERSITY ASSESSMENT

In any breeding programs, genetic diversity analysis plays a major role, and this diversity analysis allows a breeder to select parents from a germplasm accession and their use in hybridization programs. The use of AI technology helps to assess the diversity among the germplasm, crop relatives, or varieties more accurately. Through a deep learning algorithm, that is, object detection, genotypic classification and efficient genetic diversity evaluation can be achieved. The morphological characteristics can be detected using different AI algorithms like k-nearest neighbor, CNN, support vector machines, and PCA operated by eCNN, as they are an appropriate choice to reduce the time. One of the examples of the use of an AI algorithm is the experiment conducted by Sakoda and his coworkers (2019) for detecting the importance of stomatal density in the photosynthetic activity of 90 soybean accessions. To carry out the analysis, a central leaflet from each genotype was collected, and its microscopic pictures were obtained. Not only morphological character information but also molecular characterization information can be utilized to classify objects using ML techniques. Using molecular markers, the sequences of DNA and RNA can be extracted, which can be used to train CNNs for the purposes of genotypic classification or molecular biology studies. The genomic prediction and genomic selection in plant breeding, when combined with artificial neural networks in a variety of plants, resulted in high accuracy of data. Besides other advantages, the ML methods help in locating unique SNPs (Zingaretti et al., 2020).

7.10.2 ANALYSIS AND PREDICTION OF YIELD COMPONENTS

Most breeding programs always aim to increase the economic yield. On the other hand, yield is a complicated quantitative trait that is greatly impacted by the environment and typically controlled by multiple genes. Regression-based models in traditional plant breeding research have one significant flaw: nonlinear relationships within the variables cannot be analyzed. The analysis of nonlinear interactions among both independent and dependent variables is made possible by the implementation of nonlinear machine learning algorithms. ANNs are additionally utilized for predicting winter wheat and winter rapeseed yields based on meteorological information and details about the application of mineral fertilizer (Niedbala, 2019).

7.10.3 YIELD STABILITY AND GENOTYPE X ENVIRONMENT INTERACTION

Factors influencing crop productivity and yield include plant genotype, agricultural practices, and the surrounding environment. The connections and how these elements interact create a complicated scenario that affects how much a plant can potentially yield. A machine learning algorithm (ANN) provided better results when multiple dependent and independent variables were used for data analysis (Niazian et al., 2020). The prediction of yield at different locations and environments can be better analyzed through the ANN model as it considers additional variables that predict the data more quickly and precisely.

7.10.4 BIOTIC AND ABIOTIC STRESS ASSESSMENT

Plants should undergo different types of abiotic and biotic stresses during their life cycle. To avoid these stresses, various new techniques are applied to determine the superior traits and evaluate their genotypic characteristics. These techniques thus help in developing varieties that are either tolerant or resistant to different stresses. Evaluation of different omics data through machine learning techniques, in addition to digital images that stimulate and anticipate the genotypic reactions, helps in determining the genotypes that are more resilient to stress and non-stress conditions. Also, large-scale data can be interpreted through ML for testing the genotypes against different stresses. To categorize and recognize different leaf diseases in maize, an application along with the CNN algorithm was installed that provides accurate data about the stress conditions that provoked the development of diseases to farmers (Sibiya and Sumbwamyambe, 2019).

7.10.5 MATING DESIGNS AND HYBRIDIZATION PROGRAMS

A few traditional mating designs resulted from the introduction of statistics into genetics. Utilizing these techniques, genetic analysis of crops has been carried out to ascertain the type of gene actions responsible for regulating significant morphological, phenological, and yield component traits. In order to solve the problem associated with examining a large number of cross combinations in maize, matrix factorization and neural network approaches were used, and the results revealed outstanding performance over other algorithms (Khaki et al., 2020).

7.11 CONCLUSION

Debatable topics about AI's global applications have arisen because of the tremendous increase in recent years in the discourse surrounding

AI's significance. Modern plant breeding is required to capitalize on the digital revolution. To succeed in their future work, breeders and researchers will need to weigh farmer requirements against computer-generated recommendations. AI will concentrate on creating innovative, human-centered methods and evaluating how robotic technology can be applied to numerous global sectors and companies. The worldwide agricultural breeding industry will benefit from the creation of different AI systems considering the accessibility of metaphor structures, which have the capacity for deductive reasoning and the presence of wisdom. Artificial intelligence algorithms have the potential to be applied in comparative genomics studies of model plants and the transmission of the data to any crop. Breeders and farmers must provide the data collected by them during their research or fieldwork into a big data cloud AI application to accelerate the role of artificial intelligence and their functions. Although the omics data acquired through machine and deep learning methods are precise, the process of breeding is still difficult, expensive, and time-consuming, so breeding cannot be sped up solely by technological means.

KEYWORDS

- **agrobacterium**
- **artificial intelligence**
- **crop disease management**
- **data mining tools**
- **deep learning**
- **genomics**
- **greenhouse monitoring**
- ***in vitro* breeding**
- **machine learning**
- **neural networks**
- **smart sensors**

REFERENCES

Alen, A. (2016). Transformative technologies. UTS Center for Autonomous Systems. Retrieved from: www.uts.edu.au/research-and-teaching/our-research/center-autonomous-systems.

Ali, S., Shafique, O., Mahmood, T., Hanif, M. A., Ahmed, I., & Khan, B. A. (2018). A review about perspectives of nanotechnology in agriculture. *Pakistan Journal of Agricultural Research, 31*, 116–121.

Antoine, L. H., Daniel, A. J., David, K., Jonathon, C. R., Antoine, H. H., Giuseppe, S. M., Menachem, M., Gerald, A., Tuskan, J. J. B., & Arie, A. (2019). Accelerating climate-resilient plant breeding by applying next-generation artificial intelligence. *Trends in Biotechnology, 37*(11), 1217–1235.

Araus, J. L., & Cairns, J. E. (2014). Field high-throughput phenotyping: The new crop breeding frontier. *Trends in Plant Science, 19*(1), 52–61.

Bechar, A., & Vigneault, C. (2017). Agricultural robots for field operations. Part 2: Operations and systems. *Biosystems Engineering, 153*, 110–128.

Chaudhury, A., Ward, C., Talasaz, A., Ivanov, A. G., Brophy, M., & Grodzinski, B. (2017). Machine vision system for 3D plant phenotyping. *ArXiv. 1705*, 00540.

Chawade, A., Van Ham, J., Blomquist, H., Bagge, O., Alexandersson, E., & Ortiz, R. (2019). High-throughput field-phenotyping tools for plant breeding and precision agriculture. *Agronomy, 9*, 258.

Crossa, J., Perez-Rodríguez, P., Cuevas, J., Montesinos-Lopez, O., Jarquin, D., De Los Campos, G., Burgueno, J., González-Camacho, J. M., Pérez-Elizalde, S., & Beyene, Y. (2017). Genomic selection in plant breeding: Methods, models, and perspectives. *Trends in Plant Science, 22*, 961–975.

Danilevicz, M. F., Gill, M., Anderson, R., Batley, J., Bennamoun, M., Bayer, P. E., & Edwards, D. (2022). Plant genotype to phenotype prediction using machine learning. *Frontiers in Genetics, 13*, 822173.

De Souza, M. H., Pereira Junior, J. D., Steckling, S. M., Mencalha, J., Dias, F. D. S., Rocha, J. R. D. A. S. C., Carneiro, P. C. S., & Carneiro, J. E. S. (2020). Adaptability and stability analyses of plants using random regression models. *PLoS ONE, 15*(12), e0233200.

Esposito, S., Carputo, D., Cardi, T., & Tripodi, P. (2020). Applications and trends of machine learning in genomics and phenomics for next-generation breeding. *Plants, 9*, 34.

Gouldson, A., Colenbrander, S., Sudmant, A., Paparagyropoulou, E., Kerr, N., & McAnulla, F. (2016). Cities and climate change mitigation: Economic opportunities and governance challenges in Asia. *Cities, 54*, 11–19.

Hassan, M. A., Yang, M., Rasheed, A., Yang, G., Reynolds, M., & Xia, X. (2019). Rapid monitoring of NDVI across the wheat growth cycle for grain yield prediction using a multi-spectral UAV platform. *Plant Science, 282*, 95–103.

Katiyar, S., & Farhana, A. (2021). Smart agriculture: The future of agriculture using AI and IoT. *Journal of Computer Science, 17*(10), 984–999.

Khaki, S., Khalilzadeh, Z., & Wang, L. (2020). Predicting yield performance of parents in plant breeding: A neural collaborative filtering approach. *PLoS ONE, 15*, e0233382.

Khan, M. H. U., Wang, S., Wang, J., Ahmar, S., Saeed, S., Khan, S. U., Xu, X., Chen, H., Bhat, J. A., & Feng, X. (2022). Applications of artificial intelligence in climate-resilient smart-crop breeding. *International Journal of Molecular Sciences, 23*, 1–13.

Khan, M. H. U., Wang, S., Wang, J., Ahmar, S., Saeed, S., Khan, S. U., Xu, X., Chen, H., Bhat, J. A., & Feng, X. (2023). Applications of artificial intelligence in smart-crop breeding. *Encyclopedia*. Retrieved from: https://encyclopedia.pub/entry/48652 (accessed on 16 January 2025).

Khan, Z. H., Khalid, A., & Iqbal, J. (2018). Towards realizing robotic potential in future intelligent food manufacturing systems. *Innovative Food Science and Emerging Technologies, 48*, 11–24.

Kim, Y. J., Evans, R. G., & Iversen, W. M. (2008). Remote sensing and control of an irrigation system using a distributed wireless sensor network. *IEEE Transactions on Instrumentation and Measurement, 57*(7), 1379–1387.

Lu, H., Tang, L., Whitham, A. S., & Mei, Y. (2017). A robotic platform for corn seedling morphological traits characterization. *Sensors, 17*, 2082.

Mahood, E. H., Kruse, L. H., & Moghe, G. D. (2020). Machine learning: A powerful tool for gene function prediction in plants. *Applications in Plant Sciences, 8*, e11376.

Meuwissen, T. H. E., Hayes, B. J., & Goddard, M. E. (2001). Prediction of total genetic value using genome-wide dense marker maps. *Genetics, 157*, 1819–1829.

Montesinos-López, O. A., Montesinos-López, A., Hernandez-Suarez, C. M., Barrón-López, J. A., & Crossa, J. (2021). Deep-learning power and perspectives for genomic selection. *Plant Genome, 14*(3), e20122.

Muller-Sim, T., Jenkins, M., Abel, J., & Kantor, G. (2017). The Robotanist: A ground-based agricultural robot for high-throughput crop phenotyping. *IEEE International Conference on Robotics and Automation (ICRA), Singapore,* 3634–3639.

Niazian, M., Sadat-Noori, S. A., Abdipour, M., Tohidfar, M., & Mortazavian, S. M. M. (2018). Image processing and artificial neural network-based models to measure and predict physical properties of embryogenic callus and number of somatic embryos in Ajowan (*Trachyspermum ammi L. Sprague*). *In Vitro Cellular and Developmental Biology-Plant, 54,* 54–68.

Niedbala, G. (2019). Application of artificial neural networks for multi-criteria yield prediction of winter rapeseed. *Sustainability, 11,* 533.

Sakoda, K., Watanabe, T., Sukemura, S., Kobayashi, S., Nagasaki, Y., Tanaka, Y., & Shiraiwa, T. (2019). Genetic diversity in stomatal density among soybeans elucidated using high-throughput technique based on an algorithm for object detection. *Scientific Reports, 9,* 7610.

Shafiekhani, A., Kadam, S., Fritschi, B. F., & De Souza, N. G. (2017). Vinobot and Vinoculer: Two robotic platforms for high-throughput field phenotyping. *Sensors, 17,* 214.

Sibiya, M., & Sumbwanyambe, M. A. (2019). Computational procedure for the recognition and classification of maize leaf diseases out of healthy leaves using convolutional neural networks. *Agri Engineering, 1,* 9.

Singh, R., & Singh, G. S. (2017). Traditional agriculture: A climate-smart approach for sustainable food production. *Energy Ecology Environment, 2,* 296–316.

Spindel, J., Begum, H., Akdemir, D., Virk, P., Collard, B., & Redona, E. (2015). Genomic selection and association mapping in rice (*Oryza sativa*): Effect of trait genetic architecture, training population composition, marker number, and statistical model on accuracy. *PLoS Genetics, 11,* e1004982.

Srinivasarao, Ch., Srinivas, T., Rao, R. V. S., Rao, N. S., Vinayagam, S. S., & Krishnan, P. (2020). Climate change and Indian agriculture: Challenges and adaptation strategies. ICAR-National Academy of Agricultural Research Management, Hyderabad, Telangana, India, 584.

Tang, W. H., & Rollin, A. (2021). Model identification for ARMA time series through convolutional neural networks. *Decision Support Systems, 146,* 113544.

Tanha, T., Dhara, S., Nivedita, P., Hiteshri, Y., & Manan, S. (2020). Implementation of artificial intelligence in agriculture for optimization of irrigation and application of pesticides and herbicides. *Artificial Intelligence in Agriculture, 4,* 58–73.

Tomic, T., Schmid, K., Lutz, P., Domel, A., Kassecker, M., Mair, E., Grixa, I. L., Ruess, F., Suppa, M., & Burschka, D. (2012). Toward a fully autonomous UAV: Research platform for indoor and outdoor urban search and rescue. *IEEE Robotics & Automation Magazine, 19*(3), 46–56.

Tripodi, P., Massa, D., Venezia, A., & Cardi, T. (2018). Sensing technologies for precision phenotyping in vegetable crops: Current status and future challenges. *Agronomy, 8,* 57.

Van Dijk, A. D. J., Shiu, S. H., & De Ridder, D. (2022). Editorial: Artificial intelligence and machine learning applications in plant genomics and genetics. *Frontiers in Artificial Intelligence, 5,* 959470.

Wallace, J. G., Rodgers-Melnick, E., & Buckler, E. S. (2018). On the road to breeding 4.0: Unraveling the good, the bad, and the boring of crop quantitative genomics. *Annual Review of Genetics, 52,* 421–444.

Walsh, M. K., Backlund, P., Buja, L., De Gaetano, A., Melnick, R., Prokopy, L., Takle, E., Todey, D., & Ziska, L. (2020). Climate indicators for agriculture. USDA Technical Bulletin 1953, Washington, DC, 70 pages.

Walter, A., Liebisch, F., & Hund, A. (2015). Plant phenotyping: From bean weighing to image analysis. *Plant Methods, 11,* 14.

Wang, Z., Li, J., Lai, C., Wang, R. Y., Chen, X., & Lian, Y. (2018). Drying tendency dominating the global grain production area. *Global Food Security, 16,* 138–149.

Wijerathna-Yapa, A., Ramtekey, V., Ranawaka, B., & Basnet, B. R. (2022). Applications of *in vitro* tissue culture technologies in breeding and genetic improvement of wheat. *Plants, 11,* 2273.

Wright, A. (1976). The significance for breeding of linear regression analysis of genotype-environment interactions. *Heredity, 37,* 83–93.

Yoosefzadeh, N. M., Hesami, M., & Eskandari, M. (2023). Machine learning-assisted approaches in modernized plant breeding programs. *Genes (Basel), 14*(4), 777.

Yunbi, X., Xingping, Z., Li, H., Zheng, H., Zhang, J., Michael, S. O., Rajeev, K., Varshney, B., Prasanna, M., & Qian, Q. (2022). Smart breeding driven by big data, artificial intelligence, and integrated genomic-enviromic prediction. *Molecular Plant, 15*(11), 1664–1695.

Zingaretti, L. M., Gezan, S. A., Ferrao, L. F. V., Osorio, L. F., Monfort, A., Munoz, P. R., Whitaker, V. M., & Perez-Enciso, M. (2020). Exploring deep learning for complex trait genomic prediction in polyploid outcrossing species. *Frontiers in Plant Science, 11,* 1222–1230.

CHAPTER 8

Crop Cultivation Under Changing Climatic Regime of the Red-Lateritic Belt with Special Reference to Jhargram District of West Bengal – A Reappraisal

DHIMAN MUKHERJEE*

Regional Research Station (Red and Laterite Zone), Bidhan Chandra Krishi Viswavidyalaya, Jhargram, West Bengal, India

*Corresponding author

ABSTRACT

Climate change is becoming increasingly tangible, with significant impacts on the rainfed ecosystem, particularly affecting the livelihoods of marginal and poor communities from various tribal groups. Climate variability acts as a "risk multiplier," heightening these communities' vulnerability to environmental, social, political, and economic hardships. Identifying effective agronomic management techniques is essential for optimizing agricultural output amid shifting seasonal variations in the red-lateritic belt of West Bengal. Climate change can adversely affect long-term crop production in terms of yield, quality, irrigation demand, and input requirements such as fertilizers, herbicides, and insecticides. To mitigate these effects, careful management of resources like soil, water, and biodiversity is necessary. Agricultural policy implementation and scientific research must operate at local, zonal, and national levels. In the red-lateritic belt, rice remains the dominant crop, often followed by lentils, groundnuts, mustard, and wheat. Weather pattern shifts—including delayed winter onset, early monsoon withdrawal, and erratic rainfall—pose challenges to

Climate-Resilient Agriculture for Sustainable Crop Production: New Adaptation and Mitigation Strategies. Sanjay Kumar, Shivani Ranjan, Sumit Sow, Smruti Ranjan Padhan, and Dhirendra Kumar Roy (Eds.)

DOI: 10.1201/9781998511150-8

sustainable cultivation. These can be addressed by selecting crops suited to prevailing moisture levels and soil characteristics. Scientific and timely crop husbandry practices, especially under limited resources, are critical for sustaining and enhancing crop productivity in the face of ongoing climate change.

8.1 INTRODUCTION

Numerous scientific reports and empirical research have acknowledged the adverse effects of climate change and global warming on crop cultivation and its governance features. In India, agriculture plays a significant role in the economy of the whole country by ensuring food security and offering employment opportunities to a large section of the population. Approximately 70% of the Indian population relies on agriculture for their means of living. The agricultural sector alone contributes 23% to India's gross national product, signifying its crucial contribution to the country's development and its continued importance in the national economy (Mukherjee, 2015). Furthermore, the increasing global temperatures are not only leading to climate change but also impacting rainfall patterns in an unpredictable manner. According to the sixth Appraisal Document of the IPCC (Pörtner et al., 2022), the vulnerability of agriculture is very high in developing countries under the changing regime of abiotic factors, where agriculture is considered a major source of livelihood. Increases in severe instability and mixed weather phenomena alter observable seasonal patterns, and sudden changes in persistent heat during the harvest season and unexpected outbreaks of new kinds of diseases and insect appearances are major warning signs of changing weather patterns, which ultimately affect crop harvests. Uneven rainfall patterns, rising temperatures, and increased levels of carbon dioxide in the atmosphere are important climate parameters affecting crop yield and attributes. Research shows that weather parameters have a greater impact (67%) than other factors such as soil and nutrient management during the growing season (33%). The Intergovernmental Panel on Climate Change (IPCC) predicts that global average surface temperatures are likely to rise, potentially leading to uneven conditions. The increase in temperature relative to precipitation during the 20th century has reportedly played an important role in global warming. Researchers have demonstrated that crop yields decrease by 2%–6% for every 1°F increase in temperature (IPCC, 2014). In the face of a highly variable climate, food growers can influence the crop environment through enhanced adaptive management strategies (Mukherjee, 2014). In addition, the global average surface temperature is expected to be in the range of 0.3°C–0.7°C from 2016–2035 (IPCC, 2014). This not only hampers the crop microclimate but also the quality of livelihood. Our observations over the last few years, with the help of different secondary data, revealed that extreme climate events like delayed onset and early withdrawal of the monsoon/deficit rainfall patterns (drought, dry spells, etc.), extremely heavy rain within a short span of time, hailstorms, and so on, were increasing and continue to do so, which adversely affects crops and cropping patterns in the red-lateritic belt of West Bengal. This poses a serious threat not only to plant yield but also to the environment, human health, food security, and the socioeconomic livelihood of the area. Keeping the above facts in mind, the author tries to emphasize the outcomes of changing climate patterns on crops and cropping schemes in the red-lateritic zone of West Bengal, along with their remedial measures for the farming community, supported by critical observations from other researchers/workers.

8.2 PHENOMENON OF CROPS AND WEATHER

Crop information, along with its development phase and day-by-day atmospheric conditions during the cropping period, is a primary concern for decent establishment. Location-specific husbandry practices can be planned in concurrence with climatic assemblage, which reduces losses and curbs the expenditure of inputs and other miscellaneous farm dealings (Aryal et al., 2020). Various abiotic factor alterations in plant production systems may impact food security and threaten tribal people and their livelihoods through farm expenditures and food values. The problem may be even more severe for small farmers (<1.0 ha) with limited infrastructure facilities in this part of India. Greater dependency on precipitation and other crop inputs would make the impact more severe for them. Our study and sampling were confined to Jhargram and its adjoining districts, which fall under the red-lateritic belt from 2018–2023. Moreover, the agro-climatic conditions of the red and laterite zone mainly include Bankura, Jhargram, Purulia, and parts of the Birbhum region of West Bengal. The physiography is undulating, and the soil is coarse-textured and susceptible to erosion.

The Jhargram district is situated in the southwest region of West Bengal, between latitudes 21°52′ and 22°48′ North and longitudes 86°34′ and 87°20′ East. The district has a total size of 3,02,439 ha. It is bordered on the north by the districts of Bankura and Purulia; on the south by the district of Mayurbhanj in Orissa; and on the west by the district of Singbhum in Jharkhand. The temperature is tropical and quite humid. In the hot and dry months of May and June, the temperature can reach as high as 47°C, but during the cold evenings of December and January, it can drop to as low as 4°C–7°C. Over 75% of the total population resides in rural and remote forest areas and heavily depends on forest products and integrated agricultural practices (Figure 8.1).

FIGURE 8.1 Jhargram territory map.

Source: http://wbdmd.gov.in.

The territory has a varied crop ecosystem with various soil types, climatologies, vegetation kinds, and geomorphologies. Important rivers like the Subarnarekha, Dulang, Kangsabati, and their tributaries flow and improve the farming system by bringing in soil that is either sandy or fertile during the rainy season. Major crops farmed include watermelon, legumes, wheat, potatoes, mustard, sesame, groundnuts, aman-boro paddy, and maize, among others. The net cropped area is 1,68,448 hectares, with a gross cropped area of 2,27,921 hectares and a cropping intensity of 140%. The district experiences both droughts and floods, and occasionally one follows the other in the same year. This feature has been quite common for the last few years due to shifting climate patterns. Sankrail, Binpur I, Gopiballavpur I, and II have flood-prone areas of about 2120 hectares. However, the maximum area under drought-prone conditions is about 1,14,099 hectares. For the past 10 years, the average rainfall has been 1609.46 mm. However, people's experience indicates that rainfall is usually dispersed unevenly (http://wbdmd.gov.in). The monsoon showers always arrive late with an irrational distribution pattern in patches. A crucial component of the agricultural production system is contingent crop planning, which depends on location-specific short-duration crop cultivars. The goal of contingency crop planning is to offer strategies and tools for dealing with situations resulting from a variety of anomalous natural disasters.

8.2.1 MONSOON DYNAMISM

For the last several years, the pattern of monsoon onset has become very unpredictable, uncertain, and erratic. In the past, different regions of the country received the southwest monsoon at a specific and particular time, but nowadays, the commencement of the monsoon has become uncertain and unprecedented (Mani and Mukherjee, 2016). Sometimes it has hastened, and sometimes it has been delayed by a month or a fortnight, thus affecting the rainfed crops. Despite the use of supercomputers and satellite data on meteorological observations, it is still very difficult to predict the exact time of monsoon arrival, and very often, the forecasts of the monsoon by the meteorological department have been found irrelevant and wrong. This might be due to changes in the climatic multivariate, mainly affected by anthropogenic actions. During the last few years, the early onset of the monsoon in the middle of June, followed by a long gap in monsoon rain, has led to damage to crop seedlings and finally crop failure, as most farmers sow their *kharif* crops, and later, the majority of the farmers had no seeds to sow a second time. Similarly, the preparation of rice nurseries and their transplanting in early monsoon rain has led to spoilage of rice crops in rainfed areas. This may perhaps be an indication of climatic extremes and variability, mainly caused by global warming and other factors. In the last decade, the occurrence of super cyclones in Orissa, severe floods in North Bihar, Rajasthan, Gujarat, earthquakes in Uttarakhand and Gujarat, and recently the tsunami in Southeast Asia have indicated the dynamics of climatic change and aberration. This year's monsoon has changed, weakening its circulations and causing record-breaking extreme weather events in June and July across the nation. Given that the monsoon has just finished its halfway journey, July saw an increase in the amount of torrential rains throughout the nation, with 1113 stations reporting very heavy rainfall events and 205 stations reporting extremely heavy rainfall. This led to the most intense and severe rain occurrences in July during the previous five years, according to data from Climate Trends (https://economictimes.indiatimes.com). In the Binpur II block of the Jhargram district of West Bengal, a study was carried out to determine the extent of farmers' perceptions of the prevalence of climate change, as evidenced by phenomena like rising temperatures and fluctuating rainfall in this agro-ecological region. The study aimed to determine the farmers' adaptation strategies to deal with losses resulting from climate change. The findings demonstrated that farmers' attitudes on climate change had a significant impact on their willingness to modify their farming and production methods in order to adapt to the changing environment. The results showed that farmers were among the high-level respondents in terms of their perception of climate change. The findings indicated that half of the participants expressed optimism about their ability to manage the challenges posed by climate change (Biswas et al., 2020).

Paria et al. (2021) claim that changes in rainfall limit the principal crop's intensity and yield in this region of India. Since over 45% of the state's cropland is more subject to temperature shocks and rainfall, the majority of West Bengal's districts have been classified as particularly vulnerable to climate change (Birthal and Hazrana, 2019). Farmers' adaptation techniques to mitigate the negative effects of climate change are greatly influenced by the agroclimatic conditions, and it is frequently noted that intercropping and mixed cropping tactics are implemented to some degree. The improvement of climate resilience indicators was observed when efforts to enhance productivity were combined with initiatives focusing on soil health, water management, livelihood diversification, and optimum nutritional and food safety. Climate-resilient agriculture mainly encompasses the use of local and natural resources to promote crop production (Debenath et al., 2021a). Deploying both livestock and crop systems according to the location is essential. Enhancing resilience requires a focus on learning, adaptive decision-making, assessment, and observation. The farming community in this red-lateritic belt can promote climate-resilient farming practices such as making small water reservoirs, farm ponds, and intermittent check dams to capture and store rainwater, which can help recharge groundwater and supply water at critical phases of the crop life cycle (particularly during dry spells). However, this requires proper attention and funding support at local panchayat, village, and block levels.

8.2.2 ADAPTATION TO MITIGATION STRATEGY AT FIELD LEVEL

Climate change or changes in the monsoon pattern at the field level are only recognized through various field-level workers (the farming community), scientists, and various extension officials. This could be possible if farmers are able to: (i) recognize that climate change is actually happening; (ii) recognize the options for adaptation that are accessible to them in order to reduce the impacts of climate change; and (iii) respond to climate change adaptation in accordance with the production techniques that they follow in agriculture and related aspects. Therefore, it determines farmers' preparedness to modify themselves towards the mitigation of climate change by adopting certain alterations to their current cultivation and production practices. A thorough understanding of farmers' perceptions of climate change is critical. Measures should be taken at a very micro- and bottom-level so that the threat of climate change on crop production can be resolved to some degree. Appropriate and well-advanced knowledge of crops and cropping systems becomes very crucial for a sustainable crop production system.

8.3 CROPPING PATTERN

Choosing the appropriate and right kind of crop, as well as the subsequent crop in a cropping scheme, is one of the most essential risk management and income-enhancing strategies for farmers (Mukherjee, 2022). Various observations over the past five years, from February 2018 to November 2023, in different blocks of Jhargram revealed that crop choice decisions by the farming community heavily depend on rainfall patterns because this zone is more or less confined as a purely rainfed zone. Most of the farmland is barren either due to a scarcity of water or due to the encroachment of wild animals such as elephants in crop fields. Moreover, cropping patterns rely on other factors too, such as the technology applied (crop diversification, water resources, and input availability intensity), infrastructure artifacts (such as markets, roads, and warehouse keeping facilities), and environmental circumstances such as various abiotic factors (like rainfall, relative humidity, temperature, etc.). Productivity, profitability, and apparent nutrient balance under different crop sequences can be increased with proper and timely selection of location-specific crop options (Mukherjee, 2010). The dominant cropping systems in this region are mainly rice-rice, aman rice-fallow, rice-mustard, rice-wheat, blackgram-mustard, pigeonpea-fallow, rice-potato, rice-maize, rice-lentil/other pulses, and rice/blackgram/*kharif* maize-vegetable-based systems.

8.3.1 RICE-RICE AND RICE-FALLOW SYSTEMS

Rice-rice is a very common cropping pattern in the lowland area of the red-lateritic belt. This system has become unproductive, and a huge burden occurs on stored water in the soil. Puddling of soil in the rice system leads to land becoming unfit for subsequent crop cultivation in the same field, and sometimes farmers keep their land fallow for the next season. Deterioration of soil physical conditions, micronutrient deficiencies, inefficient nitrogen use, unbalanced nutrient use, and lack of suitable transplanters to address labor scarcity during the crucial transplanting period are the main problems in rice cultivation that reduce rice-rice system productivity. Several researchers' observations have shown that India's modern-bred wheat and rice cultivars are ineffective at storing iron and zinc (Debnath et al., 2021). Therefore, biofortified genotypes should be incorporated into rice to overcome malnutrition problems. In waterlogged prone areas, low nutrient use efficiency and iron toxicity become very prominent features in poorly drained submerged areas, which have become burning issues for the scientific community. Observations revealed that crop lodging due to frequent cyclone problems has become very challenging (Mukherjee, 2023). Changes in climate patterns have led to a steep rise in the problem of objectionable weeds such as *Echinochloa crusgalli, Dactyloctenium aegyptium*, and wild rice in paddy fields, and *Argemone mexicana* in mustard due to the non-availability of suitable control measures (Saha et al., 2023). Soil health deterioration and increased pests (weeds, insects, etc.), diseases, and arsenic contamination have become major constraints on low productivity in the rice-rice system (Hooi et al., 2023; Pandit and Mukherjee, 2022). Furthermore, in rice-fallow systems, monsoon season-sown rice areas remain vacant during the succeeding *rabi* season, mainly due to various biotic and abiotic constraints (Kumar et al., 2022). In this system, soil moisture conservation is a crucial aspect to consider for crop intensification. In the Sankrail and Jamboni areas of Jhargram, most farmers keep their land fallow due to the late harvesting of long-duration cultivars. However, a short-season pulse crop can be grown using the residual soil moisture and fertility after short-duration rice.

The selection of suitable cultivars and crop establishment methods must ensure proper seed germination, seedling emergence, and development. Green gram or lentil, normally established through *utera* (broadcasting seeds in the standing crop) or zero/minimum tillage, proves to be highly effective in late rice-harvesting situations. Under *utera* cropping, sowing the next crop's seeds 10–12 days before harvesting the standing rice helps utilize moisture efficiently and contributes to higher cropping intensity with increased returns.

The adoption of zero tillage, a conservation agriculture technique where crops are grown season after season without disturbing the soil through tillage, has become a highly effective practice in this region of West Bengal (Mukherjee et al., 2022). According to Moitzi et al. (2019) and Sarangi et al. (2021), zero tillage enhances system sustainability, allows for early seeding (7–10 days), reduces cultivation costs by approximately Rs. 2,500–3,000/ha, and improves the utilization of underutilized rice-fallow areas. Ploughless tillage systems are considered more energy-efficient and require less energy compared to tillage-intensive systems that involve ploughing (Moitzi et al., 2021).

8.3.2 RICE-MUSTARD SYSTEM

In this cropping system, panoramic variations in productivity and yield of mustard were recorded in different fields due to variations in agronomic management practices under a climate change scenario. However, a number of significant barriers exist in this cropping system, such as insufficient understanding of weather-based crop measures, inadequate support from improved production technologies, and the adoption of suitable high-yielding rice and mustard cultivars. A good mustard crop may generally be achieved with a medium- or short-duration, high-yielding rice variety. The mustard crop can be left in the field until March, at which point another crop can be planted to maximize system productivity during the summer.

8.3.3 RICE-GROUNDNUT SYSTEM

Rice followed by groundnut has become a very dominant cropping feature of Jhargram and its adjoining districts. Red lateritic soil provides a good soil-plant continuum and fetches higher returns per unit area for groundnut growers. Summer and winter season peanuts are rising as a crucial high-value crop under a limited water supply regime. *Rabi*/summer groundnut productivity is about twice as high as the yield produced during the *kharif* season. After the previous rice crop is harvested, groundnuts can now be grown on low-lying, productive, well-drained fields with a limited water supply regime. Achieving potential yield depends on a number of factors, including soil type, rainfall between August and September, early maturing rice and groundnut variety selection, and agro-technologies developed to fit local conditions for rice-groundnut systems.

8.3.4 RICE-PULSES/OILSEEDS SYSTEM

Pulses have become very popular for their suitability in various cropping sequences. This has become an important part of nutrition or food security issues (Mukherjee, 2021b). Rainfed crop sequences can now comprise pulse and oilseed crops due to recent advancements in their production of numerous kinds with largely different maturation durations. The rice-pulses planting system is gaining popularity nowadays; however, due to improper timing of paddy harvesting and late sowing of pulses (lentils, etc.), it has become an important constraint for the feasibility of this sequence in the red-lateritic belt. Soil microbial activity is greatly enhanced with this cropping system and is very effective for carbon sequestration as well (Mukherjee, 2019), which indirectly helps to tackle climate change threats at micro-levels. However, a few notable factors limit the productivity of this cropping scheme in the lateritic belt:

1. **Physical Factors:** Erratic dispersion of precipitation, dry spells, and high percolation and fixation of supply nutrients resulting in heavy nitrogen losses and phosphorus fixation problems in Jhargram and Purulia.
2. **Input Signal Factors:** Prolonged and delayed transplanting, insufficient coverage under high-yielding types, insufficient supply of high-quality seed, insufficient attention to timely weed control, insufficient attention to disease and insect infestation, and deficient implementation of control methods.
3. **Ethical Causes:** Low literacy rates, a higher percentage of marginalized and tribal farmers, and the practice of grazing animals on agricultural land or crop fields, along with the poor ability of the region's farmers to take risks.

8.3.5 PIGEONPEA-FALLOW SYSTEM

This system is quite common in rainfed or severely water-scarce areas. Furthermore, pigeonpea has proven to be highly useful in intercropping systems. This crop thrives in a seedbed that has been thoroughly tilled and well-drained (Mandal et al., 2015). It is advised to use a deep ploughing technique with a soil-turning plough in fallow or waste regions, zero tillage sowing under an intensive cropping strategy, and broad bed furrow/ridge planting in low-lying and intercropping areas. The suitable time for sowing varies from patch to patch based on seed and soil moisture availability. The first and second weeks of June are when early-maturing types mature, whereas medium- and late-maturing kinds mature later. Line sowing on the ridge and beds using a seed drill, desi plough, or digging; both methods are advised according to the region. The deep root structure and initial slow growth rate of pigeonpea provide an excellent opportunity for intercropping with early-maturing, fast-growing, and shallow-rooted crops (Nandhini et al., 2015). In the red-lateritic belt, this crop can be interplanted with other crops such as mungbean, urdbean, maize, soybean, and groundnut to boost system production and enhance soil health (Mukherjee, 2021a). Furthermore, compared to monocropping, intercropping has proven to be a very successful method of managing pest and disease outbreaks and is especially beneficial for resource-poor areas.

8.3.6 RICE/BLACKGRAM/KHARIF MAIZE-VEGETABLE-BASED SYSTEM

Vegetable-based cropping sequences offer a variety of choices that can be combined with climate-smart farming, machinery, genetics, agronomic options, and information application. These schemes help sequester larger quantities of CO_2 in the form of biomass and thereby improve the net greenhouse gas release balance via a reduction in methane and N_2O emissions. Crop rotation, cover cropping, and intercropping, which includes vegetables, can improve biomass production and soil carbon sequestration and help address the problem of climate change threats at the micro-level.

8.4 SIGNIFICANT EFFECTS OF CLIMATE CHANGE ON CROP YIELD

1. **Frequency of Precipitation:** The alteration in the frequency of precipitation has caused a shift in the distribution of rainfall, encompassing modifications in intensity, timing, place, and overall distribution. These changes have had significant consequences such as floods, droughts, and unpredictable rainfall patterns, all of which have a detrimental impact on the agricultural scenario and productivity. Consequently, this has resulted in a decrease in agricultural gross domestic output and an escalation in the competition for water resources. A concrete example of this occurred in 2019–2020, when India encountered delayed and insufficient monsoon rains, ultimately leading to diminished crop yields across numerous regions.

2. **Diurnal Temperature Variation:** Increased temperatures may be detrimental to the development and growth of plants. Elevated temperatures in the growing season have the potential to diminish crop yields and lower the nutritional content of crops. The well-being and productivity of livestock, wheat, mustard, as well as winter vegetables like spinach and cabbage, can be negatively influenced by heat stress.

3. **Changing Patterns of Pests and Diseases:** Changes in weather patterns have a direct impact on the spread and population of pests, weeds, and diseases, creating obstacles for the management of crop pests (Singh et al., 2004). Variation in temperature and precipitation can promote the proliferation of specific pests and diseases, which can have detrimental effects on the health of crops (Mukherjee, 2018). An example of this is the rise in occurrences of pests such as the pink bollworm, leading to negative impacts on cotton cultivation in India, as well as the invasion of locust swarms from the Somalia region caused by unpredictable rainfall patterns. Further, shifting weed flora patterns have also been observed in several crop sequences due to early or delayed shifts in the season (Mukherjee, 2021c). Huge weed biodiversity has been observed in different crops, which was not present earlier in the same field at the same time.

4. **Modifications to the Cropping Order:** The fluctuation in weather conditions can have an impact on the suitability of various crops in specific areas. As the temperature and rainfall patterns change, farmers may have to adjust their crop choices in order to maintain productivity. Certain crops may become less feasible to grow, while others may become more favorable options.

5. **A Rise in Extreme Weather Events:** Extreme weather occurrences like storms, hailstorms, and cyclones have been associated with climate change. These occurrences have the potential to seriously harm infrastructure, livestock, and crops, which could result in lower yields and difficult financial times for farmers. For example, during 2019, a heavy hailstorm reduced the production of cashew nuts in the Jhagram Block.

8.5 CROP DIVERSIFICATION

This approach improved the production of a variety of food, a strategy that broadens food sources, improves livelihoods, increases job opportunities in local agro-processing, and revitalizes the rural economy as a whole, thereby addressing the problem of micronutrient malnutrition (Douyon et al., 2022; FAO, 2021). Efficient cropping patterns refer to specialized crop cultivation involving the utmost efficient utilization of land, fertilizer, irrigation systems, and other agricultural inputs, which can positively improve land productivity and net farm income (Mukherjee, 2016a). Hence, diversification of crops is crucial for agricultural sustainability as it removes biotic and abiotic stress from the soil and sustains good productivity of cropland (Kamraju et al., 2017). Agriculture in West Bengal used to be characterized by the wide diversification of different crops cultivated in different districts over various periods. Conservation practices are an essential component of managing soil and water in agriculture for the maintenance of plant agro-ecosystems (Mukherjee and Mandal, 2017). This should be an integral part of crop diversification to reduce the cost of cultivation and be eco-friendly as well. Clearly, there was a progressive rise in the crop diversification metrics, such as the Simpson index, Herfindahl index, Modified Entropy index, and Entropy index, in West Bengal from 1980–1981 to 2007–2008, as indicated by Majumdar (2014). Nevertheless, a spatiotemporal examination demonstrated that in Gopiballavpur-II, the level of crop diversification had steadily declined from moderate to low between 2007–2008 and 2010–2011, according to Bisai et al. (2016). Food growers can combat a variety of abnormal climate conditions, such as early rains or mid-season droughts, recurrent dry spells during the crop season, heat stress issues, and so on, by diversifying their cropping patterns. Crop production managers must adapt cropping patterns appropriately in response to shifting market and climatic conditions in order to maintain the sector's viability and include farmers with limited land in the process of intensification (Figure 8.2). Legume-based crop diversification becomes very imperative for aberrant weather situations. Legumes, in association with bacteria, can fix atmospheric nitrogen in root nodules. Globally, industrial nitrogen (N) output is 1.1 times less than land-based biological nitrogen fixation. When legumes and cereals are grown in rotation, soil carbon (C) sequestration can rise by 9–45 Mg C ha^{-1}, and greenhouse gas emissions can be reduced by 5–7 times when compared to non-legumes.

Legumes can be grown in areas with little water availability and require less nitrogen. After grain legumes, non-legumes could see about a 200 kg N ha^{-1} reduction in the rate of N. By using biological nitrogen fixation to save nitrogen, carbon dioxide emissions might be decreased by about 200 kg yearly. Legume residue mineralization increases soil fertility by supplying N and other nutrients in addition to fixing nitrogen. Pulses can improve soil aggregates, modify soil water retention, and have a significant reconciling capacity to adverse conditions (Rahaman et al., 2022).

(a) Study on weed flora distribution pattern in rice-rice system.

(b) Rice cultivation with effective nutrient management in limited water supply regime.

(c) Cultivation of millet (*Eleusine coracana*) under rainfed agro-ecosystem of red-lateritic belt.

(d) Millets (*Barnyard millet*) cultivation – New option under shifting climate condition.

FIGURE 8.2 Suitable crop cultivation under the changing climate scenario of Jhargram.

8.5.1 CROP DIVERSIFICATION AND IMPORTANCE

This is a crucial element in sustaining sustainability and increasing agricultural yield. Crop diversification techniques such as crop rotation, mixed cropping, and intensive cropping not only increase output but also lessen erosion and aid food producers during unusual weather conditions (Mukherjee and Moktan, 2008). Crop rotation acts as an alternate measure because in Jhargram and other parts of the red-lateritic belt, farmers usually grow only rice-rice or rice-mustard systems (Figure 8.3). A change in crop sequence pattern with suitable non-traditional crops gives significant benefits to the farming community (Mukherjee and Hedayetullaha, 2018). Due to shifting climate and less availability of water, crop failure occurs most of the time. This can be improved by including millet or pulse-based systems in the cropping system instead of traditional cropping practices as stated above. These measures are beneficial for controlling weeds, pests, and diseases in addition to being crucial for managing soil fertility. Leguminous crops are necessary for adding nitrogen to the soil in any cycle. Mixed cropping with various pulses enhances the productivity of non-leguminous cereals, and so on; hence, it is necessary for the success of crop production in this stressed environmental area. In a few pockets of Binpur II and Nayagram block of Jhargram (West Bengal), a very intensive cropping system is followed. Relay and multiple cropping are very common in assured irrigation areas or in lowland systems. The addition of pulse-based crops to the cropping sequence increased environmental sustainability and productivity. Using vegetation or plants that are climate-ready is a good way to achieve high productivity (Haefelea et al., 2014). In order to maintain crop viability under any sequence or system, it is essential to incorporate new plant cultivars that are both economically stable and resistant to a variety of biotic and abiotic challenges. In one case study during 2021–2023 at Kenduana village, Tapsia Gram Panchayat, Gopiballavpur-II Block, Jhargram in West Bengal, observations revealed that crop diversification became very effective for higher productivity. In general, agricultural practices are actually location-specific. The strategy we support and the advice we offered were region-specific. The two main crops grown in the area, mustard in the winter (*rabi*) and rice in the *kharif* (monsoon) seasons, were the main focus. Recommendations were made for crop varieties, irrigation, nutrients, pest and disease control, and cultural methods based on various weather scenarios. Our strategy places a strong emphasis on using materials that are readily available locally and environmentally friendly methods in order to achieve two key goals: making farming both profitable and sustainable. In addition to increasing productivity and farm incomes, timely weather-based agricultural advice and training on sustainable agricultural practices will strengthen the farming sector's resilience to climate change (Table 8.1). Our critical observation revealed that it has come to our attention that it is essential to transition rice-fallow lands to alternative and lucrative crops or cropping systems. In this transition, priority should be given to oilseed and pulse crops. This diversification in cropping patterns will not only enhance soil quality but also aid in mitigating water stress-related issues.

(a) Rice cultivation under rice-groundnut system with limited water supply regime.

(b) Groundnut cultivation under rice-groundnut system with limited water supply regime.

(c) Weed shifting under rice-rice eco-system.

(d) Maize cultivation under rainfed agro-ecosystem of Jhargram.

FIGURE 8.3 Performance of different crops under different crop sequences in rainfed agroecosystems.

TABLE 8.1 Factors Affecting Climate Change on Farming Communities and Its Measures to Enhance Socio-Ecological Resiliency

	Factor	Cause	Assessment Level	Major Impact
Climate Change	Threat (weather event)	Elevate the severity, occurrence, strength, and dispersion of damage caused by insects, pests, diseases, and so on, in agricultural fields.	Decrease in economic yield and productivity of land.	Gross impact to farming community.
	Vulnerability	Characteristics of food growers and feature of agro-ecosystem that determines by severity of threat based on exposure, sensitivity, and adaptive capacity.	Change in behavioral pattern of farmer's and need to aware challenges due to climatic variability and aberrant weather condition on crop.	Threat to food security.
	Time bound measurement	Natural resource degradation.	Training and on-site demonstration tailored to specific location for maximum effectiveness.	Increase efficiency of farming community against various challenges of climate change.

8.5.2 CONSTRAINTS FOR CROP DIVERSIFICATION

Land degradation problem: In this location, it is widely recognized that specific canal command sites have concerns related to the acidity of the soil and hindered natural drainage during the monsoon season. Unusual groundwater exploitation: In some areas of the Jhargram district, specifically Binpur I and Binpur II, the groundwater table has lowered due to over-pumping of groundwater for agriculture. Due to excessive dependency on monocropping of rice-fallow or rice-rice cropping systems, this leads to the exhaustion of groundwater resources (Shah, 2019). In addition to increasing production costs because deeper well pumping requires more energy, declining water tables pose severe concerns about the long-term viability of the rice-rice or rice-vegetable/mustard systems themselves in these regions which include:

1. **Change of Land Use Pattern:** The conversion of extremely productive land to non-agricultural uses, such as business hubs,

housing, and industry, particularly along the rural-urban interface, needs to be carefully considered in various parts of Jhargram, Purulia, and Kharagpur division.

2. **Decrease in Exhibition Potentiality of the Field:** An unbalanced approach to fertilizer application, pervasive deficiencies in micro- and secondary nutrients (Boron, Molybdenum, Ca, S, etc.), and decreased levels of organic matter in farmed areas have resulted in a declining trend in major crop sequence responses to nutrients, particularly nitrogen and phosphorus. In other words, farmers must apply larger doses of chemical loading to the field in order to maintain earlier output levels.
3. **Imbalance in Nutrient Use Practice:** The problem of imbalance in primary or secondary nutrient use has been emphasized on three points. With extensive cropping, nutrient loss by crops from the soil has far overtaken replacement through manures and fertilizers. This leads to a negative equilibrium of plant nutrients in the soil system. In addition, if this pattern continues, there is still a huge risk to the long-term viability of the main cropping systems in this area. Food growers have acquired tendencies to utilize more urea or nitrogen-based fertilizers instead of phosphorus and potassium due to ease and lower cost of availability. This, consequently, has resulted in expanding ratios of nitrogen to phosphate and nitrogen to potassium to harmful levels.
4. **Build-Up of Diseases/Pests and Weed Resistance Problem:** Due to increased crop intensification and high input consumption (fertilizers, pesticides, etc.), some unwanted pests and illnesses have become more dangerous and have begun to accumulate (Mahapatra et al., 2020; Mukherjee, 2017).

8.6 CONTINGENCE CROP DESIGN

Planning for action in climate change scenarios should focus on four central arenas: (i) crop management strategies that are adaptable to changing weather conditions; (ii) accelerative farm fecundity; (iii) preserving the health of ecosystems; and (iv) intensifying soil ecosystems. To manage weather-related hazards or maximize the benefits of favorable weather, these must be led by trustworthy local weather forecasts. The final two elements determine farming sustainability. A crucial component of agricultural production planning is dependent crop planning, since agriculture is heavily reliant on the natural world. The goal of contingency crop planning is to offer strategies for dealing with the effects of a flood, a drought, or any other natural disaster (Mukherjee et al., 2019). To fight the difficulties that may arise due to drought or flood, the following steps, as suggested by scientists and extension officers from the agriculture department, should be followed:

1. **Mitigating the Negative Impact of Flooding:**

    i. **In the Case of Mid or Early Season Flood:**

    - Paddy seed should be sown in the seedbed at 10-day intervals.
    - A higher elevation should be selected for the raised seedbed.
    - It is recommended to cultivate the flood-tolerant Swaran Sub-I cultivar.
    - Enough organic matter must be added to farmed land in order to lessen the negative effects of sand deposition.
    - Sand deposition must be eliminated manually or mechanically if it is high.
    - If you plan to seed late, use a drum seeder.
    - Planting paddy twice will be highly beneficial, particularly when the floodwaters from the early and mid-season floods recede.
    - If you are transplanting late, you should increase the basal dose of nitrogen and transplant five to six seedlings per hill.

    ii. **Belated Flood:** In low-lying areas affected by late flooding, when rice farming is not feasible, post-*kharif* season *kalai* (varieties: *Sarada*, *Goutam*, *Sulata*, etc.) may be cultivated. It is possible to cultivate toria, yellow sarson, lentil, and other crops early in the *rabi* season. In very late conditions, green gram or French bean cultivation can be profitable.

2. **Overcoming the Adverse Effects of Drought and Scanty Rainfall:**

- Each plot needs to have earthen bunds built around it in order to collect rainwater or runoff.

- Applying organic manure can enhance the soil's ability to retain water.
- It is recommended to use closer spacing and a higher number of seedlings per hill.
- Rice is directly seeded using a zero-tillage or minimal-tillage machine.
- Pigeonpea, black gram, groundnut, and maize, or a combination of maize and arhar, are grown in highland areas.
- In rice fields, mechanical weeding is employed along with the application of soil mulch made from uprooted weed biomass or alternative mulching materials such as straw.
- Avoid using nitrogenous fertilizers in the soil to prevent the growth of unusual plants.

3. **Adaptation Capability:** This can be defined as "a system's (human or natural) ability to adapt to climate change, including its variability and extremes, in order to mitigate potential harm, seize opportunities, or deal with its aftermath" (IPCC, 2001). The adaptive capability of respective units and assemblages is molded by their access to and control over social, natural, physical, human, and financial resources (CARE, 2010; Table 8.2).

TABLE 8.2 Illustration of Resources Affecting Adaptive Capability

Resource	Factors Influencing Adaptive Capability
Human	Knowledge of climate risks, conservation agriculture skills, good health to enable labor.
Social	Women's savings and loan groups, farmer-based organizations, traditional welfare and social support institutions.
Natural	Reliable water resources, productive land, vegetation, and trees.
Financial	Micro-insurance, diversified income sources.

Source: CARE (2010, p. 11).

8.7 FUTURE LINE OF WORK TO TACKLE CLIMATE CHANGE PROBLEM

1. **Agroforestry Forestry:** Reducing soil erosion, increasing climate resilience, improving biodiversity, and helping to sequester carbon can all be achieved by growing different crops in the same area or

beside trees (Mukherjee, 2013a). For example, due to more forest area, in Jhargram district, effective agri-silviculture has become an effective option for high carbon sequestration and mitigating climate change threats. This has become quite acceptable to the Lodha and Santhal communities due to their close association with different flora and tree species in the forest region.

2. **Cultivation of Non-Traditional Crops:** Promotion of non-traditional crops (millets, faba beans, horse gram, medicinal plants, etc.) that are more adaptive to the vagaries of the monsoon can reduce dependence on mono-cropping and lower risk (Hedayetullaha et al., 2018; Mukherjee, 2016b).

3. **Climate-Smart Water Management:** Implementing efficient climate-smart water management practices in the red-lateritic belt can increase crop production with more efficient water use.

4. **Time-Bound Forecasting Measures:** Crop productivity and production are heavily reliant on climate, and climate variability is increasingly manifested in unpredictable weather events that are out of human control. Despite this, farmers can partially prepare for and adapt to unpredictable weather when they receive timely weather forecasts, which give them the foresight to take appropriate precautions against extreme weather events.

5. **Reduce the Amount of Land Planted:** To water-hungry crops and encourage those that require less water. Nearly 80% of irrigation water is used by three water-intensive crops: sugarcane, wheat, and rice (Mukherjee, 2013). Particularly in rain-fed and drought-prone areas of the nation, a decrease in the area planted to these crops, combined with crop diversification and micro-irrigation, will greatly help to alleviate water shortages and free up additional water resources for other uses. Planting of "water-lite" crops like lentils, oilseeds, and millets should be preferred instead of "water-guzzling crops" such as rice and wheat. Locally suitable crops, along with diverse pulses and millets such as barnyard or African millets, should be incorporated into the cropping pattern.

6. **Ecosystem-based Approaches:** Approaches to crop cultivation such as conservation agriculture are one of the key strategies for building climate-resilient agriculture. By utilizing integrated

soil and water management, eco-friendly inputs, promoting agro-biodiversity, and implementing multi-layer farming, the ecosystem-based approach helps to align farming practices with specific agro-ecological zones (location-specific), ultimately increasing yields sustainably, conserving natural resources, and improving overall agricultural productivity.

8.8 CONCLUSION

All things considered, more robust and long-term agricultural systems must be constructed in a sustainable way. Promoting and putting into practice climate-resilient agricultural techniques, however, is fraught with difficulties, including a lack of knowledge and ability on the part of farmers, restricted access to markets and funding, and institutional and governmental hurdles. Increased attempts to expand extension services, such as advocating for universal education in farming communities and funding the construction of rural infrastructure, are necessary. Increasing the farmers' capacities and abilities to adjust farming patterns in accordance with local advantages and disadvantages will also help them better adapt to the changing climate. Additionally, it is critical to invest in capacity building, research, and innovation, as well as to strengthen collaborations between farmers, researchers, extension officers, and policymakers.

KEYWORDS

- **biodiversity conservation**
- **crop diversification**
- **crop yield**
- **food security**
- **global warming**
- **irrigation water consumption**
- **soil management**
- **unpredictable rainfall**

REFERENCES

Amare, M., Mavrotas, G., & Edeh, H. (2018). Farmers' crop choice decisions: Trends and determinants in Nigeria and Uganda. *IFPRI-Discussion Papers,* 65–74.

Aryal, J. P., Sapkota, T. B., Khurana, R., Khatri-Chhetri, A., Rahut, D. B., & Jat, M. L. (2020). Climate change and agriculture in South Asia: Adaptation options in smallholder production systems. *Environment, Development and Sustainability, 22*(6), 5045–5075.

Birthal, P. S., & Hazrana, J. (2019). Crop diversification and resilience of agriculture to climatic shocks: Evidence from India. *Agricultural Systems, 173,* 345–354.

Bisai, D., Chatterjee, S., Sau, S., Samanta, S., & Saha, S. (2016). Analysis of crop diversification with spatio-temporal concept over Paschim Medinipur district, West Bengal, India. *International Journal of Innovative Science, Engineering and Technology, 3,* 363–369.

Biswas, S., Chatterjee, S., & Roy, D. C. (2020). Understanding farmers' perception of climate change and adaptation strategies: A case study in Jhargram district of West Bengal, India. *Journal of Applied and Natural Science, 12*(2), 207–212.

CARE. (2010). Toolkit for integrating climate change adaptation into development projects–Digital toolkit–Version 1.0. CARE International, with technical input by the International Institute for Sustainable Development (IISD), July 2010.

Debnath, S., Mandal, B., Saha, S., Sarkar, D., Batabyal, K., Murmu, S., Patra, B. C., Mukherjee, D., & Biswas, T. (2021). Are the modern-bred rice and wheat cultivars in India inefficient in zinc and iron sequestration? *Environmental and Experimental Botany, 189,* 1–7.

Debnath, S., Saha, S., Mandal, B., Sarkar, D., Chattopadhyay, A., Mukherjee, D., Batabyal, K., Murmu, S., Nath, R., Mishra, D. K., & Sinha, K. (2021a). Zinc and iron profiling in some commonly consumed food crops uncovers inter- and intra-crop variation. *Journal of Soil Science and Plant Nutrition.* https://doi.org/10.1007/s42729-022-00770-7.

Duncan, E., Ashton, L., Abdulai, A. R., Sawadogo-Lewis, T., King, S. E., & Fraser, E. D. G. (2022). Connecting the food and agriculture sector to nutrition interventions for improved health outcomes. *Food Security, 14,* 657–675.

FAO, IFAD, UNICEF, WFP, & WHO. (2021). The state of food security and nutrition in the world 2021. *Transforming Food Systems for Food Security, Improved Nutrition and Affordable Healthy Diets for All.* Rome: FAO.

Hedayetullaha, M., Zaman, P., & Mukherjee, D. (2018). Setaria grasses (African grass). In M. Hedayetullaha & P. Zaman (Eds.), *Forage Crop of the World (Vol. 2)* (pp. 3–12). Apple Academic Press, Taylor and Francis Group USA.

Hooi, A., Mahapatra, S., Chakraborty, S., Mukherjee, D., & Maji, A. (2023). Disease scenario and virulence pattern of major wheat pathogens occurring in Indo-Gangetic plains of West Bengal, India. *Indian Journal of Ecology, 50*(6), 2088–2095.

IPCC (Intergovernmental Panel on Climate Change). (2001). *Climate Change 2001: Synthesis Report.* A contribution of working groups I, II, and III to the third assessment report of the Intergovernmental Panel on Climate Change (R. T. Watson & the Core Writing Team, Eds.). Cambridge University Press.

IPCC (Intergovernmental Panel on Climate Change). (2014). *Climate Change 2014: Synthesis Report.* Contribution of working groups I, II, and III to the fifth assessment report of the Intergovernmental Panel on Climate Change. Geneva, Switzerland.

Kamraju, M., Vani, M., & Anuradha, T. (2017). Crop diversification pattern: A case study of Telangana state. *International Journal of Innovative Science and Research Technology, 2*(5), 366–371.

Kumar, R., Choudhary, J. S., Mishra, J. S., Mondal, S., Poonia, S. P., Monobrullah, M., Hans, H., Verma, M., Kumar, U., Bhatt, B. P., Malik, R. K., Kumar, V., & McDonald, A. (2022). Outburst of pest populations in rice-based cropping systems under conservation agricultural practices in the middle Indo-Gangetic Plains of South Asia. *Scientific Reports, 12,* 3753. https://doi.org/10.1038/s41598-022-07760-w.

Mahapatra, S., Devi, H. M., Kumar, S., Mukherjee, D., & Das, S. (2020). Performance and stability of different zone-specific genotypes evaluated against spot blotch of wheat. *International Journal of Current Microbiology and Applied Sciences, 9*(6), 2968–2981.

Majumdar, K. (2014). Nature and pattern of crop diversification in West Bengal. *International Journal for Research in Management and Pharmacy, 3*(2), 33–41.

Mandal, B., Mukherjee, D., & Ray, D. (2015). Ground reality assessment and possible strategies in improvement of red gram (*Cajanus cajan* (L.) Mill sp.) in old alluvial plains of West Bengal. *International Journal of Bioresource Sciences, 2*(2), 187–194.

Mani, J. K., & Mukherjee, D. (2016). Accuracy of weather forecast for hill zone of West Bengal for better agriculture management practices. *Indian Journal of Research, 5*(10), 325–328.

Moitzi, G., Neugschwandtner, R. W., Kaul, H. P., & Wagentristl, H. (2019). Energy efficiency of winter wheat in a long-term tillage experiment under Pannonian climate conditions. *European Journal of Agronomy, 103,* 24–31.

Moitzi, G., Neugschwandtner, R. W., Kaul, H., & Wagentristl, H. (2021). Effect of tillage systems on energy input and energy efficiency for sugar beet and soybean under Pannonian climate conditions. *Plant, Soil and Environment, 67*(3), 137–146.

Mukherjee, D. (2010). Productivity, profitability and apparent nutrient balance under different crop sequences in mid-hill conditions. *Indian Journal of Agricultural Sciences, 80*(5), 420–422.

Mukherjee, D. (2013). Sustainable uses of water resources and quality with respect to agriculture. In S. R. Maloo & L. L. Somani (Eds.), *Advances in Experimental Agriculture* (Vol. 3, pp. 5–48). Agrotech Publishing Academy.

Mukherjee, D. (2013a). Prospect and need of agroforestry in the hills of Darjeeling. *Indian Agriculturist, 57*(1), 21–27.

Mukherjee, D. (2014). Climate change and its impact on Indian agriculture. In S. Nehra (Ed.), *Plant Disease Management and Microbes* (pp. 193–206). Aavishkar Publishers.

Mukherjee, D. (2015). Food security: A worldwide challenge. *Research and Review: Journal of Agriculture and Allied Sciences, 4*(1), 3–5.

Mukherjee, D. (2016a). Evaluation of different crop sequence productivity potential, economics and nutrient balance under new alluvial situations of NEPZ. *International Journal of Horticulture and Agriculture, 1*(1), 51–56.

Mukherjee, D. (2016b). Medicinal and aromatic plants: Wealth of India. In A. Hemantaranjan (Ed.), *Advances in Plant Physiology* (Vol. 17, pp. 425–456). Scientific Publishers.

Mukherjee, D. (2017). Influence of crop geometry and different nutrient management practices on growth and yield of Aloe vera (*Aloe barbadensis*) under sub-temperate zone of Darjeeling hills. *Indian Horticulture Journal, 7*(1), 67–71.

Mukherjee, D. (2018). Tackling climate change impact on wheat production and effective adaptation strategies for the state. In A. Rakshit, V. K. Tripathi, A. Singh, S. Shekhar, & D. R. Sarkar (Eds.), *Innovative Approach of Integrated Resource management* (pp. 17–22). New Delhi Publishers.

Mukherjee, D. (2019). Microbial intervention in soil and plant health for improving crop efficiency. In D. P. Singh & R. Prabha (Eds.), *Microbial Interventions in Agriculture and Environment* (Vol. 3, pp. 17–47). Springer Nature. https://doi.org/10.1007/978-981-32-9084-6_2.

Mukherjee, D. (2021a). Production potential of green gram (*Vigna radiata*) under various sowing dates and weed control measures. *Annals of Agricultural Research, 42*(1), 46–53.

Mukherjee, D. (2021b). Food security under the era of climate change threat. *Journal of Advanced Agriculture & Horticulture Research, 1,* 1–4. https://www.sciforce.org/index.php/JAHR/article/view/78 (accessed on 16 January 2025).

Mukherjee, D. (2021c). Weeds biodiversity: Challenges and opportunities in current context. In V. Kumar, S. Kumar, N. Kamboj, T. Payum, P. Kumar, & S. Kumari (Eds.), *Biological Diversity: Current Status and Conservation Policies* (Vol. 1, pp. 46–66).

Mukherjee, D. (2022). Various approaches for mitigation of abiotic stresses for higher crop productivity. In A. Hemantaranjan (Ed.), *Advances in Plant Physiology* (Vol. 19, pp. 227–265). Scientific Publishers.

Mukherjee, D. (2023). Management of lodging and yield maximization of wheat (*Triticum aestivum* L.) under Indo-Gangetic plains. *Annals of Agricultural Research, 44*(2), 158–163.

Mukherjee, D., & Hedayetullaha, M. (2018). Non-conventional legume forage crops. In M. Hedayetullaha & P. Zaman (Eds.), *Forage Crop of the World* (Vol. 2, pp. 287–308). Apple Academic Press.

Mukherjee, D., & Mandal, B. (2017). Practice of conservation agriculture for sustaining agro-ecosystem: A review. *International Journal of Bioresource Science, 4*(2), 89–99.

Mukherjee, D., & Moktan, M. W. (2008). Studies on landholding and crop diversification pattern in hill region of West Bengal. *Journal of Crop and Weed, 4*(1), 41–45.

Mukherjee, D., Mahapatra, S., & Singh, D. P. (2019). Rainfed wheat production through efficient water management and disease control measures in new alluvial zones. *Indian Journal of Ecology, 46*(7), 11–14.

Mukherjee, D., Maji, A., Mahapatra, S., & Mandal, B. (2022). Response of various sowing dates and tillage options on the performance of wheat under late sown conditions. *International Journal of Bioresource Sciences, 9*(2), 115–121.

Pandit, K. T., & Mukherjee, D. (2022). Arsenic profanity in agro-ecosystem and its effect on plant metabolism. In A. Hemantaranjan (Ed.), *Advances in Plant Physiology* (Vol. 19, pp. 439–491). Scientific Publishers.

Paria, B., Pani, A., Mishra, P., & Behera, B. (2021). Irrigation-based agricultural intensification and future groundwater potentiality: Experience of Indian states. *SN Applied Sciences, 3*(4), 1–22.

Pörtner, H. O., Roberts, D. C., Adams, H., Adler, C., Aldunce, P., Ali, E., Begum, R. A., Betts, R., Kerr, R. B., & Biesbroek, R. (2022). Climate change 2022: Impacts, adaptation and vulnerability. Contribution of Working Group II to the Sixth Assessment Report of the Intergovernmental Panel on Climate Change. Cambridge University Press. https://doi.org/10.1017/9781009325844.

Rahman, M. M., Alam, M. S., Islam, M., & Biswas, C. J. (2022). Potential of legume-based cropping systems for climate change adaptation and mitigation. In *Advances in Legumes for Sustainable Intensification* (pp. 381–402).

Sarangi, S. K., Maji, B., Sharma, P. C., Digar, S., Mahanta, K. K., Burman, D., Mandal, U. K., Mandal, S., & Mainuddin, M. (2021). Potato (*Solanum tuberosum* L.) cultivation by zero tillage and paddy straw mulching in the saline soils of the Ganges Delta. *Potato Research, 64*(2), 277–305.

Shah, M. (2019). India's water problem has a simple solution. *Indian Express.* https://bit.ly/39d61Bt (accessed on 16 January 2025).

Shah, M. H., Mandal, B., Mukherjee, D., & Kundu, S. (2023). Effect of weed management in transplanted rice under new alluvial zone of West Bengal. *The Pharma Innovation Journal, 12*(8), 1834–1838.

Singh, R. P., Singh, R. K., & Mukherjee, D. (2004). Effect of weed interference on efficiency of crop. *Agronomica,* 24–26.

CHAPTER 9

Integration of Next-Generation Breeding Approaches to Enhance Crop Adaptability Under Adverse Climatic Conditions

RAJ KUMAR,[1] ABHISHEK KUMAR,[2] SHAMBHAVI KUMARI,[2] MUKUL KUMAR GANDHI,[1] YOGESH DASHRATH NAIK,[1,*] NAVNIT KUMAR,[3] and SANJAY KUMAR[4]

[1]*Department of Agricultural Biotechnology and Molecular Biology, Dr. Rajendra Prasad Central Agricultural University, Pusa, Bihar, India*

[2]*Department of Genetics and Plant Breeding, Dr. Rajendra Prasad Central Agricultural University, Pusa, Bihar, India*

[3]*Sugarcane Research Institute, Dr. Rajendra Prasad Central Agricultural University, Pusa, Bihar, India*

[4]*Department of Agronomy, Bihar Agricultural University, Sabour, Bihar, India*

*Corresponding author

ABSTRACT

Climate change adversely impacts the quantity and quality of agricultural products, posing a serious threat to global food security, sustainable development, and poverty reduction. If left unaddressed, it could lead to a 20% increase in hunger and malnutrition by 2050. Enhancing agricultural adaptation to extreme climatic conditions is therefore essential. However, conventional breeding, which depends on existing genetic diversity and

Climate-Resilient Agriculture for Sustainable Crop Production: New Adaptation and Mitigation Strategies. Sanjay Kumar, Shivani Ranjan, Sumit Sow, Smruti Ranjan Padhan, and Dhirendra Kumar Roy (Eds.)

DOI: 10.1201/9781998511150-9

phenotypic selection, can be time-consuming and costly. To accelerate crop improvement and bolster food security, integrating traditional breeding with modern approaches is essential. Genomic advancements are driving more sustainable agriculture and increasing crop resilience to climate-related stresses. Next-generation breeding techniques—such as high-throughput phenotyping, advanced sequencing, trait mapping, genomic-assisted breeding, gene editing, and speed breeding—support the development of superior crop varieties capable of withstanding biotic and abiotic stresses. Artificial intelligence further aids agricultural research by analyzing large datasets and optimizing breeding strategies, enhancing precision phenotyping and minimizing errors. This chapter reviews the major constraints posed by climate change and their effects on agriculture. It also explores the role of multi-omics approaches—including genomics, transcriptomics, and proteomics—in understanding the molecular and physiological adaptations in tolerant crop varieties, offering insights for advancing climate-resilient agriculture.

9.1 INTRODUCTION

Agriculture and climate change exhibit a significant and interconnected relationship in numerous aspects. Climate change serves as the primary catalyst for both biotic and abiotic stresses, which in turn exert adverse effects on the agricultural practices within a region. Climate change manifests its impact on the land and agriculture in various ways (Raza et al., 2019). This includes irregular rainfall, occurrences of heatwaves, alterations in average temperatures, changes in the behavior of pests, weeds, and microbes, and alterations in the concentration of atmospheric CO_2 and ozone levels. The variable nature of global climate poses a substantial threat that has brought significant attention from the scientific community. This is due to the fact that these climate variations are contributing negatively to global crop production and undermining food security on a global scale. Therefore, research aimed at addressing this concern often centers on identifying exotic genetic diversity that demonstrates significant stress tolerance. The goal is to understand and incorporate the genetic factors responsible for these traits or discover the fundamental genes that can serve as a foundation for precise genetic modifications. While these approaches may occasionally succeed in improving performance in specific stressful conditions (Snowdon et al., 2021), conventional breeding methods have gradually increased the yields of modern crops over the long term by accumulating advantageous, small-effect genetic variants. These variants not only enhance yield but also ensure stability by enabling stress adaptation.

The advancement in genomics has been revolutionizing agriculture, providing a foundation for sustainable intensification and enhancing crop resilience in the face of climate change. This change is mainly a result of the increasing number of high-quality reference genomes, made possible by the widespread adoption of advanced sequencing technologies (Scheben et al., 2016). Furthermore, improvements in population-level genotyping have added to our understanding of genetic diversity on a broader scale. As biological data becomes more readily available in the public domain, the primary challenge in genomics research has shifted from sequencing to computer analysis. This transition has rendered earlier statistical and bioinformatics tools and algorithms obsolete, paving the way for the emergence of new computational tools and algorithms that are designed to handle extensive datasets, often referred to as BIG data (Naik et al., 2023). The idea of plant pan-genomics, which provides deeper insights into the variety of genetic resources accessible for crop improvement and the production of novel cultivars, is being made possible by the abundance of genomic data. Next-generation breeding, on the other hand, involves the application of cutting-edge techniques, technologies, and methods in plant breeding, going beyond traditional practices and integrating genomics and high-throughput screening to advance the creation of improved crop varieties. This integration of genomics into breeding has yielded significant advantages, enabling the swift identification of specific genes linked to agronomic traits influenced by climate conditions, ultimately contributing to the development of crops well-suited for a changing climate, enhancing global food security and sustainability. In this chapter, we explore significant limitations affecting agricultural productivity and the role of genetic variability in enhancing crop improvement. Furthermore, we delve into next-generation breeding approaches aimed at bolstering crop adaptability. These approaches encompass high-throughput phenotyping, sequencing technologies, genomes and pan-genomes, trait mapping, gene editing, and speed breeding. We also highlight how artificial intelligence is used in crop breeding for climate resilience.

9.2 MAJOR ABIOTIC CONSTRAINTS FOR AGRICULTURAL PRODUCTIVITY

Global climate change is putting food and nutritional security at risk. Most of the crops grown in the tropical and subtropical regions of the world are seriously threatened by climate change. The impact of climate change is very comprehensive, as it depends mostly on the current local climate, how climate changes locally, and even soil conditions (Mendelsohn, 2009). It has been occurring due to anthropogenic activities leading to different kinds of biotic and abiotic stresses, and therefore it has become a serious threat to global food security. However, abiotic stresses like heat, drought, salinity, flooding, cold, CO_2, metal deficiency, and toxicity make agricultural practices more vulnerable. It impacts plants starting from morphological, physiological, and even molecular levels. These stresses are hostile to the growth and development of plants, leading to great crop yield penalties of approximately 51%–82% annual losses across the globe (Oshunsanya et al., 2019). Pollen sterility is the major problem of yield reduction caused by many abiotic stresses like heat, drought, and cold. Abiotic stresses mostly affect major crops like rice, wheat, maize, barley, soybean, and chickpeas. Heat stress mostly affects cool-season crops, but more or less warm-season crops are also affected. Therefore, significant yield loss has been observed in rice, wheat, maize, and chickpeas. Pollen sterility in rice grown in high-altitude regions, due to cold stress during the booting stage, has been linked to tapetum hypertrophy (causing disrupted nutrient supply to microspores) in anthers (Yamamori et al., 2021). In the case of drought stress, roots are unable to uptake more water through their expansion against gravity, and thus it produces symptoms like yellowing leaves, rolling (stress avoiding), and scorching, resulting in permanent wilting in plants (Corso et al., 2020). Furthermore, the yield penalty was also observed in barley, maize, and rice as a result of fewer tillers, spikes, and grains per plant. It has also been observed in pulses like chickpea, cowpea, lentil, and beans like faba bean, mung bean, and common bean, showing reduced yield due to terminal drought (Farooq et al., 2017). Interrupted seed germination and sodium (Na^+) toxicity have been observed in maize due to decreased osmotic potential of saline soils compared to the osmotic potential of maize seeds (Farooq et al., 2015). Ion toxicity was determined in cucumber planted in saline soil by grafting cucumber onto the rootstock of pumpkin, resulting in restricted uptake of Na^+ by cucumber (Chen et al., 2020).

9.3 ROLE OF GENETIC VARIABILITY IN CROP IMPROVEMENT

Genetic variability in crop improvement refers to the genetic diversity within a population of a specific crop species (Salgotra and Chauhan, 2023). This diversity arises from natural mutations, genetic recombination, and the introduction of new genetic material, and it plays a pivotal role in developing improved crop varieties over time. The emphasis on creating new crop varieties evolves in response to changing environmental conditions. Plant breeders create crop varieties that work well in extreme weather conditions and have good qualities (Bhandari et al., 2017). For example, consider a crop like rice. Breeders can utilize the genetic variability present in different rice varieties to create new strains that exhibit traits like higher grain yield, resistance to pests and diseases, tolerance to drought, or improved nutritional content. They may develop new types of rice plants that are more adapted to various environments by identifying and crossing plants that possess these desired characteristics. This diversity within rice populations not only ensures food security by adapting to various climates and resisting pests but also supports sustainable agriculture by reducing the need for chemical inputs and preserving genetic resources. In plant breeding programs, prior to selecting diverse parent plants, breeders and biotechnologists employ numerous methods to analyze the germplasm and understand its genetic diversity (Govindaraj et al., 2015). These techniques include the assessment of genetic diversity in plant genetic resources and the identification of superior plant varieties, involving the use of various methods like examining morphological traits, biochemical or allozyme markers, and applying molecular techniques. For instance, in rice diversity analysis, the extent of population genetic variation accounted for by SNPs (Single Nucleotide Polymorphisms) exceeded that explained by SSRs (Simple Sequence Repeats). In rice diversity analysis, they found that the gene flow among the 93 genotypes under investigation was greater than that typically observed in naturally self-pollinating crops. This elevated gene flow might be attributed to the prolonged selective breeding of Indica rice in the Qinba region (Zhang et al., 2022). SNPs have gained preference as the genetic marker of choice for examining partially or fully sequenced genomes because they are prevalent throughout the genome.

9.4 CONVENTIONAL BREEDING METHODS TO ENHANCE CROP ADAPTABILITY

The process of creating new varieties by natural processes and selection is known as conventional breeding. It produces crops with desired features by combining and matching pre-existing traits. This is accomplished by putting together desirable traits from several gene pools that are closely related through a process called cross-hybridization (Acquaah, 2015). Identifying superior cultivars through conventional breeding heavily relies on the phenotypic expressions of such cultivars because phenotypes are greatly influenced by genotype-environment interactions; selected cultivars are therefore not error-free. A developed variety's success is based on both its high producing potential and the consistency of its yield over time in different environments. A variety is considered widely adopted if it can be produced in different agro-climatic zones and yet provide a good yield while performing similarly to other varieties or even better. Accordingly, adaptability denotes a lower variance in performance between places, and stability denotes a lower fluctuation in performance between years.

There are different conventional methods to enhance crop adaptability, including pureline selection, mass selection, and backcross breeding for self-pollinated crops, but recurrent selection and hybridization for cross-pollinated crops. Besides, in contemporary agriculture, mutation breeding, cross-breeding, and transgenic breeding are the main crop improvement techniques. Such protracted, demanding, and aimless breeding programs are unable to meet the rising global food demand. Hybridization is used to create new cultivars by combining the necessary traits from closely related individuals. In traditional breeding, hybridization is the means of producing variability. Many early and late-season varieties of major agricultural crops are produced in India, including Jaya (rice), Sonalika (wheat), Pusa hybrid 1 and 2 (maize), Pusa Komal (chickpea), and Pusa Swarnim (mustard). Mutation breeding is the most popular traditional breeding technique. It swiftly produces large populations of better crops, which may then be further bred until the desired outcome is obtained. Genetic differences are produced at the initial stage of the plant breeding process. It has produced many mutant varieties resistant to biotic and abiotic stresses, like Pusa Basmati 1 (rice), Sharbati Sonora (wheat), and BGM-547 (chickpea), and so forth. To enrich the plant genome and create new plant types, conventional breeding methods are insufficient. In view of this, improved and modern plant breeding approaches have been adopted in the last decades.

9.5 MULTI-OMICS APPROACHES TO UNDERSTAND THE CROP ADAPTABILITY

Multi-OMICS approaches are powerful methods that combine various molecular and genetic technologies to gain a comprehensive understanding of how crops adapt to different environmental conditions (Jha et al., 2023). These approaches provide a holistic view of the genetic, genomic, transcriptomic, proteomic, and metabolomic factors that contribute to crop adaptability. However, plant genomics plays a pivotal role in enhancing crops, improving their resilience to environmental challenges, and boosting overall productivity (Varshney et al., 2021). The application of cutting-edge genomics techniques, alongside the abundance of extensive sequencing data, equips us with the ability to recognize the genetic variations or genes responsible for enhancing crop performance and streamlining breeding processes. Additionally, biotechnological methods, like the CRISPR/Cas genome editing approach, which has been used for precise genome editing, have accelerated progress in regulating genes and critical pathways for adapting to drought conditions (Zaidi et al., 2020). A thorough comprehension of how plants adapt to various drought scenarios is essential for securing future harvests and driving the necessary genetic improvements in crop development. These improvements can be facilitated through genetic diversity, harnessed via genomic breeding, more exact genetic alterations, and customized agricultural practices (Varshney et al., 2021).

The rapid progress in high-throughput sequencing technology, like NGS (Next Generation Sequencing), coupled with a significant reduction in sequencing costs, has made transcriptome profiling a highly effective method for studying the genetic expression of stress-related genes and their associated pathways (Wang et al., 2020a). Comparative transcriptome profiling between different plant genotypes has been extensively explored to uncover distinct molecular mechanisms underlying stress tolerance in various crop species (Mia et al., 2020). In general, when crop plants face adverse climatic conditions such as drought, they trigger adaptive mechanisms that encompass specific morphological, physiological, and molecular processes (Mia et al., 2020). These processes often result in an increased or decreased expression of genes related to stress responses, aiming to mitigate the damage caused by extreme conditions. In the case of drought, some of these genes, like *DREB* genes, produce "effector proteins" that directly defend the plant, while others play passive roles by encoding regulatory proteins like protein kinases and transcription factors (Manna et al., 2020). These regulatory proteins upregulate the expression of stress-related genes.

Additionally, proteomics and metabolomics play a crucial role in understanding and enhancing crop adaptability. They provide valuable insights into the intricate molecular processes and chemical compounds that influence the ability of crops to thrive in diverse environmental conditions (Eldakak et al., 2013). Although genomic studies explore theoretical possibilities, proteomic research reveals the functional elements driving specific cellular processes. The proteome is dynamic, unlike the static genome, and studying proteins, including post-translational modifications, is vital for understanding biological functions (Kav et al., 2007). Plants have evolved various strategies to overcome extreme adverse conditions. These strategies include modulation of photosynthesis and associated protein complexes and the utilization of metabolites like ascorbate and glutathione, along with enzymes such as peroxidases and superoxide dismutases (Hare et al., 1998). In response to drought conditions, plants increase turgor pressure by overproducing osmolytes like proline, glycine betaine, and trehalose, which provide secondary protection to proteins and prevent misfolding. Dehydration-responsive proteins like dehydrins and heat shock proteins are also overproduced to safeguard internal metabolic processes. Therefore, it is crucial to give equal importance to both proteomic and metabolomic studies.

9.6 NEXT-GENERATION BREEDING APPROACHES TO ENHANCE CROP ADAPTABILITY

Next-generation breeding approaches are revolutionary methods in agriculture that harness cutting-edge genetic technologies to enhance the adaptability of crops (Figure 9.1). In the realm of next-generation breeding approaches, a meticulously interconnected sequence of advanced techniques collaborates to enhance crop adaptability. The journey commences with high-throughput phenotyping, which employs automated sensors and imaging technologies to swiftly and accurately evaluate a wide array of pivotal plant traits, serving as the foundational step in the selection of superior individuals (Gill et al., 2022). This data feeds into the application of sequencing technologies, particularly next-generation sequencing (NGS), enabling breeders to decode and comprehend entire plant genomes rapidly and cost-effectively. In these genomes and the larger context of pan-genomes, breeders discover a valuable source of genetic diversity that expands the gene pool. Exploring this diversity, they can identify novel genes and variations that are invaluable for enhancing adaptability (Jha et al., 2022). Moving further along the continuum, trait mapping comes into play, encompassing various genetic mapping techniques that elucidate the specific genomic regions responsible for the expression of desirable traits. The cutting-edge capabilities of gene editing technologies, exemplified by CRISPR-Cas9, come to the forefront, enabling breeders to modify the gene. This facilitates the introduction or suppression of specific genes associated with beneficial or undesirable traits. Lastly, the process is accelerated by the innovative technique of speed breeding, which manipulates environmental conditions, such as light and temperature, to hasten the plant breeding cycle. This drastically reduces the time required for the generation of new crop varieties, expediting the development of crops primed for resilience in the face of changing environmental conditions (Rai, 2022). This integrated approach, from high-throughput phenotyping (HTP) to speed breeding, culminates in the development of crop varieties that exhibit superior adaptability, resistance to environmental stressors, and elevated nutritional quality. In concert, these techniques address the critical challenges of global food security and ensure the sustainability of agriculture amidst the uncertainties of a changing world.

9.6.1 HIGH-THROUGHPUT PHENOTYPING (HTP)

The phenotyping of a plant is the comprehensive evaluation of complex traits, including its growth and development, tolerance and resistance, physiology, architecture, yield, ecology, and the measurement of every quantitative parameter that establishes a foundation for the assessment of complex traits (Li et al., 2020a). HTP has accelerated plant breeding efficiency by screening a large number of plants at distinct phenological stages in a short duration. This cutting-edge technology uses automated trait analysis to provide phenotypic data of desired plant features. It also includes automated sensing, data collection, and analysis. With this technique, a plant may be analyzed and phenotyped from its earliest stages of

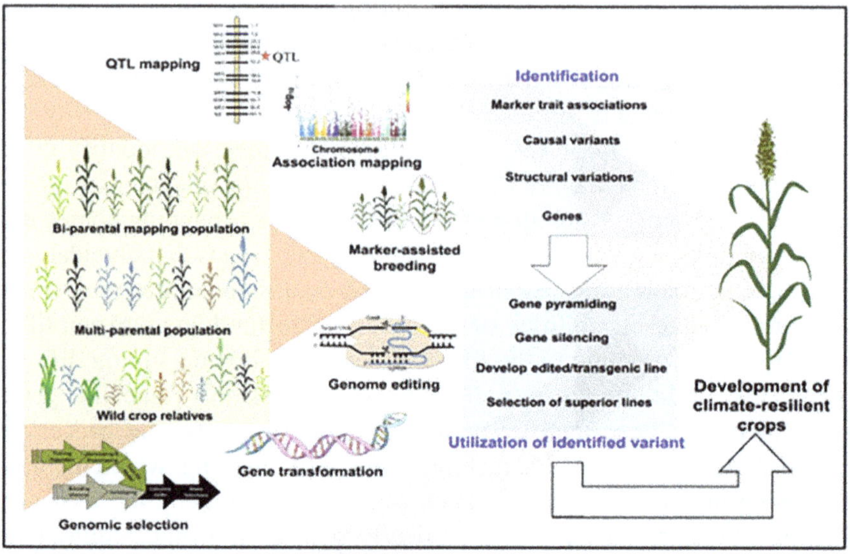

FIGURE 9.1 Leveraging genomic tools to enhance crop adaptability under adverse climatic conditions.

Note: Genomic resources, such as breeding lines and wild relatives, facilitate the identification of causal variants through approaches like QTL mapping and association mapping. Identified markers can aid marker-assisted breeding, while key genes can be targeted for gene editing to unravel complex traits linked to climate resilience.

growth, negating the need to wait for it to mature and express itself later in its life cycle (Jangra et al., 2021). There are several HTP techniques available for different crops. They involve different imaging techniques like visible light imaging, fluorescence imaging, thermal imaging, imaging spectroscopy, and some tomography-based techniques like computed tomography and positron emission tomography. Imaging technologies are basically carried out by electromagnetic radiation. The absorption, emission, transmission, reflection, and fluorescence are the properties of electromagnetic radiation that reflect the health of plants, which are usually not detected by human eyes. In that context, HTP has been used for crop stress management in field conditions during the last few decades. The biomass of plants, root architecture, growth dynamics, rate of seed germination, vegetation indices, plant height and structure, morphological characters (height, shape, and color), panicle traits, and so forth, have been recorded for Arabidopsis, barley, rice, maize, Medicago, and bean through the use of visible light for imaging. Moreover, thermal imaging with near-infrared cameras has been used in crops like wheat, maize, grapevine, rice, and barley for recording leaf area index, disease severity, seed infestation by insects, composition of seed, canopy temperature, transpiration, stomatal conductance, and plant-water status. In spite of that, fluorescence imaging has been used in crops like wheat, maize, tomato, bean, sugar beet, and barley to record photosynthetic activity, quantum yield, disease severity of leaves, non-photochemical quenching, chlorophyll conductance, pigment composition, and so forth. On the contrary, three-dimensional imaging is used in crops like soybean, maize, pepper, and rye to measure the architecture of roots, structure of shoots, plant height, and so forth. Similarly, the different morphological parameters in some crops of economic interest like soybean, wheat, maize, sugar beet, barley, and rice have been recorded using laser imaging, magnetic resonance, positron emission tomography, and computed tomography.

9.6.2 SEQUENCING

Determining the order of four nitrogenous bases of nucleotides or even the amino acid code of proteins is known as sequencing. Sanger sequencing was the first among DNA sequencing techniques to be commercialized to identify point mutations precisely in small stretches of DNA, mostly in gene editing technology. It was still costly and time-consuming, but it was used to read genomic sequences from plants like *Arabidopsis thaliana* and rice. The study of genomics and molecular biology was revolutionized by the advent of high-throughput sequencing approaches like NGS, which make it possible to generate huge quantities of sequencing data more quickly, accurately, and affordably than using Sanger sequencing, which was previously employed. NGS technology involves RNA sequencing, DNA sequencing, QTL sequencing, exome sequencing, and many more. RNA sequencing, also known as transcriptome profiling, was reported by Garg et al. (2023) to reveal the expression and regulation of genes associated with fusarium wilt resistance in chickpeas. Innovations in technology for sequencing DNA have had a significant impact on crop improvement and have changed biological research. In the context of DNA sequencing, different approaches like genotyping-by-sequencing (GBS) and whole genome resequencing (WGRS) have been applied to study the genomic regions associated with agronomic traits. GBS is an ultimate tool to accelerate plant breeding, used for discovering and genotyping SNPs in large crop genomes like maize and wheat. Large-scale plant breeding programs have successfully used GBS to carry out GWAS, genomic diversity, genetic linkage analysis, molecular marker discoveries, and genomic selection (He et al., 2014). WGRS-based QTL sequencing has been used in identifying genomic regions in rice (Takagi et al., 2013), pigeon pea (Singh et al., 2022), groundnut (Pandey et al., 2017), chickpea (Deokar et al., 2019), and so forth, so that climate-smart cultivars could be developed in the future. Exome sequencing is a significant tool for studying natural variation in plants, host-pathogen interaction, and improvement in crop production, as allelic variation associated with phenotypic characters is determined by exons (Hashmi et al., 2015). Saintenac et al. (2011) and Winfield et al. (2016) have utilized it to determine the genetic diversity and natural variation between hexaploid wheat types and wild and cultivated durum wheat accessions, respectively. The assembly of genomes is seriously impeded by a high proportion of repeat sequences. However, a foundation has been laid for many significant crops, and more studies may yield more data that can be used in crop breeding. After Arabidopsis, rice was the first crop whose genome was sequenced. The several rice genome initiatives provided a wealth of valuable material and developed a solid framework for rice genomics. The *Gn1a* significant QTL for rice grain yield was discovered using these data, and it was later determined to be a cytokinin oxidase/dehydrogenase, an enzyme that breaks down cytokinin. In order to learn more about their genome of interest, scientists are using second-generation sequencing as a result of the latest advancements in crop genome sequencing technologies, which are accelerating the speed of technological advancement. A draft consensus sequence of the grape genome has been developed using the Sanger and Roche 454 technologies, estimating 94.6% of the genome with 2093 meta-contigs (Velasco et al., 2007). Moreover, Illumina Solexa combined with Roche 454 has been applied to characterize the cotton genome (Wilkins et al., 2009). The oil palm genome has therefore been completed (1.7 Gbps) with the application of a combination of Roche 454, whole genome shotgun, and bacterial artificial chromosome. In addition to this, many economically important crops have been sequenced, and the information obtained is utilized in various crop improvement programs related to biotic and abiotic stress conditions.

9.6.3 GENOMES AND PAN-GENOMES

Plants are able to adapt to changing environments because of the genetic variation included in their dynamic plant genomes (Raza et al., 2019). The genetic diversity across the population is not adequately represented by a single reference genome of a species. A pangenome can represent a group of plants' genomic diversity, as opposed to a single reference genome. The pangenome is the complete collection of genes found in a species. It is made up of the dispensable and the core genome (approximately 80% of the genome), which contains sequences that are shared by every member of the species. In plant species, the dispensable genome is typically host to a large number of agronomically significant genes, as reported in the pangenome of 54 lines of the grass *Brachypodium distachyon* containing biotic stress response genes. Similarly, some disease-resistant genes were found among 4873 genes in the dispensable genome of tomato. It labels genes as either variable, meaning they are missing in at least one accession, or core, meaning they are present in all accessions. The first pangenome was published by Li et al. (2014), and they sequenced seven accessions of wild soybeans (*Glycine soja*), a potential source of new genes for the improvement of cultivated soybeans (*Glycine max*). Additionally, pangenomes have been constructed for a variety of crops of economic interest like maize, soybean, tomato, cotton, rice, chickpea, sorghum, sunflower, and canola. These genomes are a valuable resource for the breeding of climate-resilient varieties. Plants have been able to adapt to changing climates for millions of years, as evidenced by their ability to grow in a variety of circumstances, including cold, saline, dry, hot, or regularly changing environments. The current crops that lost their ability to withstand biotic and abiotic challenges during domestication and breeding improvement are not like the diverse crop wild relatives.

9.6.4 TRAIT MAPPING

Genomic and genetic resources have been generated by numerous crops to uncover the genetic basis of complex traits. Trait mapping is a fundamental aspect of untying the genetic basis of complex traits, and two prominent approaches employed for this purpose are quantitative trait locus (QTL) mapping and association mapping/genome-wide association studies (GWAS). QTL mapping, often conducted in controlled breeding populations like F2 or recombinant inbred lines (RILs), identifies genomic regions correlated with quantitative traits by analyzing offspring resulting from crosses with differing trait values. On the other hand, GWAS utilizes natural populations or diverse germplasm collections to uncover associations between genetic markers and phenotypic variations, capturing a broader range of natural allelic diversity. QTL mapping offers precision and control over the genetic background, while association mapping excels in capturing natural allelic diversity, making it well-suited for traits influenced by multiple genes (Gangurde et al., 2022). The realm of plant genomics stands as a pivotal accelerator for breeding programs, playing a critical role in enhancing crop performance by identifying traits and exploring genetic variations within the crop genome. Furthermore, plant genomics plays a fundamental role in elucidating gene functionality and their intricate interplay, governing plant growth and performance in diverse agro-climatic conditions (Bevan et al., 2017; Varshney, 2016). Several mapping studies on abiotic stress in various crop plants have been reported (Table 9.1). In the realm of trait mapping, especially in the context of climate change, the selection of crucial traits is paramount for understanding and enhancing adaptability in adverse climatic conditions. Traits like relative water content, chlorophyll content, and root-related traits are often integral components of trait mapping studies (Schleuning et al., 2020). Through trait mapping approaches such as QTL mapping and GWAS, researchers can precisely identify the genomic regions associated with these key traits (Xu et al., 2023). This knowledge not only aids in unraveling the genetic basis of these traits but also provides valuable insights for breeding programs aimed at developing climate-resilient crops. By integrating trait mapping methodologies with the identification of important adaptive traits, scientists can contribute significantly to the development of crops better suited to navigate the challenges posed by climate change.

9.6.5 GENOME EDITING

A variety of cutting-edge molecular biology techniques are used in genome editing to enable targeted, accurate, and efficient changes at specified genomic sites. In recent years, interest has been increasing in the Clustered Regularly Interspaced Short Palindromic Repeats-CRISPR-associated

TABLE 9.1 Summarizes the Recent Progress in Trait Mapping for Heat Stress Tolerance in Major Field Crops

Crop	Mapping Population	Traits	QTL/MTA	PVE (%)	References
Rice	256 RIL (D 71 × Zhonghui 161)	Heading date and spikelet fertility	6 QTLs	6.70–34.61	Huang et al. (2023)
	185 RIL (Cypress × LaGrue)	Heat associated traits	15 QTLs	3.33–8.27	Kumar et al. (2023)
	173 accessions	Spikelet fertility	5 QTLs	8.93–13.52	Hu et al. (2022)
Maize	21 RIL (Huangzaosi × Mo17)	Flowering Time	71 QTLs	8.34–33.68	Leng et al. (2022)
	165 $F_{3:4}$ (B 73 × Zheng 58)	Leaf angle	8 QTL	4.30–14.21	Zhang and Huang (2021)
	199 $F_{2:3}$ (SG 5 × SG 7)	Kernel size	10 QTLs	8.41–23.02	Wang et al. (2020a)
Wheat	82 RIL (TAM 112 × Duster)	Kernel traits	134 QTLs	1.51–23.30	Wang et al. (2022)
	166 DH (PBW 3438 × IC 252874)	Morpho-physiological traits	12 QTLs	11.84–21.24	Pankaj et al. (2021)
	309 RIL (Berkut ×Worrakatta)	Germination associated traits	18 QTLs	2.71–6.52	Ren ct al. (2021)
Chickpea	186 RIL (Pusa 362 × SBD 377)	Heat associated traits	9 QTLs	6.29–90.68	Yadava et al. (2023)
	187 RIL (GPF 2 × ILWC 292)	Heat associated traits	51 QTLs	5.01–13.59	Kushwah et al. (2021)
	184 RIL (DCP 92-3 × ICCV 92944)	Morpho-physiological traits	77 QTLs	7.12–43.49	Jha et al. (2021)

Abbreviations: DH: Double haploid; MTA: Marker trait associations; PVE: Phenotypic variation explained; QTL: Quantitative trait loci; and RIL: Recombinant inbred lines.

protein (CRISPR-Cas) system because of its simplicity and great effectiveness in targeted gene editing. The rise of genome editing technologies, particularly CRISPR-Cas9, has dramatically reshaped the field of crop breeding. Genome editing is widely used to precisely modify crop genomes, enhancing adaptability by targeting specific genes (Table 9.2). This technology allows researchers to improve traits associated with abiotic stress tolerance, accelerating crop improvement compared to traditional methods and contributing to global food security.

9.6.5.1 EXPANDING THE TOOLKIT FOR CRISPR-CAS TECHNOLOGY

In terms of genome editing, the CRISPR/CRISPR-associated system stands as an incredibly innovative and revolutionary technology (Jinek et al., 2012). Ongoing efforts persist in addressing and reducing the limitations of CRISPR/Cas systems in plants while striving to develop the next generation of genome editing tools. Utilizing the CRISPR/Cas system has led to significant progress in engineering plant genomes for crop enhancement (Li et al., 2013; Shan et al., 2013). This innovative technology is transforming plant breeding by offering extensive applications in creating crops resilient to diverse climates (Puchta, 2017). By using CRISPR/Cas9 genome editing to create precise point mutations in the Acetolactate Synthase gene, basmati rice has been engineered to be resistant to herbicides (Zafar et al., 2023). Through precise editing of three homologs of *TaCIPK14* (the susceptible gene) for wheat stripe rust, which is responsible for significant wheat losses worldwide, researchers have developed a disease-resistant wheat variety with the help of the CRISPR/Cas9 genome editing system (He et al., 2023). Additionally, many studies highlight its efficacy in improving agronomic traits and enhancing stress tolerance to various environmental pressures, both biotic and abiotic (Table 9.2). For instance, drought, a highly damaging abiotic stress causing significant crop losses, was addressed by using the CRISPR/Cas9 system to create knockout mutants of *SlLBD40* in tomatoes. These mutant lines showed higher water-holding capacity than the overexpressing *SlLBD40* line. These results suggest that the *SlLBD40* gene negatively regulates drought stress in tomato plants (Liu et al., 2020).

TABLE 9.2 List of Targeted Genes for Biotic and Abiotic Stress Responses Using the CRISPR System

Crop	Target Gene	Role	Transformation	References
Rice	*OsPUB7*	Abiotic stress	Agrobacterium	Kim et al. (2023)
	OsDjA2 and *OsERF104*	Blast resistance	Agrobacterium	Tavora et al. (2022)
	Xa13	Bacterial blight	Agrobacterium	Li et al. (2020b)
	OsERA1	Drought stress	Agrobacterium	Ogata et al. (2020)
	HDA710	Salinity stress	Agrobacterium	Ullah et al. (2020)
	OsPDS	Heat tolerance	Gene Gun	Nandy et al. (2019)
	Os8N3	Leaf blight	Agrobacterium	Kim et al. (2019)

TABLE 9.2 *(Continued)*

Crop	Target Gene	Role	Transformation	References
Maize	*ZmSRL5*	Drought stress	Agrobacterium	Pan et al. (2020)
	ARGOS	Drought stress	Biolistic	Shi et al. (2017)
Wheat	*aWD40-4B.1*	Drought stress	Agrobacterium	Tian et al. (2023)
	TaHSFA6e	Heat tolerance	Agrobacterium	Wen et al. (2023)
Soybean	*GmAITR*	Salinity stress	Agrobacterium	Wang et al. (2021)
	GmPLAs	Flooding and drought	Agrobacterium	Xiao et al. (2021)
	GmF3H1, GmF3H2 and *GmFNSII-1*	Soybean mosaic virus	Agrobacterium	Zhang et al. (2020)
Tomato	*SlLBD40*	Drought	Agrobacterium	Liu et al. (2020)
	PMR4	Powdery mildew	Agrobacterium	Martinez et al. (2020)
	HyPRP1 domain	Salinity	Agrobacterium	Tran et al. (2021)
	SlJAZ2	Bacterial speck	Agrobacterium	Ortigosa et al. (2019)
Cassava	*AC2, AC3*	African cassava mosaic virus	Agrobacterium	Mehta et al. (2019)
	nCBP-1 and *nCBP-2*	Cassava brown streak virus	Agrobacterium	Gomez et al. (2019)

9.6.5.2 INNOVATIVE TECHNOLOGICAL ADVANCEMENTS

The application of genome-editing technology in agriculture displays significant potential, although it grapples with limitations such as low efficiency of homologous recombination (HR), off-target effects, and restrictions due to protospacer adjacent motif (PAM) sequences. Encouragingly, continuous innovations are augmenting the genome-editing toolkit to overcome these challenges. A number of characteristics of the CRISPR/Cas toolkit have been enhanced recently. These properties include increased target specificity, increased target range, decreased off-target effects, precise nuclease activity, varied PAM sites, and more effective delivery mechanisms (Koonin et al., 2017). This technology has emerged as the preferred system for conducting diverse genetic modifications, encompassing the study of mutation patterns (Jia et al., 2018), incorporation of external genes (Collonnier et al., 2017), regulation of genes (Qi et al., 2013), and even cell imaging (Xue and Acar, 2018).

9.6.5.2.1 Base and prime editing

Utilizing CRISPR/nCas9 (Cas9 nickase) in conjunction with cytidine deaminase, base editing is a novel approach that allows for the targeted modification of single bases in plant genomic sequences. Base Editor 3 (BE3) is a very efficient technology designed to induce specific base substitutions in cotton, with mutation rates ranging from 26.67%–57.78% (Qin et al., 2020). Additionally, Li et al. (2018) innovatively designed an adenine base editor (ABE) to generate wheat and rice plants resistant to herbicides.

Recently, prime editing has emerged as a revolutionary toolkit capable of overcoming previous limitations by facilitating insertions, deletions, and base substitutions without creating double-strand breaks (Marzec et al., 2020). The Xu research group pioneered the development of the initial prime-editing system in rice, establishing the plant prime editors (pPE2) toolkit to induce mutations in various genomic regions, achieving frequencies ranging from 0%–31.3% (Xu et al., 2020). For example, a primary editor system was improved to introduce insertions, deletions, and point mutations in wheat and rice protoplasts utilizing particular promoters and codons. Additionally, Xu et al. (2020) reported that it was employed to produce single and multiple base edits in rice, exhibiting mutation rates of 21.8% and 26%, respectively.

9.6.5.3 DNA-FREE GENOME EDITING SYSTEMS

Traditional genome editing typically entails introducing DNA cassettes encoding editing components into the host genome, a process that can lead to random integration and unintended genetic alterations. However, DNA-free genome editing methods have been successfully implemented through both protoplast-mediated transformation and particle bombardment. Woo and his team achieved the initial successful application of DNA-free genome editing in various plants like Arabidopsis, tobacco, lettuce, and rice by transfecting CRISPR/Cas9 ribonucleoproteins (RNPs) into protoplasts. Additionally, DNA-free genome editing technology using particle bombardment has been developed for wheat and maize.

9.6.6 SPEED BREEDING

The escalation of global food security concerns arises as the human population expands while environmental shifts persist, challenging the pace of enhancement in several vital crops to meet future needs. One contributing factor to the slow progress is the extended generation times of crop plants. However, the introduction of speed breeding has presented a transformative solution, significantly reducing generation times along with facilitating multiple crop cycles annually and thus accelerating breeding and research programs aimed at enhancing crop varieties. It is a promising avenue toward ensuring nutritional assurance and sustainable agriculture by abbreviating breeding cycles for the enhancement of both food and industrial crops. Moreover, accelerating the development, release, and commercialization of cultivars is a remarkable capability of speed breeding. This method empowers plant breeders to elevate crop production by manipulating factors like temperature, duration of light, and its intensity to accelerate the growth of plants. Employing a continuous artificial light source triggers photosynthesis, leading to earlier growth and reproduction. An extended photoperiod of light for long-day plants leads to a reduction of the vegetative phase and acceleration of the reproductive phase, allowing researchers to obtain more generations per year (Sysoeva et al., 2010). This innovative approach aims to address the needs of an expanding global population in the future.

9.6.6.1 ENHANCEMENT OF CROP PLANTS THROUGH SPEED BREEDING

Speed breeding represents a transformative approach compared to conventional breeding approaches in agriculture, significantly expediting various crop breeding processes such as crossing, backcrossing, swift gene identification, population mapping, trait pyramiding, and the development of transgenic pipelines (Hickey et al., 2019). Traditional breeding typically achieves just 1–2 generations per season for any crop. Nevertheless, speed breeding enables the production of up to four generations of *B. napus* and six generations of *B. distachyon, Hordeum vulgare, Pisum sativum, Cicer arietinum*, and *Triticum aestivum* in the same amount of time (Watson et al., 2018). Researchers are employing gene pyramiding to introduce multiple disease resistance genes into European two-rowed barley (Scarlet). Using four donor lines to combine multiple disease resistance (leaf rust, net, and spot forms of net blotch and spot blotch), they were able to develop 87 BC1F3:4 promising lines in less than two years with the aid of speed breeding. Phenotyping of introgression lines (ILs) revealed that they had a high level of multiple disease resistance; for testing the yield of ILs, 12 ILs were selected, out of which 3 promising lines indicated significantly higher yield than the Scarlet at one location, while all 12 ILs displayed an equivalent yield to the Scarlet at another location (Hickey et al., 2017).

Moreover, it serves as an efficient, sturdy, and cost-effective approach for conducting comprehensive initiatives for crop improvement, seamlessly integrating phenomics and genomics. This inclusive approach encompasses discovering candidate genes through various genomic-assisted breeding methods like genomic selection (GS), pan-genome assembly, and genotyping-by-sequencing (GBS). It proceeds to employ multiplex gene editing or metabolic pathway editing for desired traits, followed by HTP to visualize outcomes. Additionally, combining speed breeding with advanced metabolomic tools enables swift risk estimation of gene-edited crops across many generations in a sturdy way (Razzaq et al., 2019). Consequently, the amalgamation of next-generation OMICS tools with speed breeding technology presents a promising path forward for accelerating crop breeding programs aimed at enhancement.

9.7 AI-BASED APPLICATIONS IN BREEDING FOR CLIMATE-RESILIENT CROP

Artificial intelligence (AI) or machine intelligence is a branch of computer science that deals with the development and management of technology and replicates the intellect of the human mind. It teaches us how to carry out activities and make decisions without requiring human intervention (Dwivedi et al., 2021). In the era of climate change, AI has emerged as a progressive field, offering an outstanding opportunity to explore contemporary crop breeding approaches. It has accelerated trait identification for climate adaptation and food security. Progress in agricultural phenomics, enviromics, and other "OMICS" fields is opening doors to clarify the intricate and multifaceted biological processes driving crop responses to environmental stressors. With the use of these "omics" techniques, plant scientists can now precisely assess the crucial agronomic characteristics of large-scale genetic material in a shorter time frame during the early development phases. The vast amount of data and the intricate linkages it contains, however, make it difficult to comprehend the complex processes behind the genes that shape agronomic traits. Ultimately, AI has tremendous computing capability, many advanced devices, and approaches for a sustainable agricultural future (Khan et al., 2022). In this regard, applications of AI in climate-resilient crop breeding are described briefly.

9.7.1 OVERCOMING THE CONVENTIONAL PHENOTYPING BOTTLENECK

The conventional manual phenotyping approaches were labor-intensive and time-consuming; nevertheless, they were unable to accomplish research goals for plant scientists. Keeping that in mind, there has been rapid adoption of AI-based HTP, using ground and aerial platforms with multiple sensors. It can be utilized for phenotyping various traits such as biotic and abiotic stresses, yield-related traits, phenological traits, and even morphological traits throughout the developmental stages of plants. It has been used to replace the laborious and inconsistent method of conventional manual phenotyping, thus enabling the evaluation of a larger population with improved selection intensity and accuracy simultaneously (Araus et al., 2018). In addition, deep learning (DL) and machine learning (ML) offer alternative ways of obtaining useful information. In that context, DL and ML have been used for biotic and abiotic stress management in field conditions for many crops during the last decades. The crops include rice, maize, wheat, barley, soybean, citrus, and cotton.

9.7.2 EXPLORING THE FUNCTIONAL ANALYSIS OF GENES

The advent of high-throughput sequencing technology has produced abundant genomic data, creating storage and analysis issues (Esposito et al., 2019).

In this regard, AI technology in bioinformatics has enabled scientists to measure the simultaneous expression of more than one gene or even a single gene present in the genome (Lemmon et al., 2013). Additionally, ML, one of the AI tools with algorithms, has enabled different kinds of bioinformatics analysis. It entails a variety of tasks, including the identification of cis-regulatory elements and protein-coding genes, gene expression, subcellular localization, protein-protein interactions, metabolic pathway analysis, and many more (Mahood et al., 2020).

9.7.3 INTEGRATING THE CROP GENOME AND PHENOME WITH AI

Rapid gene identification linked to crop phenotypes may be made possible by combining AI with phenomics and genomics techniques, which would ultimately speed up agricultural development initiatives. In that context, yield traits and crop performances can be predicted on time by incorporating genotypic and phenotypic data. Scientists have modulated the information flow from generic DNA to genetic-based phenotypes using AI and its models to investigate possible variances in natural populations (Mahood et al., 2020). More specifically, AI will assist breeders in discovering novel genetic loci to boost agricultural output by initiating genome algorithms and facilitating high-throughput crop phenotyping in quantitative traits for open-field and controlled environments (Razzaq et al., 2021). Recently, Deep BSA, a deep-learning method for BSA, was developed for the purpose of functional gene cloning and QTL mapping in maize.

9.8 CURRENT CHALLENGES AND FUTURE PROSPECTS

The increasing global population and the challenges posed by climate change have intensified the need for ensuring worldwide food security. Next-generation breeding technologies are emerging as a powerful solution to address these challenges. These advanced techniques go beyond traditional breeding methods, integrating genomics and high-throughput screening to develop crop varieties that are not only high-yielding but also resilient to the changing climate (Razzaq et al., 2021). The current challenges of next-generation breeding include addressing issues related to regulatory compliance, ensuring public acceptance of genetically modified crops, and managing ethical concerns surrounding gene editing. Additionally, there is a need to develop efficient delivery methods for genome editing tools, expand the understanding of gene function, and enhance data management and analysis capabilities to harness the full potential of genomic data in breeding programs. Meeting these challenges is crucial to advancing the field of next-generation breeding and achieving its potential in crop improvement.

Innovative pan-genome techniques are allowing us to explore the full range of genetic variations in crops. An increase in genomic data is anticipated as a result, especially when third-generation sequencing and the pan-genome theory are used to generate large gene repertoires from landraces, domesticated plants, and crop wild relatives. Additionally, super pan-genomes, which combine different pan-genomes, offer a holistic view of the genetic profile of specific plant genera and encourage pan-genome directed by haplotype breeding for agricultural crop development (Khan et al., 2019). AI plays a key role in the construction of super pan-genomes. This AI-driven process empowers the comprehensive investigation of genetic variations and their real-world applications in crop breeding and genomics research. DL also has potential applications in combining sequencing data to create models for genomic selection and trait prediction (Montesinos-Lopez et al., 2021). Remarkable precision in modifying plant genomes is made possible by emerging CRISPR technologies, such as tissue-specific editing, prime editing, *de novo* domestication, and fine-tuning of gene expression. It generated the possibility of editing plant organelles like mitochondria and chloroplasts using CRISPR/Cas systems (Kang et al., 2021). This requires further exploration, particularly in developing delivery methods for these systems. AI-driven phenotyping platforms, including robotics and 3D imaging sensors, when combined with other OMICS data, are revolutionizing plant science. Handling big data in a multiscale, multidimensional, and multidomain phenomics context necessitates the application of next-generation procedures like swarm intelligence, hybrid intelligence, and data fusion. These approaches facilitate the analysis of vast datasets, propelling crop breeding and research to new heights.

9.9 CONCLUSION

Plant breeders face challenges like unpredictable weather and stagnant crop yields. New tools help us understand trait genetics better. To overcome these hurdles, scientists are using innovative tools that provide deeper insights into the genetic basis of essential traits in plants. Despite the challenges posed by climate change, there's optimism about making significant advances in crop breeding. This progress is fueled by a better understanding of genetic variations within crops and the application of advanced genetic technologies. Techniques like targeted genetic recombination and genome editing are being employed to introduce novel genetic variations. One noteworthy advancement is the potential of genomic-assisted breeding, which focuses on optimizing methods that use information from a crop genome to enhance adaptability. This advancement holds promise for creating crops with improved traits. Nevertheless, continuous innovation in plant breeding methods is crucial for maximizing the benefits of these technological advancements and adapting to evolving challenges.

KEYWORDS

- **climate resilience**
- **crop adaptability**
- **crop improvement**
- **gene editing**
- **genetic diversity**
- **multi-omics approaches**
- **next-generation breeding**
- **sustainable agriculture**

REFERENCES

Acquaah, G. (2015). Conventional plant breeding principles and techniques. In *Advances in Plant Breeding Strategies: Breeding, Biotechnology and Molecular Tools* (pp. 115–158). https://doi.org/10.1007/978-3-319-22521-0_5.

Araus, J. L., Kefauver, S. C., Zaman-Allah, M., Olsen, M. S., & Cairns, J. E. (2018). Translating high-throughput phenotyping into genetic gain. *Trends in Plant Science, 23*(5), 451–466. https://doi.org/10.1016/j.tplants.2018.02.001.

Bevan, M. W., Uauy, C., Wulff, B. B., Zhou, J., Krasileva, K., & Clark, M. D. (2017). Genomic innovation for crop improvement. *Nature, 543*(7645), 346–354. https://doi.org/10.1038/nature22011.

Bhandari, H. R., Bhanu, A. N., Srivastava, K., Singh, M. N., & Shreya, H. A. (2017). Assessment of genetic diversity in crop plants: An overview. *Advances in Plants and Agriculture Research, 7*(3), 279–286. https://doi.org/10.15406/apar.2017.07.00255.

Chen, T. W., Gomez Pineda, I. M., Brand, A. M., & Stützel, H. (2020). Determining ion toxicity in cucumber under salinity stress. *Agronomy, 10*(5), 677. https://doi.org/10.3390/agronomy10050677.

Collonnier, C., Guyon-Debast, A., Maclot, F., Mara, K., Charlot, F., & Nogué, F. (2017). Towards mastering CRISPR-induced gene knock-in plants: Survey of key features and focus on the model *Physcomitrella patens. Methods, 121, 122*, 103–117. https://doi.org/10.1016/j.ymeth.2017.04.024.

Corso, D., Delzon, S., Lamarque, L. J., Cochard, H., Torres-Ruiz, J. M., King, A., & Brodribb, T. (2020). Neither xylem collapse, cavitation, nor changing leaf conductance drive stomatal closure in wheat. *Plant, Cell and Environment, 43*(4), 854–865. https://doi.org/10.1111/pce.13722.

Deokar, A., Sagi, M., Daba, K., & Tar'an, B. (2019). QTL sequencing strategy to map genomic regions associated with resistance to Ascochyta blight in chickpea. *Plant Biotechnology Journal, 17*(1), 275–288. https://doi.org/10.1111/pbi.12964.

Dwivedi, Y. K., Hughes, L., Ismagilova, E., Aarts, G., Coombs, C., Crick, T., Duan, Y., Dwivedi, R., Edwards, J., Eirug, A., Galanos, V., Ilavarasan, P. V., Janssen, M., Jones, P., Kar, A. K., Kizgin, H., Kronemann, B., Lal, B., Lucini, B., & Williams, M. D. (2021). Artificial intelligence (AI): Multidisciplinary perspectives on emerging challenges, opportunities, and agenda for research, practice, and policy. *International Journal of Information Management, 57,* 101994. https://doi.org/10.1016/j.ijinfomgt.2019.08.002.

Eldakak, M., Milad, S. I., Nawar, A. I., & Rohila, J. S. (2013). Proteomics: A biotechnology tool for crop improvement. *Frontiers in Plant Science, 4,* 35. https://doi.org/10.3389/fpls.2013.00035.

Esposito, S., Carputo, D., Cardi, T., & Tripodi, P. (2019). Applications and trends of machine learning in genomics and phenomics for next-generation breeding. *Plants, 9*(1), 34. https://doi.org/10.3390/plants9010034.

Farooq, M., Gogoi, N., Barthakur, S., Baroowa, B., Bharadwaj, N., Alghamdi, S. S., & Siddique, K. H. M. (2017). Drought stress in grain legumes during reproduction and grain filling. *Journal of Agronomy and Crop Science, 203*(2), 81–102. https://doi.org/10.1111/jac.12169.

Farooq, M., Hussain, M., Wakeel, A., & Siddique, K. H. M. (2015). Salt stress in maize: Effects, resistance mechanisms, and management. A review. *Agronomy for Sustainable Development, 35*(2), 461–481. https://doi.org/10.1007/s13593-015-0287-0.

Gangurde, S. S., Xavier, A., Naik, Y. D., Jha, U. C., Rangari, S. K., Kumar, R., Reddy, M. S., Channale, S., Elango, D., Mir, R. R., Zwart, R., Laxuman, C., Sudini, H. K., Pandey, M. K., Punnuri, S., Mendu, V., Reddy, U. K., Guo, B., Gangarao, N. V. P. R., Sharma, V. K., Wang, X., Zhao, C., & Thudi, M. (2022). Two decades of association mapping: Insights on disease resistance in major crops. *Frontiers in Plant Science, 13,* 1064059. https://doi.org/10.3389/fpls.2022.1064059.

Garg, V., Chitikineni, A., Sharma, M., Ghosh, R., Samineni, S., Varshney, R. K., & Kudapa, H. (2023). Transcriptome profiling reveals the expression and regulation of genes associated with Fusarium wilt resistance in chickpea (*Cicer arietinum* L.). *Plant Genome, 16*(3), e20340. https://doi.org/10.1002/tpg2.20340.

Gill, T., Gill, S. K., Saini, D. K., Chopra, Y., de Koff, J. P., & Sandhu, K. S. (2022). A comprehensive review of high throughput phenotyping and machine learning for plant stress phenotyping. *Phenomics, 2*(3), 156–183. https://doi.org/10.1007/s43657-022-00048-z.

Gomez, M. A., Lin, Z. D., Moll, T., Chauhan, R. D., Hayden, L., Renninger, K., Beyene, G., Taylor, N. J., Carrington, J. C., Staskawicz, B. J., & Bart, R. S. (2019). Simultaneous CRISPR/Cas9-mediated editing of cassava eIF 4E isoforms nCBP-1 and nCBP-2 reduces cassava brown streak disease symptom severity and incidence. *Plant Biotechnology Journal, 17*(2), 421–434. https://doi.org/10.1111/pbi.12987.

Govindaraj, M., Vetriventhan, M., & Srinivasan, M. (2015). Importance of genetic diversity assessment in crop plants and its recent advances: An overview of its analytical perspectives. *Genetics Research International, 2015*, 431487. https://doi.org/10.1155/2015/431487.

Hare, P. D., Cress, W. A., & Van Staden, J. (1998). Dissecting the roles of osmolyte accumulation during stress. *Plant, Cell and Environment, 21*(6), 535–553. https://doi.org/10.1046/j.1365-3040.1998.00309.x.

Hashmi, U., Shafqat, S., Khan, F., Majid, M., Hussain, H., Kazi, A. G., John, R., & Ahmad, P. (2015). Plant exomics: Concepts, applications and methodologies in crop improvement. *Plant Signaling and Behavior, 10*(1), e976152. https://doi.org/10.4161/15592324.2014.976152.

He, F., Wang, C., Sun, H., Tian, S., Zhao, G., Liu, C., Wan, C., Guo, J., Huang, X., Zhan, G., Yu, X., Kang, Z., & Guo, J. (2023). Simultaneous editing of three homoeologues of TaCIPK14 confers broad-spectrum resistance to stripe rust in wheat. *Plant Biotechnology Journal, 21*(2), 354–368. https://doi.org/10.1111/pbi.13956.

He, J., Zhao, X., Laroche, A., Lu, Z. X., Liu, H., & Li, Z. (2014). Genotyping-by-sequencing (GBS), an ultimate marker-assisted selection (MAS) tool to accelerate plant breeding. *Frontiers in Plant Science, 5*, 484. https://doi.org/10.3389/fpls.2014.00484.

Hickey, L. T., Germán, S. E., Pereyra, S. A., Diaz, J. E., Ziems, L. A., Fowler, R. A., & Dieters, M. J. (2017). Speed breeding for multiple disease resistance in barley. *Euphytica, 213*, 1–14.

Hickey, L. T., N Hafeez, A., Robinson, H., Jackson, S. A., Leal-Bertioli, S. C. M., Tester, M., Gao, C., Godwin, I. D., Hayes, B. J., & Wulff, B. B. H. (2019). Breeding crops to feed 10 billion. *Nature Biotechnology, 37*(7), 744–754. https://doi.org/10.1038/s41587-019-0152-9.

Hu, C., Jiang, J., Li, Y., Song, S., Zou, Y., Jing, C., Zhang, Y., Wang, D., He, Q., & Dang, X. (2022). QTL mapping and identification of candidate genes using a genome-wide association study for heat tolerance at anthesis in rice (*Oryza sativa* L.). *Frontiers in Genetics, 13*, 983525. https://doi.org/10.3389/fgene.2022.983525.

Huang, D., Zhang, Z., Fan, Y., Tang, S., Zhuang, J., & Zhu, Y. (2023). Detection of QTL for high-temperature tolerance in rice using a high-density bin map. *Agronomy, 13*(6), 1582. https://doi.org/10.3390/agronomy13061582.

Jangra, S., Chaudhary, V., Yadav, R. C., & Yadav, N. R. (2021). High-throughput phenotyping: A platform to accelerate crop improvement. *Phenomics, 1*(2), 31–53. https://doi.org/10.1007/s43657-020-00007-6.

Jha, U. C., Nayyar, H., Chattopadhyay, A., Beena, R., Lone, A. A., Naik, Y. D., & Siddique, K. H. (2023). Major viral diseases in grain legumes: Designing disease resistant legumes from plant breeding and OMICS integration. *Frontiers in Plant Science, 14*, 1183505. https://doi.org/10.3389/fpls.2023.1183505.

Jha, U. C., Nayyar, H., Palakurthi, R., Jha, R., Valluri, V., Bajaj, P., Chitikineni, A., Singh, N. P., Varshney, R. K., & Thudi, M. (2021). Major QTLs and potential candidate genes for heat stress tolerance identified in chickpea (*Cicer arietinum* L.). *Frontiers in Plant Science, 12*, 655103. https://doi.org/10.3389/fpls.2021.655103.

Jha, U. C., Nayyar, H., von Wettberg, E. J. B., Naik, Y. D., Thudi, M., & Siddique, K. H. M. (2022). Legume pangenome: Status and scope for crop improvement. *Plants, 11*(22), 3041. https://doi.org/10.3390/plants11223041.

Jia, C., Huai, C., Ding, J., Hu, L., Su, B., Chen, H., & Lu, D. (2018). New applications of CRISPR/Cas9 system on mutant DNA detection. *Gene, 641*, 55–62. https://doi.org/10.1016/j.gene.2017.10.023.

Jinek, M., Chylinski, K., Fonfara, I., Hauer, M., Doudna, J. A., & Charpentier, E. (2012). A programmable dual-RNA–guided DNA endonuclease in adaptive bacterial immunity. *Science, 337*(6096), 816–821. https://doi.org/10.1126/science.1225829.

Kang, B. C., Bae, S. J., Lee, S., Lee, J. S., Kim, A., Lee, H., Baek, G., Seo, H., Kim, J., & Kim, J. S. (2021). Chloroplast and mitochondrial DNA editing in plants. *Nature Plants, 7*(7), 899–905. https://doi.org/10.1038/s41477-021-00943-9.

Kav, N. N., Srivastava, S., Yajima, W., & Sharma, N. (2007). Application of proteomics to investigate plant-microbe interactions. *Current Proteomics, 4*(1), 28–43. https://doi.org/10.2174/157016407781387357.

Khan, A. W., Garg, V., Roorkiwal, M., Golicz, A. A., Edwards, D., & Varshney, R. K. (2019). Super-pangenome by integrating the wild side of a species for accelerated crop improvement. *Trends in Plant Science, 25*(2), 148–158. https://doi.org/10.1016/j.tplants.2019.10.012.

Khan, M. H. U., Wang, S., Wang, J., Ahmar, S., Saeed, S., Khan, S. U., Xu, X., Chen, H., Bhat, J. A., & Feng, X. (2022). Applications of artificial intelligence in climate-resilient smart-crop breeding. *International Journal of Molecular Sciences, 23*(19), 11156. https://doi.org/10.3390/ijms231911156.

Kim, M. S., Ko, S. R., Jung, Y. J., Kang, K. K., Lee, Y. J., & Cho, Y. G. (2023). Knockout mutants of OsPUB7 generated using CRISPR/Cas9 revealed abiotic stress tolerance in rice. *International Journal of Molecular Sciences, 24*(6), 5338. https://doi.org/10.3390/ijms24065338.

Kim, Y. A., Moon, H., & Park, C. J. (2019). CRISPR/Cas9-targeted mutagenesis of Os8N3 in rice to confer resistance to *Xanthomonas oryzae* pv. *oryzae*. *Rice, 12*(1), 67. https://doi.org/10.1186/s12284-019-0325-7.

Koonin, E. V., Makarova, K. S., & Zhang, F. (2017). Diversity, classification and evolution of CRISPR-Cas systems. *Current Opinion in Microbiology, 37*, 67–78. https://doi.org/10.1016/j.mib.2017.05.008.

Kumar, A., Thomas, J., Gill, N., Dwiningsih, Y., Ruiz, C., Famoso, A., & Pereira, A. (2023). Molecular mapping and characterization of QTLs for grain quality traits in a RIL population of US rice under high nighttime temperature stress. *Scientific Reports, 13*(1), 4880. https://doi.org/10.1038/s41598-023-31399-w.

Kushwah, A., Bhatia, D., Singh, I., Thudi, M., Singh, G., Bindra, S., Vij, S., Gill, B. S., Bharadwaj, C., Singh, S., & Varshney, R. K. (2021). Identification of stable heat tolerance QTLs using inter-specific recombinant inbred line population derived from

GPF 2 and ILWC 292. *PLOS ONE, 16*(8), e0254957. https://doi.org/10.1371/journal.pone.0254957.

Lemmon, E. M., & Lemmon, A. R. (2013). High-throughput genomic data in systematics and phylogenetics. *Annual Review of Ecology, Evolution, and Systematics, 44*(1), 99–121. https://doi.org/10.1146/annurev-ecolsys-110512-135822.

Leng, P., Khan, S. U., Zhang, D., Zhou, G., Zhang, X., Zheng, Y., Wang, T., & Zhao, J. (2022). Linkage mapping reveals QTL for flowering time-related traits under multiple abiotic stress conditions in maize. *International Journal of Molecular Sciences, 23*(15), 8410. https://doi.org/10.3390/ijms23158410.

Li, C., Li, W., Zhou, Z., Chen, H., Xie, C., & Lin, Y. (2020b). A new rice breeding method: CRISPR/Cas9 system editing of the Xa13 promoter to cultivate transgene-free bacterial blight-resistant rice. *Plant Biotechnology Journal, 18*(2), 313–315. https://doi.org/10.1111/pbi.13217.

Li, C., Zong, Y., Wang, Y., Jin, S., Zhang, D., Song, Q., Zhang, R., & Gao, C. (2018). Expanded base editing in rice and wheat using a Cas9-adenosine deaminase fusion. *Genome Biology, 19*(1), 59. https://doi.org/10.1186/s13059-018-1443-z.

Li, D., Quan, C., Song, Z., Li, X., Yu, G., Li, C., & Muhammad, A. (2020a). High-throughput plant phenotyping platform (HT3P) as a novel tool for estimating agronomic traits from the lab to the field. *Frontiers in Bioengineering and Biotechnology, 8*, 623705. https://doi.org/10.3389/fbioe.2020.623705.

Li, J. F., Norville, J. E., Aach, J., McCormack, M., Zhang, D., Bush, J., Church, G. M., & Sheen, J. (2013). Multiplex and homologous recombination–mediated genome editing in Arabidopsis and *Nicotiana benthamiana* using guide RNA and Cas9. *Nature Biotechnology, 31*(8), 688–691. https://doi.org/10.1038/nbt.2654.

Li, Y. H., Zhou, G., Ma, J., Jiang, W., Jin, L. G., Zhang, Z., Guo, Y., Zhang, J., Sui, Y., Zheng, L., Zhang, S. S., Zuo, Q., Shi, X. H., Li, Y. F., Zhang, W. K., Hu, Y., Kong, G., Hong, H. L., Tan, B., & Qiu, L. J. (2014). De novo assembly of soybean wild relatives for pan-genome analysis of diversity and agronomic traits. *Nature Biotechnology, 32*(10), 1045–1052. https://doi.org/10.1038/nbt.2979.

Liu, L., Zhang, J., Xu, J., Li, Y., Guo, L., Wang, Z., Zhang, X., Zhao, B., Guo, Y. D., & Zhang, N. (2020). CRISPR/Cas9 targeted mutagenesis of SlLBD40, a lateral organ boundaries domain transcription factor, enhances drought tolerance in tomato. *Plant Science, 301*, 110683. https://doi.org/10.1016/j.plantsci.2020.110683.

Mahood, E. H., Kruse, L. H., & Moghe, G. D. (2020). Machine learning: A powerful tool for gene function prediction in plants. *Applications in Plant Sciences, 8*(7), e11376. https://doi.org/10.1002/aps3.11376.

Manna, M., Thakur, T., Chirom, O., Mandlik, R., Deshmukh, R., & Salvi, P. (2020). Transcription factors as key molecular target to strengthen the drought stress tolerance in plants. *Physiologia Plantarum, 172*(2), 847–868. https://doi.org/10.1111/ppl.13268.

Martinez, S. M. I., Bracuto, V., Koseoglou, E., Appiano, M., Jacobsen, E., Visser, R. G., Wolters, A. M. A., & Bai, Y. (2020). CRISPR/Cas9-targeted mutagenesis of the tomato susceptibility gene PMR4 for resistance against powdery mildew. *BMC Plant Biology, 20*(1), 1–13. https://doi.org/10.1186/s12870-020-02497-y.

Marzec, M., Brąszewska-Zalewska, A., & Hensel, G. (2020). Prime editing: A new way for genome editing. *Trends in Cell Biology, 30*(4), 257–259. https://doi.org/10.1016/j.tcb.2020.01.004.

Mehta, D., Stürchler, A., Anjanappa, R. B., Zaidi, S. S. E. A., Hirsch-Hoffmann, M., Gruissem, W., & Vanderschuren, H. (2019). Linking CRISPR-Cas9 interference in cassava to the evolution of editing-resistant geminiviruses. *Genome Biology, 20*(1), 80. https://doi.org/10.1186/s13059-019-1678-3.

Mendelsohn, R. (2009). The impact of climate change on agriculture in developing countries. *Journal of Natural Resources Policy Research, 1*(1), 5–19. https://doi.org/10.1080/19390450802495882.

Mia, M. S., Liu, H., Wang, X., Zhang, C., & Yan, G. (2020). Root transcriptome profiling of contrasting wheat genotypes provides an insight to their adaptive strategies to water deficit. *Scientific Reports, 10*(1), 4854. https://doi.org/10.1038/s41598-020-61680-1.

Montesinos-López, O. A., Montesinos-López, A., Pérez-Rodríguez, P., Barrón-López, J. A., Martini, J. W. R., Fajardo-Flores, S. B., Gaytan-Lugo, L. S., Santana-Mancilla, P. C., & Crossa, J. (2021). A review of deep learning applications for genomic selection. *BMC Genomics, 22*(1), 19. https://doi.org/10.1186/s12864-020-07319-x.

Naik, Y. D., Zaho, C., Channale, S., Nayak, S. N., Bhutia, K. L., Wallace, J., et al. (2023). Bioinformatics for plant genetic and breeding research. In M. K. Pandey (Ed.), *Frontier Technologies for Crop Improvement, Sustainable* (pp. 71–100). Springer. https://doi.org/10.1007/978-981-99-4673-0_3.

Nandy, S., Pathak, B., Zhao, S., & Srivastava, V. (2019). Heat-shock-inducible CRISPR/Cas9 system generates heritable mutations in rice. *Plant Direct, 3*(5), e00145. https://doi.org/10.1002/pld3.145.

Ogata, T., Ishizaki, T., Fujita, M., & Fujita, Y. (2020). CRISPR/Cas9-targeted mutagenesis of OsERA1 confers enhanced responses to abscisic acid and drought stress and increased primary root growth under nonstressed conditions in rice. *PLOS ONE, 15*(12), e0243376. https://doi.org/10.1371/journal.pone.0243376.

Ortigosa, A., Gimenez-Ibanez, S., Leonhardt, N., & Solano, R. (2019). Design of a bacterial speck resistant tomato by CRISPR/Cas9-mediated editing of sl JAZ 2. *Plant Biotechnology Journal, 17*(3), 665–673. https://doi.org/10.1111/pbi.13006.

Oshunsanya, S. O., Nwosu, N. J., & Li, Y. (2019). Abiotic stress in agricultural crops under climatic conditions. In *Sustainable Agriculture, Forest and Environmental Management* (pp. 71–100).

Pan, Z., Liu, M., Zhao, H., Tan, Z., Liang, K., Sun, Q., Gong, D., He, H., Zhou, W., & Qiu, F. (2020). ZmSRL5 is involved in drought tolerance by maintaining cuticular wax structure in maize. *Journal of Integrative Plant Biology, 62*(12), 1895–1909. https://doi.org/10.1111/jipb.12982.

Pandey, M. K., Khan, A. W., Singh, V. K., Vishwakarma, M. K., Shasidhar, Y., Kumar, V., Garg, V., Bhat, R. S., Chitikineni, A., Janila, P., Guo, B., & Varshney, R. K. (2017). QTL-seq approach identified genomic regions and diagnostic markers for rust and late leaf spot resistance in groundnut (*Arachis hypogaea* L.). *Plant Biotechnology Journal, 15*(8), 927–941. https://doi.org/10.1111/pbi.12686.

Pankaj, Y. K., Pal, L., Nagarajan, R., Gill, K. S., Kumar, V., Sangwan, S., Panigrahi, S., & Kumar, R. (2021). Mapping QTLs for phenotypic and morphophysiological traits related to grain yield under late sown conditions in wheat (*Triticum aestivum* L.). *bioRxiv, 2021*(June 17), 2021–2006. https://doi.org/10.1101/2021.06.17.448834.

Puchta, H. (2017). Applying CRISPR/Cas for genome engineering in plants: The best is yet to come. *Current Opinion in Plant Biology, 36*, 1–8. https://doi.org/10.1016/j.pbi.2016.11.011.

Qi, L. S., Larson, M. H., Gilbert, L. A., Doudna, J. A., Weissman, J. S., Arkin, A. P., & Lim, W. A. (2013). Repurposing CRISPR as an RNA-guided platform for sequence-specific control of gene expression. *Cell, 152*(5), 1173–1183. https://doi.org/10.1016/j.cell.2013.02.022.

Qin, L., Li, J., Wang, Q., Xu, Z., Sun, L., Alariqi, M., Manghwar, H., Wang, G., Li, B., Ding, X., Rui, H., Huang, H., Lu, T., Lindsey, K., Daniell, H., Zhang, X., & Jin, S. (2020). High-efficient and precise base editing of C·G to T·A in the allotetraploid cotton (*Gossypium hirsutum*) genome using a modified CRISPR/Cas9 system. *Plant Biotechnology Journal, 18*(1), 45–56. https://doi.org/10.1111/pbi.13168.

Rai, K. K. (2022). Integrating speed breeding with artificial intelligence for developing climate-smart crops. *Molecular Biology Reports, 49*(12), 11385–11402. https://doi.org/10.1007/s11033-022-07769-4.

Raza, A., Razzaq, A., Mehmood, S. S., Zou, X., Zhang, X., Lv, Y., & Xu, J. (2019). Impact of climate change on crops adaptation and strategies to tackle its outcome: A review. *Plants, 8*(2), 34. https://doi.org/10.3390/plants8020034.

Razzaq, A., Kaur, P., Akhter, N., Wani, S. H., & Saleem, F. (2021). Next-generation breeding strategies for climate-ready crops. *Frontiers in Plant Science, 12*, 620420. https://doi.org/10.3389/fpls.2021.620420.

Razzaq, A., Sadia, B., Raza, A., Khalid Hameed, M., & Saleem, F. (2019). Metabolomics: A way forward for crop improvement. *Metabolites, 9*(12), 303. https://doi.org/10.3390/metabo9120303.

Ren, Y., Liu, J., Zhang, J., Dreisigacker, S., Xia, X., & Geng, H. (2021). QTL mapping of drought tolerance at germination stage in wheat using the 50 K SNP array. *Plant Genetic Resources: Characterization and Utilization, 19*(5), 453–460. https://doi.org/10.1017/S1479262121000551.

Saintenac, C., Jiang, D., & Akhunov, E. D. (2011). Targeted analysis of nucleotide and copy number variation by exon capture in allotetraploid wheat genome. *Genome Biology, 12*(9), R88. https://doi.org/10.1186/gb-2011-12-9-r88.

Salgotra, R. K., & Chauhan, B. S. (2023). Genetic diversity, conservation and utilization of plant genetic resources. *Genes, 14*(1), 174. https://doi.org/10.3390/genes14010174.

Scheben, A., Yuan, Y., & Edwards, D. (2016). Advances in genomics for adapting crops to climate change. *Current Plant Biology, 6*, 2–10. https://doi.org/10.1016/j.cpb.2016.09.001.

Schleuning, M., Neuschulz, E. L., Albrecht, J., Bender, I. M., Bowler, D. E., Dehling, D. M., Fritz, S. A., Hof, C., Mueller, T., Nowak, L., & Sorensen, M. C. (2020). Trait-based assessments of climate-change impacts on interacting species. *Trends in Ecology & Evolution, 35*(4), 319–328. https://doi.org/10.1016/j.tree.2019.12.010.

Shan, Q., Wang, Y., Li, J., Zhang, Y., Chen, K., Liang, Z., Zhang, K., Liu, J., Xi, J. J., Qiu, J. L., & Gao, C. (2013). Targeted genome modification of crop plants using a CRISPR-Cas system. *Nature Biotechnology, 31*(8), 686–688. https://doi.org/10.1038/nbt.2650.

Shi, J., Gao, H., Wang, H., Lafitte, H. R., Archibald, R. L., Yang, M., Hakimi, S. M., Mo, H., & Habben, J. E. (2017). ARGOS 8 variants generated by CRISPR-Cas9 improve maize grain yield under field drought stress conditions. *Plant Biotechnology Journal, 15*(2), 207–216. https://doi.org/10.1111/pbi.12603.

Singh, V., Sinha, P., Obala, J., Khan, A. W., Chitikineni, A., Saxena, R. K., & Varshney, R. K. (2022). QTL-seq for the identification of candidate genes for days to flowering and leaf shape in pigeonpea. *Heredity, 128*(6), 411–419. https://doi.org/10.1038/s41437-021-00486-x.

Snowdon, R. J., Wittkop, B., Chen, T. W., & Stahl, A. (2021). Crop adaptation to climate change as a consequence of long-term breeding. *Theoretical and Applied Genetics, 134*(6), 1613–1623. https://doi.org/10.1007/s00122-020-03729-3.

Sysoeva, M. I., Markovskaya, E. F., & Shibaeva, T. G. (2010). Plants under continuous light: A review. *Plant Stress, 4*(1), 5–17.

Takagi, H., Abe, A., Yoshida, K., Kosugi, S., Natsume, S., Mitsuoka, C., Uemura, A., Utsushi, H., Tamiru, M., Takuno, S., Innan, H., Cano, L. M., Kamoun, S., & Terauchi, R. (2013). QTL-seq: Rapid mapping of quantitative trait loci in rice by whole genome resequencing of DNA from two bulked populations. *Plant Journal: For Cell and Molecular Biology, 74*(1), 174–183. https://doi.org/10.1111/tpj.12105.

Távora, F. T. P. K., Meunier, A. C., Vernet, A., Portefaix, M., Milazzo, J., Adreit, H., Tharreau, D., Franco, O. L., & Mehta, A. (2022). CRISPR/Cas9-targeted knockout of rice susceptibility genes OsDjA2 and OsERF104 reveals alternative sources of resistance to *Pyricularia oryzae*. *Rice Science, 29*(6), 535–544. https://doi.org/10.1016/j.rsci.2022.04.001.

Tian, G., Wang, S., Wu, J., Wang, Y., Wang, X., Liu, S., Han, D., Xia, G., & Wang, M. (2023). Allelic variation of TaWD40-4B.1 contributes to drought tolerance by modulating catalase activity in wheat. *Nature Communications, 14*(1), 1200. https://doi.org/10.1038/s41467-023-36901-6.

Tran, M. T., Doan, D. T. H., Kim, J., Song, Y. J., Sung, Y. W., Das, S., Kim, E. J., Son, G. H., Kim, S. H., Van Vu, T., & Kim, J. Y. (2021). CRISPR/Cas9-based precise excision of SlHyPRP1 domain(s) to obtain salt stress-tolerant tomato. *Plant Cell Reports, 40*, 999–1011.

Ullah, F., Xu, Q., Zhao, Y., & Zhou, D. X. (2020). Histone deacetylase HDA710 controls salt tolerance by regulating ABA signaling in rice. *Journal of Integrative Plant Biology, 63*(3), 451–467. https://doi.org/10.1111/jipb.13042.

Varshney, R. K. (2016). Exciting journey of 10 years from genomes to fields and markets: Some success stories of genomics-assisted breeding in chickpea, pigeonpea, and groundnut. *Plant Science, 242*, 98–107. https://doi.org/10.1016/j.plantsci.2015.09.009.

Varshney, R. K., Bohra, A., Yu, J., Graner, A., Zhang, Q., & Sorrells, M. E. (2021). Designing future crops: Genomics-assisted breeding comes of age. *Trends in Plant Science, 26*(6), 631–649. https://doi.org/10.1016/j.tplants.2021.03.010.

Velasco, R., Zharkikh, A., Troggio, M., Cartwright, D. A., Cestaro, A., Pruss, D., Pindo, M., Fitzgerald, L. M., Vezzulli, S., Reid, J., Malacarne, G., Iliev, D., Coppola, G., Wardell, B., Micheletti, D., Macalma, T., Facci, M., Mitchell, J. T., Perazzolli, M., & Viola, R. (2007). A high-quality draft consensus sequence of the genome of a heterozygous grapevine variety. *PLOS ONE, 2*(12), e1326. https://doi.org/10.1371/journal.pone.0001326.

Wang, T., Xun, H., Wang, W., Ding, X., Tian, H., Hussain, S., Dong, Q., Li, Y., Cheng, Y., Wang, C., Lin, R., Li, G., Qian, X., Pang, J., Feng, X., Dong, Y., Liu, B., & Wang, S. (2021). Mutation of GmAITR genes by CRISPR/Cas9 genome editing results in enhanced salinity stress tolerance in soybean. *Frontiers in Plant Science, 12*, 779598. https://doi.org/10.3389/fpls.2021.779598.

Wang, X., Li, N., Li, W., Gao, X., Cha, M., Qin, L., & Liu, L. (2020). Advances in transcriptomics in the response to stress in plants. *Global Medical Genetics, 7*(2), 30–34. https://doi.org/10.1055/s-0040-1714414.

Wang, Z., Dhakal, S., Cerit, M., Wang, S., Rauf, Y., Yu, S., Maulana, F., Huang, W., Anderson, J. D., Ma, X. F., Rudd, J. C., Ibrahim, A. M. H., Xue, Q., Hays, D. B., Bernardo, A., St Amand, P., Bai, G., Baker, J., Baker, S., & Liu, S. (2022). QTL mapping of yield components and kernel traits in wheat cultivars TAM 112 and Duster. *Frontiers in Plant Science, 13*, 1057701. https://doi.org/10.3389/fpls.2022.1057701.

Watson, A., Ghosh, S., Williams, M. J., Cuddy, W. S., Simmonds, J., Rey, M. D., Asyraf Md Hatta, M., Hinchliffe, A., Steed, A., Reynolds, D., Adamski, N. M., Breakspear, A., Korolev, A., Rayner, T., Dixon, L. E., Riaz, A., Martin, W., Ryan, M., Edwards, D., & Hickey, L. T. (2018). Speed breeding is a powerful tool to accelerate crop research and breeding. *Nature Plants, 4*(1), 23–29. https://doi.org/10.1038/s41477-017-0083-8.

Wen, J., Qin, Z., Sun, L., Zhang, Y., Wang, D., Peng, H., Yao, Y., Hu, Z., Ni, Z., Sun, Q., & Xin, M. (2023). Alternative splicing of TaHSFA6e modulates heat shock protein–mediated translational regulation in response to heat stress in wheat. *New Phytologist, 239*(6), 2235–2247. https://doi.org/10.1111/nph.19100.

Wilkins, T. A., Mudge, J., Abidi, N., Allen, R., & Auld, D. (2009). The sequencing and resequencing of cotton. In *Plant and Animal Genomes XVII*. San Diego, CA.

Winfield, M. O., Allen, A. M., Burridge, A. J., Barker, G. L., Benbow, H. R., Wilkinson, P. A., Coghill, J., Waterfall, C., Davassi, A., Scopes, G., Pirani, A., Webster, T., Brew, F., Bloor, C., King, J., West, C., Griffiths, S., King, I., Bentley, A. R., & Edwards, K. J. (2016). High-density SNP genotyping array for hexaploid wheat and its secondary and tertiary gene pool. *Plant Biotechnology Journal, 14*(5), 1195–1206. https://doi.org/10.1111/pbi.12485.

Xiao, Y., Karikari, B., Wang, L., Chang, F., & Zhao, T. (2021). Structure characterization and potential role of soybean phospholipases A multigene family in response to multiple abiotic stress uncovered by CRISPR/Cas9 technology. *Environmental and Experimental Botany, 188*, 104521. https://doi.org/10.1016/j.envexpbot.2021.104521.

Xu, R., Li, J., Liu, X., Shan, T., Qin, R., & Wei, P. (2020). Development of plant prime-editing systems for precise genome editing. *Plant Communications, 1*(3), 100043. https://doi.org/10.1016/j.xplc.2020.100043.

Xu, Y., Yang, Y., Wu, S., Liu, D., & Ren, Y. (2023). QTL mapping for root traits and their effects on nutrient uptake and yield performance in common wheat (*Triticum aestivum* L.). *Agriculture, 13*(1), 210. https://doi.org/10.3390/agriculture13010210.

Xue, Y., & Acar, M. (2018). Live-cell imaging of chromatin condensation dynamics by CRISPR. *Science, 4*, 216–235. https://doi.org/10.1016/j.isci.2018.06.001.

Yadava, Y. K., Chaudhary, P., Yadav, S., Rizvi, A. H., Kumar, T., Srivastava, R., Soren, K. R., Bharadwaj, C., Srinivasan, R., Singh, N. K., & Jain, P. K. (2023). Genetic mapping of quantitative trait loci associated with drought tolerance in chickpea (*Cicer arietinum* L.). *Scientific Reports, 13*(1), 17623. https://doi.org/10.1038/s41598-023-44990-y.

Yamamori, K., Ogasawara, K., Ishiguro, S., Koide, Y., Takamure, I., Fujino, K., Sato, Y., & Kishima, Y. (2021). Revision of the relationship between anther morphology and pollen sterility by cold stress at the booting stage in rice. *Annals of Botany, 128*(5), 559–575. https://doi.org/10.1093/aob/mcab091.

Zafar, K., Khan, M. Z., Amin, I., Mukhtar, Z., Zafar, M., & Mansoor, S. (2023). Employing template-directed CRISPR-based editing of the OsALS gene to create herbicide tolerance in Basmati rice. *AoB Plants, 15*(2), 059. https://doi.org/10.1093/aobpla/plac059.

Zaidi, S. S., Mahas, A., Vanderschuren, H., & Mahfouz, M. M. (2020). Engineering crops of the future: CRISPR approaches to develop climate-resilient and disease-resistant plants. *Genome Biology, 21*(1), 289. https://doi.org/10.1186/s13059-020-02204-y.

Zhang, N., & Huang, X. (2021). Mapping quantitative trait loci and predicting candidate genes for leaf angle in maize. *PLOS ONE, 16*(1), e0245129. https://doi.org/10.1371/journal.pone.0245129.

Zhang, P., Du, H., Wang, J., Pu, Y., Yang, C., Yan, R., Yang, H., Cheng, H., & Yu, D. (2020). Multiplex CRISPR/Cas9-mediated metabolic engineering increases soya bean isoflavone content and resistance to soya bean mosaic virus. *Plant Biotechnology Journal, 18*(6), 1384–1395. https://doi.org/10.1111/pbi.13302.

CHAPTER 10

Economic Viability and Market Prospects of Organic and Natural Farming in Climate-Resilient Agriculture

HARSHIT MISHRA,[1,*] SHRIPATI DWIVEDI,[2] RAJAN KUMAR,[3] LALITA RANA,[3] HARUN GITARI,[4] and DHIRENDRA KUMAR ROY[3]

[1]*Department of Agricultural Economics, College of Agriculture, Acharya Narendra Deva University of Agriculture and Technology, Kumarganj, Ayodhya, Uttar Pradesh, India*

[2]*Department of Agricultural Economics, Dr. Rajendra Prasad Central Agricultural University, Pusa, Samastipur, Bihar, India*

[3]*Department of Agronomy, Dr. Rajendra Prasad Central Agricultural University, Pusa, Samastipur, Bihar, India*

[4]*Department of Agricultural Sciences and Technology, School of Agriculture and Environmental Sciences, Kenyatta University, Nairobi, Kenya*

*Corresponding author

ABSTRACT

India is a significant producer of organic crops. According to the Agricultural and Processed Food Products Export Development Authority (APEDA), India's organic crop production reached 2.66 million tons in the 2022–2023 marketing year, up from 2.35 million tons the previous year. Organic farming is an integral component of climate-resilient agriculture,

Climate-Resilient Agriculture for Sustainable Crop Production: New Adaptation and Mitigation Strategies. Sanjay Kumar, Shivani Ranjan, Sumit Sow, Smruti Ranjan Padhan, and Dhirendra Kumar Roy (Eds.)

DOI: 10.1201/9781998511150-10

rooted in sustainability and environmental stewardship. Based on principles of biodiversity, ecological balance, and reduced external inputs, it promotes regenerative practices that improve soil health, enhance water conservation, and support carbon sequestration. The benefits of natural farming extend beyond environmental outcomes to include improved product quality, nutrition, and consumer trust due to the absence of chemical residues. Economic evaluations provide insights into the viability of organic farming, highlighting input costs, yield expectations, and financial indicators such as return on investment, break-even analysis, and long-term profitability. Market prospects are shaped by shifting consumer preferences and growing demand, supported by certification and labeling standards that ensure transparency. However, challenges such as market access, regulatory complexity, and technical knowledge gaps hinder broader adoption. Addressing these barriers through strategic incentives, subsidies, and supportive policy frameworks is essential to catalyze a robust transition to organic farming and promote sustainable agriculture.

10.1 INTRODUCTION

Conventional farming techniques have been the foundation of global agricultural practices, catering to the ever-growing demands of food production. In India, and in many parts of the world, these practices have undergone significant transformations over the years to meet the needs of an increasing population (Mathukia et al., 2016; Patel and Kumar, 2022). However, the environmental and sustainability challenges posed by conventional farming techniques have become increasingly evident. It is against this backdrop that the principles of organic and natural farming in climate-resilient agriculture have gained prominence (Sarangthem et al., 2023).

The organic food market in India is an emerging and attractive market. The organic food market in India is on the rise, with an estimated market size of $138 million by 2024. This growth is driven by a compound annual growth rate (CAGR) of 13% between 2019 and 2024 (Manavalan, 2022). India is indeed a significant producer of organic crops (Mishra et al., 2023; Rai et al., 2018). According to the Agricultural and Processed Food Products Export Development Authority (APEDA), India's organic crop production reached 2.66 million tons in the marketing year 2022–2023.

This represents a significant increase from the 2.35 million tons produced in the previous year. Major organic crops produced in India include oilseeds, fiber crops, sugar, cereals, and millets (Statista, 2023). In India, the organic farming sector has devoted 2.3 million hectares of land to organic farming, making it the world's largest producer of organic farming products. In terms of exports, India has also shown growth in the organic farming sector. In the marketing year 2022–2023, India's total export value of organic products was USD 708.33 million (APEDA, 2023). The organic food market in India has witnessed increased consumption due to economic growth, higher purchasing power, and a growing interest in health and wellness (Manavalan, 2022; Saldanha, 2018). Globally, the organic farming market has been expanding. The total area under organic production reached 69.8 million hectares in 2017, demonstrating the increased growth and adoption of organic farming practices worldwide (Beni et al., 2021; Nishad et al., 2023).

Natural farming is gaining prominence as a pivotal component of climate-resilient agriculture. Organic and natural farming practices, rooted in sustainable agriculture principles, are instrumental in enhancing climate resilience, conserving resources, and safeguarding the environment (Tiwari et al., 2023). Organic farming, as a subset of these practices, plays a significant role in bolstering resistance to the changing climate. Its core principles emphasize the elimination of synthetic chemicals and the promotion of natural soil health, thereby reducing the carbon footprint of agricultural activities (Kumar et al., 2022; Srivastav et al., 2021).

As agriculture evolves, organic and natural farming face a distinct set of challenges and barriers. These include gaps in farming knowledge, obstacles in market access and distribution, regulatory and certification hurdles, and risk management issues specific to organic farming (Sinclair et al., 2019; van der Pol et al., 2021). Policies and government support emerge as pivotal tools for fostering the growth of organic and natural farming.

The chapter also investigates sustainable farming systems, discussing the integration of organic and natural practices, the importance of crop diversification, and the critical role of soil health and nutrient management in ensuring long-term sustainability (Reddy et al., 2019). Lastly, this chapter paves the way for future exploration by highlighting emerging trends in organic and natural farming, shedding light on research gaps, and identifying areas for further study. It is a comprehensive exploration of the intricate web that is organic and natural farming within the broader framework of climate-resilient agriculture (Dev et al., 2023; Li and Siddique, 2020).

10.2 ORGANIC AND NATURAL FARMING IN CLIMATE-RESILIENT AGRICULTURE

10.2.1 DEFINITION AND PRINCIPLES OF ORGANIC AND NATURAL FARMING IN CLIMATE-RESILIENT AGRICULTURE

10.2.1.1 ORGANIC FARMING

Organic farming is an agricultural approach that emphasizes sustainability, environmental stewardship, and the avoidance of synthetic chemicals and genetically modified organisms (GMOs) (Singh and Mishra, 2023). The principles of organic farming include:

1. **Health:** Organic farming practices should promote the health of soils, plants, animals, and humans as one interconnected system. This means avoiding the use of synthetic pesticides, fertilizers, and GMOs, which can harm the environment and human health. Instead, organic farmers use natural methods to control pests and weeds, and they focus on building healthy soil that can support strong crops.

2. **Ecology:** Organic farming should work with nature, not against it. This means protecting biodiversity, conserving natural resources, and minimizing pollution. Organic farmers use practices such as crop rotation, cover cropping, and composting to improve soil health and fertility, and they raise animals in humane conditions that allow them to express their natural behaviors.

3. **Fairness:** Organic farming should promote fairness and equity in all aspects of the food system, from production to consumption. This means treating workers fairly, paying fair prices for products, and ensuring that everyone has access to healthy, nutritious food. Organic farmers often participate in fair trade agreements and other initiatives that promote social justice.

4. **Care:** Organic farming should be based on care and respect for all living things. This means treating animals humanely, protecting

the environment, and ensuring that the food we eat is safe and healthy. Organic farmers take a holistic approach to their work, and they are committed to producing food that is good for the planet and for people.

10.2.1.2 NATURAL FARMING

Natural farming, sometimes referred to as "do-nothing farming" or "regenerative farming," is a system that strives to mimic natural ecological processes. The principles of natural farming include:

1. **Minimal Intervention:** Natural farming practices involve minimal human intervention. Farmers aim to disturb the ecosystem as little as possible, promoting self-regulation.
2. **Biodiversity Promotion:** Like organic farming, natural farming encourages biodiversity through the cultivation of multiple crops and minimal use of monoculture systems.
3. **Soil Health:** It is a fundamental principle of natural farming. Farmers strive to enhance soil structure and fertility by incorporating organic matter and avoiding synthetic chemicals.

TABLE 10.1 Comparison between Organic and Natural Farming

Feature	Organic Farming	Natural Farming
Practices	Compost and manure, green manure, vermicompost, biofertilizers, biopesticides.	Natural composting, crop rotation, cover cropping, intercropping, integrated pest management.
Certification	Requires certification from a third-party organization.	Does not require certification.
Benefits	Produces food that is free of synthetic chemicals, improves soil health, reduces environmental impact.	Produces food that is even more nutritious and healthy than organic food, improves soil health, reduces environmental impact, promotes biodiversity.
Challenges	Can be more expensive than conventional farming, may require more labor.	Can be more difficult to control pests and diseases.
Examples	Rodale Institute, Organic Valley, Stonyfield Farm.	Masanobu Fukuoka, Hitoshi Kushi.

10.2.2 THE ROLE OF ORGANIC FARMING IN CLIMATE RESILIENCE

Organic farming plays a crucial role in climate-resilient agriculture for several reasons:

1. **Improved Soil Health:** Organic farming enhances the nutrient and water retention capacity of soil. This, in turn, helps crops withstand drought and erratic rainfall patterns.
2. **Carbon Sequestration:** Organic farming practices, such as reduced tillage and cover cropping, sequester more carbon in the soil, mitigating climate change.
3. **Reduced Environmental Impact:** By avoiding synthetic chemicals, organic farming reduces the risk of chemical pollution, protecting ecosystems and biodiversity.
4. **Enhanced Pest and Disease Resistance:** Organic farms often rely on natural pest control and integrated pest management techniques, reducing the need for synthetic pesticides. This helps ecosystems remain balanced and adaptable.
5. **Promotes Crop Diversity:** Organic farming encourages crop diversification, reducing the vulnerability of monoculture systems to climate-related crop failures.

10.2.3 BENEFITS OF NATURAL FARMING PRACTICES

Natural farming practices offer several benefits in the context of climate-resilient agriculture:

1. **Reduced Resource Dependency:** Natural farming requires fewer external inputs like synthetic fertilizers and pesticides, making it economically viable for resource-limited farmers.
2. **Enhanced Ecosystem Resilience:** Minimal intervention and the promotion of biodiversity allow ecosystems to adapt to changing climate conditions more effectively.
3. **Water Management:** Natural farming practices such as mulching and cover cropping reduce water runoff and evaporation, improving water use efficiency in agriculture.

4. **Soil Regeneration:** By focusing on improving soil health, natural farming aids in soil regeneration and fertility, which is essential for long-term agricultural sustainability.
5. **Reduced Risk:** The avoidance of synthetic chemicals reduces the risk of crop loss due to unpredictable climate events and pests.
6. **Market Demand:** As consumer interest in sustainable and environmentally friendly products grows, the market for natural and organic produce is expanding, potentially offering better economic prospects for farmers.

Organic and natural farming practices align well with the principles of climate-resilient agriculture. They promote soil health, biodiversity, and sustainability while reducing the environmental impact of agriculture (Kumar et al., 2023). These practices offer economic viability and better market prospects as consumers increasingly value sustainable and natural food production methods (Amoak et al., 2022; Chandel et al., 2021).

10.3 ECONOMIC ASPECTS OF ORGANIC FARMING

10.3.1 COST-BENEFIT ANALYSIS (CBA)

CBA is a systematic process for calculating and comparing the benefits and costs of a project, decision, or government policy. In the context of organic farming, it helps in evaluating the financial viability of adopting organic farming practices over conventional methods (Hansika and Wijerathna, 2021).

10.3.1.1 INPUT COSTS IN ORGANIC FARMING

In organic farming, the input costs are associated with the expenses incurred in acquiring organic inputs such as organic fertilizers, organic pesticides, and other organic farming materials. These inputs are generally more expensive than their conventional counterparts (Singh and Mishra, 2023). However, the long-term benefits of reduced environmental impact and improved soil quality should also be factored into the analysis. The input cost can be calculated using the following formula:

$$\textit{Input Costs} = \textit{Cost of organic fertilizer} + \textit{Cost of organic pesticides} \\ + \textit{Other organic farming materials}$$

Higher input costs in organic farming can affect the overall profitability of the farming operation. However, these costs can be offset by the potential benefits of organic farming in terms of improved soil fertility, reduced health risks, and enhanced biodiversity.

10.3.1.2 OUTPUT AND YIELD EXPECTATIONS

Output and yield expectations in organic farming refer to the estimated amount of produce that can be obtained using organic farming practices. Organic farming may lead to slightly lower yields in the short term compared to conventional farming due to the absence of synthetic inputs. However, it is essential to consider the long-term effects on soil health and productivity, which can lead to sustained or improved yields over time (Khurana and Kumar, 2020). To determine the expected yield of a crop grown using organic farming methods, the following formula is used:

$$Yield\ Expectations = Total\ quantity\ of\ produce\ obtained\ using\ organic\ farming\ pra$$

While the initial yield might be slightly lower in organic farming, the long-term benefits include improved soil health, reduced environmental degradation, and potentially increased yields in the future. Moreover, the premium prices often associated with organic produce can also offset the lower initial yields. By thoroughly understanding the costs and benefits associated with organic farming, farmers, and policymakers can make informed decisions that promote sustainable agriculture and environmental conservation.

10.3.2 ECONOMIC VIABILITY METRICS

Evaluating the economic viability of organic farming involves many metrics, including Return on Investment (ROI), Break-Even Analysis, and Profitability Indicators.

10.3.2.1 RETURN ON INVESTMENT (ROI)

ROI is a major financial metric that assesses the profitability of an investment. In the context of organic farming, ROI helps farmers understand the efficiency of their investment in terms of costs and revenue.

$$ROI = \left[\frac{Net\ Profit - Investment\ Cost}{Initial\ Investment}\right] \times 100$$

Example: Let's say a farmer invests $10,000 in transitioning their conventional farm to organic practices. After one year, the net profit from organic farming is $4000.

$$ROI = \left[\frac{\$4{,}000 - \$10{,}000}{\$10{,}000}\right] \times 100 = -60\%$$

A negative ROI in this example indicates that the initial investment in transitioning to organic farming has not yet been recovered. The farmer would need to continue for a few more years to achieve a positive ROI. Suppose in the second year, the net profit from organic farming is $8000. Now, let's calculate the ROI for the combined two-year period:

$$Total\ Net\ Profit = \$4000 + \$8000 = \$12{,}000$$

$$Total\ Investment = \$10{,}000$$

$$ROI = \left[\frac{\$12{,}000 - \$10{,}000}{\$10{,}000}\right] \times 100 = 20\%$$

Farmer achieved a positive ROI of 20% over the two-year period. This indicates that the investment in transitioning to organic farming has not only been recovered but also generated a return of 20%. This positive ROI suggests that the decision to transition to organic farming has become financially beneficial for the farmer.

10.3.2.2 BREAK-EVEN ANALYSIS

Break-even analysis helps farmers determine the point at which total revenue equals total costs, resulting in zero profit. This is a critical milestone for organic farming, as it indicates the minimum level of sales required to cover all expenses.

$$BEP = \frac{Fixed\ costs}{Selling\ price\ per\ unit - Variable\ costs\ per\ unit}$$

Example: Suppose the fixed costs (e.g., land, equipment) amount to $20,000, the selling price of organic produce is $5 per unit, and variable costs (e.g., seeds, labor) per unit are $3.

$$BEP = \frac{\$20{,}000}{5-3} = 10{,}000\ units$$

In this example, the farmer needs to sell 10,000 units of organic produce to cover all costs and reach the break-even point. Selling more units will result in profits, while selling fewer units will result in losses.

10.3.3 PROFITABILITY INDICATORS

Various profitability indicators can be used to assess the economic viability of organic farming included here.

10.3.3.1 GROSS MARGIN

Example: If total revenue from organic farming is $50,000, and total variable costs are $30,000, the gross margin is $20,000.

$$\text{Gross margin} = \text{Total revenue} - \text{Total variable costs}$$

A higher gross margin suggests better profitability in organic farming.

10.3.3.2 NET PROFIT MARGIN

$$\text{Net profit margin} = \frac{\text{Net profit}}{\text{Total revenue}} \times 100$$

Example: If the net profit from organic farming is $10,000, and total revenue is $50,000, the net profit margin is 20%.

$$\text{Net profit margin} = \frac{\$10{,}000}{\$50{,}000} \times 100 = 20\%$$

A higher net profit margin indicates a higher level of profitability.

10.3.3.3 RETURN ON ASSETS (ROA)

$$ROA = \frac{\text{Net profit}}{\text{Total assets}} \times 100$$

Example: If the net profit is $10,000, and the total assets employed in organic farming are $1,00,000, the ROA is 10%.

$$ROA = \frac{\$10{,}000}{\$1{,}00{,}000} \times 100 = 10\%$$

ROA measures how efficiently the assets are utilized to generate profit. A higher ROA suggests better asset utilization. Evaluating the economic viability of organic farming involves assessing ROI, conducting a break-even analysis, and considering profitability indicators like gross margin, net profit margin, and return on assets. These metrics help farmers make informed decisions about the financial aspects of organic farming and its potential for success in climate-resilient agriculture (Amjath-Babu et al., 2019).

10.4 MARKET PROSPECTS FOR ORGANIC AND NATURAL FARMING PRODUCTS

10.4.1 CONSUMER TRENDS AND DEMAND FOR ORGANIC PRODUCE

Consumer trends and the demand for organic produce have seen significant growth in recent years, driven by a variety of factors related to health and environmental concerns. Understanding these trends is crucial for assessing the market prospects of organic and natural farming in the context of climate-resilient agriculture (Liu et al., 2016; Mishra, 2021).

1. **Health and Wellness:** One of the primary drivers of the demand for organic produce is the perception that it is healthier and safer for consumption. Consumers are becoming increasingly health-conscious and are willing to pay a premium for organic products. They believe that organic farming practices, which avoid synthetic pesticides, herbicides, and genetically modified organisms, are likely to contain fewer harmful residues and are more nutritious.

2. **Environmental Concerns:** Consumers are increasingly aware of the environmental impacts of conventional agriculture, including soil degradation, water pollution, and greenhouse gas emissions. This has led to a growing interest in organic and natural farming, which emphasizes sustainable and regenerative practices. Many consumers believe that by choosing organic products, they are supporting agriculture that is less harmful to the environment (Wilson and Lovell, 2016).

3. **Ethical Considerations:** The ethical aspect of food production is gaining prominence. Consumers are concerned about the welfare of farm animals, fair treatment of farmworkers, and the overall ethics

of food production systems (Sarmah and Deka, 1993). Organic and natural farming often aligns with these values, as they typically involve fair treatment of farm animals and labor, thus attracting consumers who prioritize ethical considerations (Dev, 2013).

4. **Transparency and Traceability:** Consumers are increasingly interested in knowing the origins of their food and how it was produced. Organic and natural farming systems often offer greater transparency and traceability, as they require rigorous certification standards and documentation of farming practices. This transparency appeals to consumers who want to make informed choices about the food they purchase.
5. **Local and Sustainable Food Movements:** The trend towards buying locally produced and sustainable food is closely related to the organic and natural farming movement. Many consumers prefer to support local farmers and food producers who use eco-friendly and climate-resilient practices. This trend has resulted in increased demand for organic products from local sources (Shilomboleni et al., 2023).
6. **Market Expansion and Accessibility:** The availability of organic and natural products has expanded significantly, with a wider range of options in grocery stores, markets, and online platforms. This increased accessibility has contributed to the growth in consumer demand, as consumers now have more choices when it comes to purchasing organic produce (Mishra and Mishra, 2023).
7. **Price Sensitivity:** While demand for organic products continues to grow, price sensitivity remains a factor. Organic products often come with a price premium due to the higher costs associated with organic farming practices. However, as the market matures and economies of scale are achieved, the price gap between organic and conventional products may narrow, making organic produce more affordable to a broader range of consumers.

10.4.2 CERTIFICATION AND LABELING

Certification and labeling play a pivotal role in the organic and natural farming sector in India. These mechanisms help build consumer trust, ensure product authenticity, and promote sustainable agricultural practices. In this section, we will identify the significance of certification and labeling, the various standards and regulations governing organic and natural products, and their impact on market prospects (Tran et al., 2021).

10.4.2.1 CERTIFICATION FOR ORGANIC FARMING

Organic farming in India is governed by the National Program for Organic Production (NPOP) under the Ministry of Commerce and Industry. The NPOP regulates the organic sector by setting standards and accreditation procedures. Elements of organic farming certification in India include:

1. **Organic Standards:** The NPOP lays down the standards and regulations that organic farmers and producers must adhere to. These standards cover aspects such as soil health, crop management, pest control, and livestock care.
2. **Certifying Bodies:** Organic certification is carried out by accredited certifying bodies. These organizations assess farms and processing units to ensure compliance with NPOP standards. Some well-known certifying agencies in India include the Organic Farmers and Growers (OF&G), Ecocert, and Control Union.
3. **Inspection and Documentation:** Organic farmers and processors are subject to regular inspections. They must maintain detailed records of farming practices, inputs used, and sales. This documentation is critical for the certification process.
4. **Labeling:** Once certified, products can carry the "India Organic" logo. This label assures consumers that the product has been produced according to organic standards. Additionally, organic products may also bear the certification mark of the certifying agency.
5. **Costs and Duration:** Obtaining organic certification involves costs for inspections, documentation, and annual renewal. The certification process can take several months to complete.

10.4.2.2 SIGNIFICANCE OF CERTIFICATION AND LABELING

1. **Consumer Trust:** Organic and natural products are often priced higher than their conventional counterparts. Consumers pay this premium with the expectation that the products are indeed produced

according to specific standards. Certification and labeling provide consumers with the assurance that they are getting what they pay for, fostering trust in the industry.

2. **Market Access:** Many countries and regions have strict regulations governing the sale of organic and natural products. Certification allows producers to access these markets, which can be more lucrative due to higher demand and better prices.
3. **Environmental and Health Benefits:** Certification and labeling also reflect the commitment of organic and natural farming to sustainable and eco-friendly practices. This is an attractive selling point to consumers who prioritize environmental and health considerations in their purchases.

10.4.2.3 CERTIFICATION STANDARDS

1. **USDA Organic:** In the United States, the USDA National Organic Program sets standards for organic agriculture. Products meeting these standards can be labeled as "USDA Organic." This certification is recognized in many countries, making it crucial for international trade.
2. **European Union (EU) Organic:** In the EU, organic farming is regulated by strict guidelines. Products meeting these standards receive the "EU Organic" label. This certification is one of the most recognized and trusted globally.
3. **International Federation of Organic Agriculture Movements (IFOAM) Organics International:** The IFOAM sets global standards for organic agriculture. Many countries rely on the regulations based on IFOAM's guidelines, making this certification widely recognized worldwide.
4. **Natural Product Certification:** Some products may not meet the strict criteria for organic certification but can still be labeled as "natural" or "chemical-free." These certifications vary widely and are often more flexible.

10.4.2.4 LABELING AND ITS IMPACT

1. **Consumer Awareness:** Clear and easily understandable labeling is essential. The labels should highlight the certification logos and provide information about the origin, ingredients, and farming practices involved in the production process. Effective labeling enables consumers to make informed choices.
2. **Market Differentiation:** Proper labeling allows organic and natural products to be differentiated from conventionally produced items. It not only attracts consumers who prioritize eco-friendly and health-conscious options but also allows producers to command premium prices for their products.
3. **Regulatory Compliance:** Labels must adhere to specific regulations to avoid misrepresentation and fraud. Violations can lead to fines and damage to the reputation of the producer.
4. **Global Trade:** Labeling in compliance with international standards is crucial for export markets. It ensures that the products meet the criteria of importing countries, facilitating trade and access to a broader consumer base.

10.4.2.5 CHALLENGES IN CERTIFICATION AND LABELING

1. **Costs:** Obtaining certification can be expensive, especially for small-scale farmers. This can create barriers to entry for some producers and limit the diversity of organic and natural products in the market.
2. **Complexity:** Transitioning from conventional to organic or natural farming can pose challenges in meeting certification criteria. Proper training and support are crucial for successful certification.
3. **Fraud:** In some cases, products are mislabeled or falsely advertised as organic or natural. Regulatory bodies must invest in monitoring and enforcement to maintain the integrity of certification.

Certification and labeling are essential components of organic and natural farming, providing assurance to consumers, facilitating market access, and differentiating products. Clear and accurate labeling not only serves as a marketing tool but also upholds the values of sustainable and climate-resilient agriculture. However, challenges such as costs and fraud prevention need to be addressed to ensure the continued growth and success of organic and natural farming in the global market (Babalad and Navali, 2021).

10.4.3 MARKET CHALLENGES AND OPPORTUNITIES FOR ORGANIC AND NATURAL FARMING

Organic and natural farming is gaining traction worldwide due to increasing consumer awareness of health and environmental concerns. However, it also faces several challenges and opportunities in the market (Eeswaran et al., 2021).

1. **Challenges:**
   - i. **Cost of Production:** Organic and natural farming methods often demand more labor and require initial investments that are higher compared to conventional agriculture. This can dissuade some farmers from transitioning to organic practices.
   - ii. **Yield Variability:** Organic farming can sometimes result in lower yields compared to conventional methods, which can be a financial challenge for farmers.
   - iii. **Pest and Disease Management:** Organic farming relies on non-synthetic pesticides and herbicides, which may be less effective in pest and disease management compared to their chemical counterparts, leading to crop losses.
   - iv. **Consumer Awareness and Education:** There is a need to educate consumers about the benefits of organic and natural products, as many are unaware of the differences and potential health and environmental advantages.
   - v. **Certification and Regulation:** Meeting organic certification standards can be costly and challenging for small-scale farmers, which may limit their market access.
   - vi. **Competition:** Conventional farming still dominates the market, and organic products often face competition from more affordable conventionally produced goods.

2. **Opportunities:**

   i. **Consumer Demand:** There is a growing consumer preference for organic and natural products due to health and environmental concerns, providing a substantial market opportunity.

   ii. **Price Premium:** Consumers are generally willing to pay more for products labeled as organic or natural, which can offset some of the higher production costs.

   iii. **Climate Resilience:** Organic farming employs diversified cropping systems and fosters improved soil health, which can make farms more resilient to climate change, reducing risks associated with extreme weather events.

   iv. **Local and Niche Markets:** Small-scale organic farmers thrive in local and niche markets, where consumers are willing to pay a premium for sustainable and locally produced goods.

   v. **Government Support:** Many governments offer incentives, subsidies, and support for organic and sustainable farming practices, which can help reduce the financial burden on farmers.

   vi. **Investor Interest:** The interest of investors in sustainable and ethical agriculture can provide funding and support for organic and natural farming ventures.

10.4.4 EXPORT POTENTIAL AND GLOBAL MARKETS FOR ORGANIC AND NATURAL FARMING

The global market for organic food and beverages is expected to reach USD 435.54 billion by 2027, growing at a CAGR of 10.4% from 2022–2027. This growth is being driven by increasing consumer awareness of the health and environmental benefits of organic products. Consumers are also increasingly willing to pay a premium for organic products, which is supporting the growth of the market (Jaganathan et al., 2021).

Figure 10.1 shows the percentage of organic food market segments in 2022. Organic fruits and vegetables account for the largest share at 40.2%, followed by organic meat, fish, and poultry at 19.8%, organic dairy products at 18.1%, organic frozen and processed foods at 10.6%, and others at 11.3%. The market for natural products is also growing rapidly. The global market for natural personal care products is expected to reach USD 54.5 billion by 2025, growing at a CAGR of 9.4% from 2020–2025. This growth is being driven by increasing consumer demand for products that are free from synthetic chemicals and preservatives (APEDA, 2023).

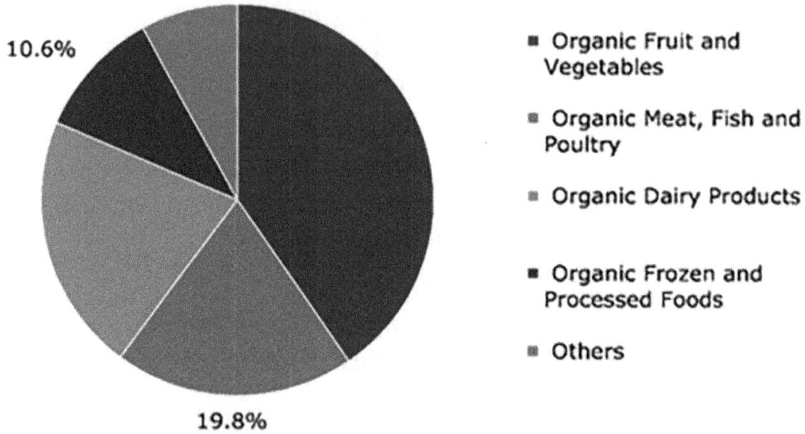

FIGURE 10.1 Global organic food market share, by food type (2021).

Source: Adapted from Research Dive Analysis.

10.4.4.1 EXPORT POTENTIAL AND GLOBAL MARKETS FOR ORGANIC AND NATURAL FARMING IN INDIA

India has a significant export potential for organic and natural products. The country has a large and growing organic farming sector, and a wide variety of organic and natural products are produced in India. The country also has a strong export infrastructure, and the government is supportive of the export of organic and natural products (Mishra and Mishra, 2023; Munawar et al., 2020).

The following are some of the global markets for organic and natural products from India:

1. **North America:** The United States is the largest market for organic food and beverages in the world. Canada is also a significant market for organic products.
2. **Europe:** The European Union is the second-largest market for organic food and beverages in the world. Germany, France, and

the United Kingdom are the largest markets for organic products in Europe.

3. **Asia:** Japan is the largest market for organic food and beverages in Asia. China is also a growing market for organic products.

The following are some of the major challenges and opportunities for exporting organic and natural products from India:

- **Challenges:**
  - High cost of certification;
  - Lack of awareness about Indian organic products in global markets;
  - Competition from other countries.

- **Opportunities:**
  - Growing demand for organic and natural products in global markets;
  - Government support for the export of organic and natural products;
  - Increasing awareness about the health and environmental benefits of organic and natural products.

India has the potential to become a major exporter of organic and natural products. The country has a strong agricultural base, a large and growing organic farming sector, and a supportive government. With the right strategies in place, India can capitalize on the growing global demand for organic and natural products (Asokan et al., 2020; Bagheri et al., 2019). Here are some specific recommendations for promoting the export of organic and natural products from India:

1. **Increase Awareness About Indian Organic Products in Global Markets:** This can be done through participation in international trade shows and exhibitions, and by conducting marketing campaigns in major export markets.

2. **Strengthen the Certification Process:** This will help to ensure that Indian organic products meet international standards and are accepted in global markets.

3. **Develop Value-Added Products:** This will help to increase the export value of Indian organic products.

4. **Promote Organic Farming Clusters:** This will help to improve the efficiency of the organic farming sector and make it more competitive.

By implementing these recommendations, India can significantly increase its export of organic and natural products and become a major player in the global market (Bonilla-Cedrez et al., 2023).

The export potential for organic and natural products is significant, as global demand for these goods continues to rise. Various factors contribute to their success in international markets:

1. **Growing Global Demand:** As consumers worldwide become more health and environmentally conscious, the demand for organic and natural products is increasing. This trend opens up opportunities for exports.
2. **Quality and Safety Standards:** Many countries have established stringent standards and regulations for organic products, ensuring that certified organic goods meet high-quality and safety criteria.
3. **Organic Certification:** Organic certification is often recognized internationally, making it easier for certified producers to access global markets.
4. **Diverse Product Range:** Organic and natural farming can produce a wide range of products, including fruits, vegetables, grains, dairy, meat, and more. This diversity allows for exports of various products to cater to different markets.
5. **Market Research:** In-depth market research is essential to identify which products have the most export potential and in which countries or regions.
6. **Trade Agreements and Partnerships:** Government trade agreements and partnerships can ease export restrictions and offer favorable terms for organic and natural product exports.

However, challenges exist as well:

1. **Competition:** The global organic market is highly competitive, with many established players. New entrants need to differentiate themselves to succeed.

2. **Logistics and Transportation:** Ensuring the freshness and quality of organic products during long-distance transportation can be challenging.

3. **Export Regulations:** Different countries have various rules and regulations governing the import of organic products. Understanding and complying with these regulations is crucial.

The market prospects for organic and natural farming are promising due to consumer demand, certification, and labeling standards, and growing interest in sustainable and climate-resilient agriculture. However, the sector also faces challenges, particularly related to price premiums and competition. Export potential and global markets offer opportunities for expansion, but exporters must navigate regulatory requirements and effectively market their products to succeed on a global scale (Mishra and Singh, 2023; Singh, 2012).

10.5 CHALLENGES AND STRATEGIES IN PROMOTING ORGANIC AND NATURAL FARMING FOR CLIMATE-RESILIENT AGRICULTURE

10.5.1 FARMING PRACTICES AND KNOWLEDGE GAPS

One of the significant challenges in promoting organic and natural farming in climate-resilient agriculture is the need for a paradigm shift in farming practices and addressing knowledge gaps. Traditional and conventional farming methods have been deeply ingrained in agricultural practices for decades, making it challenging for farmers to transition to organic and natural farming methods (Meena and Jha, 2018; Reddy, 2015).

1. **Knowledge and Training:** Farmers may lack the necessary knowledge and training in organic and natural farming techniques. They may not understand the intricacies of soil health, crop rotation, organic pest control, and the use of natural fertilizers. Bridging this knowledge gap is crucial for the successful adoption of these practices.

2. **Resistance to Change:** Transitioning to organic and natural farming often requires a fundamental change in mindset and approach.

Farmers need to understand the benefits of these methods, which include improved soil health, reduced chemical inputs, and long-term sustainability.

3. **Research and Development:** Continuous research and development efforts are necessary to create and disseminate innovative organic farming practices that are both sustainable and economically viable. Addressing these knowledge gaps requires investment in agricultural research and extension services.

10.5.2 MARKET ACCESS AND DISTRIBUTION CHALLENGES

Another significant challenge for organic and natural farming in climate-resilient agriculture is ensuring that the produce reaches the market efficiently and profitably. Access to markets and effective distribution are major issues that need to be addressed (Lal, 2013).

1. **Limited Market Opportunities:** While the demand for organic and natural produce is growing, it still represents a niche market compared to conventional agriculture. Finding profitable market opportunities can be challenging for small and medium-sized organic farmers.
2. **Distribution and Logistics:** Maintaining the integrity of organic and natural products throughout the supply chain can be difficult. Organic products require special handling and certification to ensure that they remain organic. This can add complexity and cost to distribution.
3. **Price Premiums:** While organic products often command higher prices, they can also be costlier to produce. Balancing the price premium with the increased cost of production is a continual challenge for farmers.

10.5.3 REGULATORY AND CERTIFICATION HURDLES

The regulatory framework for organic and natural farming is essential to maintain the integrity of these practices and products. However, navigating the regulatory and certification process can be a barrier for many farmers (Marke, 2014).

1. **Certification Costs:** Obtaining organic and natural certifications can be expensive, especially for small-scale farmers. These costs may include inspections, documentation, and the development of organic management plans.
2. **Compliance Challenges:** Meeting the strict standards required for certification can be challenging, especially for farmers who have limited resources or are in regions with unique environmental conditions.
3. **Market Access:** Organic and natural farmers may face difficulties accessing international markets due to varying global standards and regulations. Harmonizing standards and ensuring compliance with international requirements is a continuous challenge.

10.5.4 RISK MANAGEMENT IN ORGANIC FARMING

Organic and natural farming may involve higher risks compared to conventional farming due to reduced chemical inputs and potential yield fluctuations (Mishra and Singh, 2023). Addressing these risks is crucial for the economic viability of these practices.

1. **Pest and Disease Control:** Organic farming relies on natural methods for pest and disease control, which can be less predictable and effective than chemical alternatives. Farmers need to develop expertise in integrated pest management and resilience to cope with potential losses.
2. **Market Volatility:** The organic market can experience price volatility, and demand for organic products may fluctuate. Farmers need to develop marketing strategies and diversify their product offerings to manage this risk effectively.
3. **Climate Variability:** Climate change can lead to unpredictable weather patterns, affecting crop yields. Implementing climate-resilient agricultural practices is crucial for mitigating the impact of extreme weather events.

Addressing the challenges and barriers associated with organic and natural farming in climate-resilient agriculture requires a multi-faceted approach. This includes providing education and training, improving market access and distribution networks, streamlining regulatory processes, and developing risk management strategies. Overcoming these obstacles is essential to ensure the economic viability and long-term sustainability of organic and natural farming practices (Huyer et al., 2021; Taylor and Bhasme, 2021).

10.6 POLICY AND GOVERNMENT SUPPORT

10.6.1 GOVERNMENT INITIATIVES AND PROGRAMS

Government initiatives and programs play a pivotal role in shaping the economic viability and market prospects of organic and natural farming within the context of climate-resilient agriculture (Dorin, 2022). These initiatives encompass various aspects, including research, funding, education, and regulatory frameworks.

1. **Research and Development:** Government agencies often allocate funds for research and development in the field of organic and natural farming. These initiatives aim to enhance the understanding of sustainable agricultural practices, promote climate resilience, and improve crop yields without the excessive use of synthetic chemicals (Samsudin et al., 2021; Singh et al., 2014). Researchers work on the development of organic farming techniques, disease-resistant varieties, and sustainable crop management strategies.
2. **Education and Training:** One of the major components of government support is the provision of educational programs and training for farmers. These programs are designed to educate farmers about the benefits of organic and natural farming practices. They provide guidance on soil health management, pest control, and the use of organic fertilizers. Farmers are taught how to adapt to climate change and use environmentally friendly methods for agriculture (Mishra and Singh, 2023).
3. **Certification and Standards:** Governments often establish and enforce standards and certification procedures for organic and natural farming products. These standards ensure the authenticity and quality of organic products, providing assurance to consumers. Government agencies also regulate the use of organic labels, preventing fraudulent claims in the market.

10.6.2 INCENTIVES AND SUBSIDIES

To encourage the adoption of organic and natural farming practices, governments frequently provide financial incentives and subsidies to farmers (Singh and Kumar, 2022). These incentives are essential in making such practices economically viable.

1. **Subsidies for Transition:** Many governments offer subsidies for farmers who are transitioning from conventional to organic or natural farming methods. These subsidies may cover the cost of organic inputs, organic certification, or even financial support during the initial years of transition when yields might be lower (Shehrawat et al., 2023).
2. **Tax Benefits:** Tax incentives, such as reduced property taxes or income tax credits, are often provided to farmers who engage in sustainable and climate-resilient farming practices. These financial incentives can significantly reduce the operating costs of organic and natural farming.
3. **Marketing Support:** To expand the market prospects for organic and natural farming, governments can assist in marketing and distribution. This might include funding for markets, creating online platforms for selling organic products, or assisting with export and trade agreements, thereby increasing the economic viability of organic farming (Attri and Mohapatra, 2021).

10.7 ADOPTION STRATEGIES FOR FARMERS

10.7.1 TRANSITIONING TO ORGANIC FARMING

Transitioning from conventional farming to organic and natural farming is a significant step for farmers seeking climate-resilient agriculture. The following strategies can help them navigate this transition successfully:

1. **Assessment and Planning:** Before transitioning, farmers should assess their current farming practices, soil health, and the potential impact on their income. This involves understanding the initial costs and expected returns.

2. **Gradual Transition:** Sudden shifts can be risky and overwhelming. Farmers can consider a phased transition approach, converting a portion of their land to organic methods in the beginning. This approach helps manage risks and learn as they go.
3. **Soil Health Improvement:** Organic farming relies heavily on soil health. Farmers should focus on enhancing soil fertility through techniques such as cover cropping, composting, and reduced tillage. Soil testing can guide them in making necessary amendments.
4. **Crop Rotation and Diversity:** Diversifying crops and implementing crop rotation can reduce pest attacks and enhance soil health. This transition may require acquiring knowledge about suitable crop combinations.
5. **Certification and Compliance:** Farmers must understand the certification requirements for organic farming and adhere to them. This may involve keeping meticulous records and working with certifying agencies.
6. **Market Research:** Before transitioning, farmers should research the market demand for organic produce. Understanding which crops are in demand and at what price can guide their crop choices.
7. **Access to Resources:** Farmers may need financial support or access to organic farming expertise. Government grants, agricultural extensions, or partnerships with NGOs can provide valuable assistance.

10.7.2 EDUCATION AND TRAINING

Education and training are pivotal for farmers looking to adopt organic and natural farming. These strategies can be put into action through the following steps:

1. **Workshops and Seminars:** Farmers should attend workshops and seminars on organic farming practices. These events can provide insights into the latest techniques and connect farmers with experts and other like-minded individuals.

2. **Agricultural Extension Services:** Government agricultural agencies and universities often offer extension services that provide technical guidance on organic farming. These services can help farmers access up-to-date information and resources.

3. **Online Resources:** In the digital age, numerous online resources, including webinars, forums, and instructional videos, are available. These can be particularly useful for farmers who may not have easy access to physical training events.

4. **Farm Visits and Demonstrations:** Learning by doing is a powerful method. Farmers can visit successful organic farms or participate in on-site demonstrations to witness best practices firsthand.

5. **Local Farming Communities:** Joining or forming local farming communities and cooperatives can foster knowledge sharing and peer-to-peer learning. These networks can offer valuable support and mentorship.

10.7.3 FARMER COOPERATIVES AND NETWORKS

Collaboration through farmer cooperatives and networks can significantly enhance the success and sustainability of organic and natural farming. These strategies can be effectively employed in the following ways:

1. **Resource Sharing:** Farmers in cooperatives can pool resources like equipment, labor, and inputs, reducing individual costs and improving efficiency.

2. **Market Access:** Cooperatives often have better bargaining power in the market. They can negotiate higher prices for their organic produce and collectively explore market opportunities.

3. **Knowledge Exchange:** Cooperatives and networks provide platforms for farmers to share their experiences and insights. This fosters continuous learning and adaptation to changing conditions.

4. **Risk Mitigation:** In the face of climate uncertainties, cooperative farming can help diversify risks. If one farmer faces a setback, the group can provide support and buffer the impact.

5. **Advocacy and Policy Influence:** Large, organized groups of farmers have a stronger voice in advocating for policies that favor organic and natural farming practices. They can lobby for subsidies, incentives, and regulations that support climate-resilient agriculture.

The adoption of organic and natural farming in climate-resilient agriculture requires a multifaceted approach that encompasses transitioning strategies, education, and training, and the establishment of farmer cooperatives and networks. These strategies can empower farmers to make the shift successfully, improving their economic viability and market prospects while contributing to a more sustainable and climate-resilient agricultural system (Gugissa et al., 2022; Mishra and Singh, 2023).

10.8 SUSTAINABLE FARMING SYSTEMS

10.8.1 INTEGRATING ORGANIC AND NATURAL FARMING PRACTICES

Organic and natural farming practices are integral components of sustainable agriculture. These systems prioritize the health of the ecosystem, minimize synthetic inputs, and aim to create a balanced and self-regulating farming environment. The integration of these practices involves a shift away from conventional methods and a commitment to regenerative and eco-friendly approaches (Bezabih et al., 2016).

1. **Benefits:**
   a. **Reduced Chemical Dependency:** One of the primary goals is to reduce the reliance on synthetic pesticides, herbicides, and fertilizers. By using organic alternatives like neem oil, compost, and cover cropping, farmers can significantly decrease their chemical inputs.
   b. **Biodiversity Conservation:** Integrating natural practices such as creating habitats for beneficial insects or planting cover crops can enhance biodiversity on the farm. This can improve pest control, pollination, and overall ecosystem resilience.

c. **Soil Health:** Organic farming emphasizes building and maintaining healthy soil through techniques like composting and reduced tillage. Healthy soils promote better water retention, root development, and nutrient availability.

2. **Challenges:**

a. **Transition Period:** The transition from conventional to organic farming may involve initial yield reductions and increased labor demands, which can be financially challenging for farmers.

b. **Certification Costs:** To access organic markets, farmers may need to bear the cost of organic certification, which can be prohibitive for small-scale farmers.

c. **Market Demand:** The integration of organic and natural practices must align with market demand for such products. Consumer education and marketing strategies are essential.

10.8.2 CROP DIVERSIFICATION

Diversification in agriculture involves growing a variety of crops rather than monoculture. It is a fundamental practice in sustainable farming systems, as it can reduce vulnerabilities to climate change, pests, and market fluctuations (Koner and Laha, 2020).

1. **Benefits:**

a. **Risk Mitigation:** Different crops have different environmental requirements and vulnerabilities. By growing a variety of crops, farmers can spread risk and reduce the impact of adverse weather conditions or disease outbreaks on their income.

b. **Enhanced Soil Health:** Crop rotation and diversification can improve soil health by preventing the depletion of specific nutrients and reducing pest and disease pressure.

c. **Market Access:** Diversification allows farmers to access a broader range of markets, including those for specialty or niche crops.

2. **Challenges:**

   a. **Market Knowledge:** Farmers need to understand market demand for various crops and adapt their production accordingly. This may require investment in training and research.

   b. **Infrastructure:** Diversifying may necessitate changes in equipment, storage, and processing infrastructure.

   c. **Knowledge and Skills:** Farmers must acquire knowledge about the cultivation and management of various crops, which can be time-consuming.

10.8.3 SOIL HEALTH AND NUTRIENT MANAGEMENT

Soil health is the cornerstone of sustainable agriculture. It is crucial for crop productivity, nutrient cycling, and overall ecosystem stability. Effective soil health management involves practices that enhance the biological, physical, and chemical properties of the soil (Iñiguez-Gallardo et al., 2022).

1. **Benefits:**

   a. **Nutrient Retention:** A healthy soil structure with abundant organic matter retains nutrients better, reducing the need for synthetic fertilizers and preventing nutrient runoff, which can harm water bodies.

   b. **Enhanced Water Management:** Soils rich in organic matter have improved water-holding capacity, reducing the risk of drought stress for crops.

   c. **Carbon Sequestration:** Sustainable soil management contributes to carbon sequestration, which can help mitigate climate change.

2. **Challenges:**

   a. **Transition Period:** Transitioning to soil health-improving practices can be challenging, as it often involves changing established farming methods.

   b. **Balancing Nutrients:** Managing nutrient levels in the soil can be complex, as it requires a good understanding of the nutrient requirements of various crops.

c. **Economic Viability:** Initially, investment in soil health may not yield immediate returns, which can be a barrier for some farmers.

Sustainable farming systems that integrate organic and natural practices, promote crop diversification, and prioritize soil health and nutrient management are essential components of climate-resilient agriculture. While these practices offer numerous benefits, they also present challenges, particularly in terms of transitioning from conventional methods and addressing economic considerations (Haddad et al., 2021). A supportive policy environment, access to training and resources, and alignment with market demands are major factors in the successful adoption and economic viability of these practices (Rijsberman, 2017).

10.9 FUTURE OUTLOOK AND RESEARCH NEEDS

10.9.1 EMERGING TRENDS IN ORGANIC AND NATURAL FARMING

Organic and natural farming methods have been gaining momentum in recent years due to increased awareness of environmental sustainability, health concerns, and the need for climate-resilient agriculture (Kong et al., 2016). The following emerging trends highlight the direction in which these practices are heading:

1. **Soil Health and Regeneration:**
   a. One of the primary focuses in organic and natural farming is improving soil health and promoting soil regeneration. This trend emphasizes the importance of healthy soils for sustainable agriculture and climate resilience.
   b. Practices such as no-till farming, cover cropping, and organic matter incorporation are becoming more common to enhance soil structure, water retention, and carbon sequestration.

2. **Agroecology and Biodiversity:**
   a. The adoption of agroecological principles is on the rise. Organic and natural farming methods encourage the creation of diverse agroecosystems, incorporating companion planting, beneficial insect habitats, and agroforestry.

   b. These practices not only enhance biodiversity but also contribute to pest control, reducing the reliance on synthetic pesticides.

3. **Precision Agriculture and Technology:**
   a. Technology is being integrated into organic and natural farming to improve efficiency and yields. Precision agriculture tools, data analytics, and IoT devices are used to monitor and manage resources more effectively.
   b. This helps in optimizing resource use, reducing waste, and achieving higher economic returns.

4. **Local and Direct-to-Consumer Sales:**
   a. Many organic and natural farmers are shifting towards local and direct-to-consumer sales channels, such as farmers' markets, community-supported agriculture (CSA), and online platforms.
   b. These channels provide farmers with higher profit margins and direct relationships with consumers.

5. **Climate-Resilient Varieties:**
   a. The development and adoption of crop varieties bred for climate resilience are becoming increasingly important. Organic and natural farmers are seeking crops that can withstand extreme weather conditions and pests without the use of synthetic chemicals.
   b. Breeding programs focusing on these traits are gaining attention.

10.9.2 RESEARCH GAPS AND AREAS FOR FURTHER STUDY

While organic and natural farming show great promise for climate-resilient agriculture, several research gaps and areas for further study must be addressed to ensure their long-term viability and success such as:

1. **Long-Term Economic Viability:** Comprehensive, long-term economic assessments of organic and natural farming systems

are needed. These studies should consider the costs, benefits, and returns over extended periods to determine the true economic sustainability of these practices.

2. **Market Demand and Consumer Behavior:** Understanding consumer preferences and behaviors related to organic and natural products is crucial. Research should focus on market demand, pricing dynamics, and the willingness of consumers to pay premiums for such products.
3. **Ecosystem Services and Climate Resilience:** In-depth research on the ecosystem services provided by organic and natural farming systems, such as carbon sequestration, water quality improvement, and biodiversity conservation, is essential (Zougmore et al., 2021).
4. **Pest and Disease Management:** Investigating effective, sustainable pest and disease management strategies for organic and natural farming is necessary. Research should target innovative techniques, biopesticides, and biological controls to reduce dependency on synthetic chemicals.
5. **Scaling Up and Policy Support:** Research should explore ways to scale up organic and natural farming, both in terms of acreage and the number of farmers adopting these practices. Policy frameworks that support sustainable agriculture and provide incentives for transitioning to organic and natural methods need to be examined.
6. **Socioeconomic Impacts:** The social and economic impacts of organic and natural farming on rural communities, labor, and farmer livelihoods require in-depth analysis. Research should investigate income generation, employment opportunities, and social benefits.

Ensuring the future of organic and natural farming in climate-resilient agriculture hinges on addressing critical research gaps, embracing emerging trends, and fostering an enabling environment that nurtures sustainable farming practices (Singh et al., 2021).

10.10 CONCLUSION

Natural farming, based on its core principles of minimal intervention and reliance on natural processes, stands as a pivotal component of climate-resilient agriculture. This holistic approach prioritizes sustainability, soil health, and biodiversity, thereby contributing significantly to mitigating the adverse effects of climate change. Organic farming plays an instrumental role in promoting climate resilience by reducing greenhouse gas emissions, conserving water resources, and enhancing ecosystem services. The economic viability of organic farming is a critical factor for its widespread adoption. Our analysis of cost-benefit considerations demonstrates that while input costs may initially be higher, organic farming offers substantial long-term benefits, including improved soil fertility and reduced dependency on external inputs. ROI, break-even analysis, and profitability indicators underscore the profitability of this practice, highlighting organic farming's economic sustainability. The market prospects for organic and natural products are encouraging, as consumer trends reflect an increasing demand for healthier, sustainable alternatives. Certification and labeling systems have been pivotal in building consumer trust, and though market challenges persist, the export potential and global opportunities are substantial. Nevertheless, the challenges and barriers in organic farming are undeniable. Addressing these issues requires a multifaceted approach, involving enhanced farming practices, improved market access, streamlined regulations, and comprehensive risk management strategies. Government support is essential for fostering the growth of organic farming. Numerous initiatives, incentives, and subsidies can play a pivotal role in promoting the transition to organic practices, alongside the formulation of sound policy recommendations. Farmers, as the backbone of agriculture, must be equipped with the knowledge and resources necessary for the successful adoption of organic farming. Transitioning to this system, education and training, and the development of farmer cooperatives and networks are integral to this process. Sustainable farming systems that integrate organic and natural practices, emphasize crop diversification, and prioritize soil health and nutrient management are the way forward in ensuring the long-term sustainability of agriculture.

KEYWORDS

- **agricultural sustainability**
- **biodiversity**
- **carbon sequestration**
- **climate-resilient agriculture**
- **crop diversification**
- **environmental stewardship**
- **organic farming**
- **policy frameworks**

REFERENCES

Amjath-Babu, T. S., Aggarwal, P. K., & Vermeulen, S. (2019). Climate action for food security in South Asia? Analyzing the role of agriculture in nationally determined contributions to the Paris agreement. *Climate Policy, 19*(3), 283–298.

Amoak, D., Luginaah, I., & McBean, G. (2022). Climate change, food security, and health: Harnessing agroecology to build climate-resilient communities. *Sustainability, 14*(21), 13954.

APEDA. (2023). Statistics for the year 2022-23. https://www.apeda.gov.in/apedawebsite/organic/data.htm.

Asokan, R., Murugan, D., & Sathiyaraj, S. (2020). Environmental sustainability through zero budget natural farming. *Studies in Indian Place Names, 40*(12), 1876–1885.

Attri, S. D., & Mohapatra, M. (2021). Agrometeorological services for climate resilient agriculture. In *Climate Resilience and Environmental Sustainability Approaches: Global Lessons and Local Challenges* (pp. 127–139).

Babalad, H. B., & Navali, G. V. (2021). Comparative economics of zero budget natural farming with conventional farming systems in northern dry zone (Zone-3) of Karnataka. *Economic Affairs, 66*(2), 355–361.

Bagheri, M., Delbari, S. H., Pakzadmanesh, M., & Kennedy, C. A. (2019). City-integrated renewable energy design for low-carbon and climate-resilient communities. *Applied Energy, 239*, 1212–1225.

Beni, C., Neri, U., Papetti, P., & Altimari, A. (2021). Natural horticultural systems in organic farming as a tool for resilience: Improvement of economic performance and prevention of soil erosion. *Agroecology and Sustainable Food Systems, 45*(9), 1375–1398.

Bezabih, M., Lovo, S., Singer, G., & McLaren, C. (2016). Climate-resilient development in agrarian economies. In *The Economics of Climate-Resilient Development* (p. 164).

Bonilla-Cedrez, C., Steward, P., Rosenstock, T. S., Thornton, P., Arango, J., Kropff, M., & Ramirez-Villegas, J. (2023). Priority areas for investment in more sustainable and climate-resilient livestock systems. *Nature Sustainability*, 1–8.

Chandel, R. S., Gupta, M., Sharma, S., Sharma, P. L., Verma, S., & Chandel, A. (2021). Impact of Palekar's natural farming on farmers' economy in Himachal Pradesh. *Indian Journal of Ecology, 48*(3), 872–877.

Dev, P. K. (2013). Waterlogging through soil-less agriculture as a climate resilient adaptation option. In *Climate Change and Disaster Risk Management* (pp. 681–692).

Dev, P., Khandelwal, S., Yadav, S. C., Arya, V., & Mali, H. R. (2023). Climate-based smart agriculture: Need for food security and sustainability. *International Journal of Environment and Climate Change, 13*(3), 224–231.

Dorin, B. (2022). Theory, practice and challenges of agroecology in India. *International Journal of Agricultural Sustainability, 20*(2), 153–167.

Eeswaran, R., Nejadhashemi, A. P., & Miller, S. R. (2021). Evaluating the climate resilience in terms of profitability and risk for a long-term corn-soybean-wheat rotation under different treatment systems. *Climate Risk Management, 32*, 100284.

Gorripati, R., Thakur, M., & Kolagani, N. (2023). Promoting climate-resilient sustainable agriculture through participatory system dynamics with crop-water-income dynamics. *Water Resources Management, 37*(10), 1–17.

Gugissa, D. A., Abro, Z., & Tefera, T. (2022). Achieving a climate-change resilient farming system through push–pull technology: Evidence from maize farming systems in Ethiopia. *Sustainability, 14*(5), 2648.

Haddad, F. F., Ariza, C., & Malmer, A. (2021). Building climate-resilient dryland forests and agrosilvopastoral production systems: An approach for context-dependent economic, social and environmentally sustainable transformations. *Food & Agriculture Organization.*

Hansika, S., & Wijerathna, M. (2021). Evaluation of short organic food supply chains with special reference to climate smartness—The case of direct farmers' market, Kurunegala, Sri Lanka. *Journal of Agricultural Sciences (Sri Lanka), 16*(2).

Huyer, S., Simelton, E., Chanana, N., Mulema, A. A., & Marty, E. (2021). Expanding opportunities: A framework for gender and socially-inclusive climate resilient agriculture. *Frontiers in Climate, 3*, 718240.

Iñiguez-Gallardo, V., Loján Córdova, J., Ordoñez-León, A., & Reyes-Bueno, F. (2022). Food markets and free fairs as contributors for designing climate resilient cities: A study case in southern Ecuador. *Sustainability, 14*(12), 7214.

Jaganathan, D., Suja, G., & Byju, G. (2021). Enhancing farm income by coconut-based cropping systems involving climate resilient tuber crops. *Indian Coconut Journal, 17.*

Khurana, A., & Kumar, V. (2020). State of organic and natural farming: Challenges and possibilities. New Delhi.

Koner, N., & Laha, A. (2020). Economics of zero budget natural farming in Purulia district of West Bengal: Is it economically viable? *Studies in Agricultural Economics, 122*(1), 22–28.

Kong, R., Sar, V., Leng, V., Trang, S., Boulakia, S., Tivet, F., & Séguy, L. (2016). Conservation agriculture for climate-resilient rainfed uplands in the Western regions of Cambodia: Challenges, opportunities, and lessons from a 10-year R&D program. In P.

E. Sajise, M. C. H. Cadiz, & R. B. Bantayan (Eds.), *Learning and Coping with Change: Case Stories of Climate Change Adaptation in Southeast Asia* (pp. 55–81). SEARCA.

Kumar, N., Kushwaha, R. R., Meena, N. R., Mishra, H., & Yadav, A. P. (2023). A study on costs and returns of paddy cultivation in Ambedkar Nagar district of Uttar Pradesh. *International Journal of Statistics and Applied Mathematics, SP-8*(3), 107–111.

Kumar, R., Krishna, B., Sundaram, P. K., Kumawat, N., Jeet, P., & Singh, A. K. (2022). Crop diversification: An approach for productive and climate-resilient production system. In *Sustainable Agriculture Systems and Technologies* (pp. 63–80).

Lal, R. (2013). Climate-resilient agriculture and soil organic carbon. *Indian Journal of Agronomy, 58*(4), 440–450.

Li, X., & Siddique, K. H. (2020). Future smart food: Harnessing the potential of neglected and underutilized species for zero hunger. *Maternal & Child Nutrition, 16*, e13008.

Liu, S., Connor, J., Butler, J. R. A., Jaya, I. K. D., & Nikmatullah, A. (2016). Evaluating economic costs and benefits of climate resilient livelihood strategies. *Climate Risk Management, 12*, 115–129.

Manavalan. (2022). Organic farming sector of India–A statistical review and recommendations. In *Springer eBooks* (pp. 169–198). https://doi.org/10.1007/978-3-031-10406-0_11.

Marke, A. T. (2014). Climate-resilient agriculture and food security. *Global Solutions Network*, 1–43.

Mathukia, R. K., Ponkia, H. P., Mathukia, P. R., & Savalia, N. V. (2016). Organic farming: Climate resilient agriculture. *Journal of Eco-Friendly Agriculture, 11*(2), 95–105.

Meena, R. P., & Jha, A. (2018). Conservation agriculture for climate change resilience: A microbiological perspective. In *Microbes for Climate Resilient Agriculture* (pp. 165–190).

Mishra, H. (2021). A new era for Indian farming. *Times of Agriculture, 17*, 80–81. www.timesofagriculture.in.

Mishra, H., & Mishra, D. (2023). Artificial intelligence and machine learning in agriculture: Transforming farming systems. In *Research Trends in Agriculture Science* (Volume I, pp. 1–16). Bhumi Publishing.

Mishra, H., & Mishra, D. (2023). From comparative advantage to protectionism: Economic effects of trade wars on agricultural markets. *Agriallis, 5*(7), 29–36. www.agriallis.com (accessed on 16 January 2025).

Mishra, H., & Singh, M. (2023). Agritech in India: An overview. *Agriallis*, 28–38. www.agriallis.com (accessed on 16 January 2025).

Mishra, H., & Singh, M. (2023). Market liberalization and agricultural sector transformation: Paving the path to a dynamic future. *Times of Agriculture, 3*(7), 67–68.

Mishra, H., & Singh, M. (2023). Revolutionizing agriculture: Empowering farmers through technological adoption for increased productivity and profitability. *The Agriculture Magazine, 2*(8), 17–22. www.theagricultureonline.com (accessed on 16 January 2025).

Mishra, H., & Singh, M. (2023). The significance of wildlife sanctuaries for Indian agriculture. *Indian Farmer, 10*(05), 243–246.

Mishra, H., Supriya, & Mishra, D. (2023). Farm management and agribusiness strategies (pp. 110–128). In *Research and Reviews in Agriculture Science* (Volume I). Bhumi Publishing.

Munawar, S., ul Qamar, M. T., Mustafa, G., Khan, M. S., & Joyia, F. A. (2020). Role of biotechnology in climate resilient agriculture. In *Environment, Climate, Plant and Vegetation Growth* (pp. 339–365).

Nishad, D. C., Mishra, H., Tiwari, A. K., & Pandey, A. (2023). Towards sustainable agriculture: Mitigating the adverse effects of stubble burning in India. In *Research Trends in Environmental Science* (Volume I, pp. 42–48). Bhumi Publishing.

Patel, J., & Kumar, G. S. (2022). Climate resilient agriculture practices—The future of farming. *Plant Science Today, 9*(sp3), 63–70.

Rai, R. K., Bhatta, L. D., Acharya, U., & Bhatta, A. P. (2018). Assessing climate-resilient agriculture for smallholders. *Environmental Development, 27*, 26–33.

Reddy, B., Reddy, V., & Reddy, M. S. (2019). Potential and constraints of zero budgeted natural farming (ZBNF): A study of Andhra Pradesh. *Indian Journal of Agricultural Economics, 74*(3), 321–332.

Reddy, P. P. (2015). Climate resilient agriculture for ensuring food security (Vol. 373). New Delhi: Springer India.

Research Dive. (2022). Organic food market leading segment over the forecast period 2022–2030. *Research Dive*. https://www.researchdive.com/346/organic-food-market (accessed on 16 January 2025).

Rijsberman, F. (2017). The key role of the meat industry in transformation to a low-carbon, climate resilient, sustainable economy. *Meat Science, 132*, 2–5.

Saldanha, L. F. (2018). A review of Andhra Pradesh's climate resilient zero budget natural farming programme. *Environment Support Group, Bangalore*.

Samsudin, Y. B., Adzani, T., Ramadian, M. A., Naito, D., & Baral, H. (2021). Potential of agarwood as a climate-resilient livelihood option in Indonesia. *Bogor Regency, Indonesia: Center for International Forestry Research*.

Sarangthem, I., Haldhar, S. M., Mishra, L. K., & Thakuria, D. (2023). The book of abstract: International conference on natural farming for revitalizing environment and resilient agriculture (NF-RERA, 2023). *College of Agriculture, CAU, Imphal*, 374.

Sarmah, A., & Deka, P. C. (1993). Zero budget natural farming in Assam. *Health, 2*, 5–8.

Shehrawat, P. S., Bhakar, S., & Arulmanikandan, B. (2023). Extent of awareness and adoption of climate resilient crop production practices among farmers in Haryana state of India. *International Journal of Environment and Climate Change, 13*(10), 773–783.

Shilomboleni, H., Recha, J., Radeny, M., & Osumba, J. (2023). Scaling climate resilient seed systems through SMEs in Eastern and Southern Africa: Challenges and opportunities. *Climate and Development, 15*(3), 177–187.

Sinclair, F., Wezel, A., Mbow, C., Chomba, S., Robiglio, V., & Harrison, R. (2019). The contribution of agroecological approaches to realizing climate-resilient agriculture. *GCA: Rotterdam, The Netherlands*.

Singh, J., & Kumar, R. (2022). Concepts, importance and scope of natural farming in India. *RASSA Journal of Science for Society, 4*(2 and 3), 128–131.

Singh, M., & Mishra, H. (2023). Medicinal plant farming to improve farmers' economy. *Agriallis*, 30–36. www.agriallis.com (accessed on 16 January 2025).

Singh, M., & Mishra, H. (2023). Trade development and value chains in agriculture: Unleashing the potential for sustainable growth. *Times of Agriculture*, 56–57. www.timesofagriculture.in.

Singh, R. B. (2012). Climate change and food security. In *Improving Crop Productivity in Sustainable Agriculture* (pp. 1–22).

Singh, R. K., Kumar, S., Jat, H. S., Singh, A., Raju, R., & Sharma, D. K. (2014). Adaptation in rice-wheat based sodic agroecosystems: A case study on climate resilient farmers' practices. *Indian Journal of Traditional Knowledge, 13*(2), 377–389.

Singh, R., Machanuru, R., Singh, B., & Shrivastava, M. (2021). Climate-resilient agriculture: Enhance resilience toward climate change. In *Global Climate Change* (pp. 45–61). Elsevier.

Srivastav, A. L., Dhyani, R., Ranjan, M., Madhav, S., & Sillanpää, M. (2021). Climate-resilient strategies for sustainable management of water resources and agriculture. *Environmental Science and Pollution Research, 28*(31), 41576–41595.

Statista. (2023). Organic food market in India. https://www.statista.com/topics/10397/organic-food-market-in-india/ (accessed on 16 January 2025).

Taylor, M., & Bhasme, S. (2021). Between deficit rains and surplus populations: The political ecology of a climate-resilient village in South India. *Geoforum, 126*, 431–440.

Tiwari, A. K., Mishra, H., Nishad, D. C., & Pandey, A. (2023). Sustainable water management in agriculture: Irrigation techniques and water conservation (pp. 53–68). In *Research Trends in Agriculture Science* (Volume II). Bhumi Publishing.

Tran, D. D., Huu, L. H., Hoang, L. P., Pham, T. D., & Nguyen, A. H. (2021). Sustainability of rice-based livelihoods in the upper floodplains of Vietnamese Mekong Delta: Prospects and challenges. *Agricultural Water Management, 243*, 106495.

van der Pol, L. K., Tibbetts, C. A., & Lin, D. E. (2021). Removing barriers and creating opportunities for climate-resilient agriculture by optimizing federal crop insurance. *J. Sci. Policy. Govern, 18*, 0213.

Wilson, M. H., & Lovell, S. T. (2016). Agroforestry—The next step in sustainable and resilient agriculture. *Sustainability, 8*(6), 574.

Zougmoré, R. B., Läderach, P., & Campbell, B. M. (2021). Transforming food systems in Africa under climate change pressure: Role of climate-smart agriculture. *Sustainability, 13*(8), 4305.

CHAPTER 11

Innovative Extension Programs and Approaches for Climate Change Education in India

DIKSHA SRIVASTAVA,[1,*] GYAN SHUKLA,[2] SUDHANAND PRASAD LAL,[3] RATNESH KUMAR JHA,[4] DIBYAJYOTI NATH,[5] and SANJAY KUMAR[6]

[1]*Department of Extension Education and Communication Management, Dr. Rajendra Prasad Central Agricultural University, Pusa, Samastipur, Bihar, India*

[2]*Department of Extension Education, Institute of Agricultural Sciences, Banaras Hindu University, Varanasi, Uttar Pradesh, India*

[3]*Department of Agricultural Extension Education (PGCA), Dr. Rajendra Prasad Central Agricultural University, Pusa, Samastipur, Bihar, India*

[4]*Department of Agronomy, Center for Advance Studied on Climate Change (CASCC), Dr. Rajendra Prasad Central Agricultural University, Pusa, Samastipur, Bihar, India*

[5]*Department of Soil Science, Dr. Rajendra Prasad Central Agricultural University, Pusa, Samastipur, Bihar, India*

[6]*Department of Agronomy, Bihar Agricultural University, Sabour, Bhagalpur, Bihar, India*

*Corresponding author

ABSTRACT

Climate change poses significant challenges to agriculture, threatening the livelihoods and food security of farmers worldwide. To address these

Climate-Resilient Agriculture for Sustainable Crop Production: New Adaptation and Mitigation Strategies. Sanjay Kumar, Shivani Ranjan, Sumit Sow, Smruti Ranjan Padhan, and Dhirendra Kumar Roy (Eds.)

DOI: 10.1201/9781998511150-11

challenges, various innovative and community-based adaptation programs have been implemented to equip farmers with climate-smart strategies through Farmer Field Schools (FFS) under the NICRA Project and the ClimaAdapt Project. In addition to digital platforms such as the Harita Priya Project mobile application and the e-Arik Project, these initiatives promote climate adaptation and sustainable agriculture. Through the FFS approach, farmers are empowered with knowledge and skills to adopt climate-resilient agricultural practices. Digital innovation and mobile technology have become powerful tools for updating farmers on weather variabilities and providing real-time weather forecasts, crop advisories, and market intelligence. These tools support informed decision-making and optimized resource use. The e-Arik Project, an online platform, enables the exchange of best practices and knowledge among farmers, extension workers, and researchers, effectively bridging information gaps. It also promotes climate resilience across diverse farming communities. By equipping farmers with knowledge, digital tools, and community support, these initiatives have strengthened resilience to climate change impacts and offer valuable models for scaling up climate adaptation efforts in agriculture and securing sustainable livelihoods for farmers globally.

11.1 INTRODUCTION

Climate change stands out as one of the most prevailing issues challenging mankind nowadays. It pertains to the extent to which we are able to predict the weather for long periods of time (such as 30 years). Any systematic change in the long-term statistical patterns of climate variables over the course of several decades or longer is considered to be climate change (Malhi et al., 2021). Climate variables include temperatures, precipitation, pressure, and wind, and can be measured over a broad range of years. Moreover, different research on climate change has defined the concept of climate change as given in Table 11.1.

11.2 FACTORS CAUSING CLIMATE CHANGE

Climate change is being caused primarily by human (Srivastava et al., 2022) activities, including the combustion of fossil fuels, deforestation, and agricultural practices, such as livestock (Dube and Jaiswal, 2023)

TABLE 11.1 Selected Definitions of Climate Change

SL. No.	Definition	References
1.	"Climate change means a change in climate which is attributed directly or indirectly to human activity that alters the composition of the global atmosphere and which is in addition to natural climate variability is observed over comparable time periods."	Bodansky (1993)
2.	"Climate change refers to a change in the state of the climate that can be identified (e.g., by using statistical tests) by changes in the mean and/or the variability of its properties and that persists for an extended period, typically decades or longer."	IPCC (2018)
3.	"Climate change is generally defined as a significant variation of average weather conditions—say, conditions becoming warmer, wetter, or drier—over several decades or more."	NRDC (2021)
4.	"Global temperatures rose about 1.98°F offsite link (1.1°C) from 1901–2020, but climate change refers to more than an increase in temperature."	NOAA (2021)
5.	"Climate change refers to a statistically significant variation in either the mean state of the climate or in its variability, persisting for an extended period (typically decades or longer)."	WMO (2022)
6.	"Climate change refers to long-term shifts in temperatures and weather patterns. Such shifts can be natural, due to changes in the sun's activity or large volcanic eruptions."	UN (2023)
7.	"Climate change is a long-term shift in global or regional climate patterns. Often climate change refers specifically to the rise in global temperatures from the mid-20th century to present."	National Geographic (2023)
8.	"Weather conditions have changed significantly over multiple decades or longer as a result of climate change, for example, they have become hotter, wetter, or drier."	World Bank (2023)
9.	"Climate change refers to the increasing changes in the measures of climate over a long period of time – including precipitation, temperature, and wind patterns."	USGS (2023)
10.	"Climate change refers to the long-term changes in the Earth's climate that are heating up the atmosphere, ocean, and land."	UNDP (2023)

farming. A substantial quantity of greenhouse gases is produced by these activities, which results in an increase in these gases, which, in turn, leads to an increase in global temperatures (Kapri and Raj, 2020). The climate of the Earth has undergone noticeable and widespread changes over the last few decades due to human-induced changes in atmospheric composition. Since 1750, the concentration of greenhouse gases has increased manifold (Malhi et al., 2021), as shown in Table 11.2.

TABLE 11.2 Percentage Increment in Concentration of Greenhouses Gases (1750–2020)

SL. No.	Greenhouse Gases	Percentage Increase in Concentration (%)
1.	Methane (CH_4)	150
2.	Carbon dioxide (CO_2)	40
3.	Nitrous oxide (N_2O)	20

11.3 IMPACT OF CLIMATE CHANGE ON AGRICULTURE

The agricultural sector is more reliant on weather and climate conditions than any other economic sector, demonstrating a marked sensitivity to the effects of climate change, as well as a remarkable ability to adapt to it as it occurs (Shukla et al., 2023). Climate change and increased environmental instability are already having a significant impact on local communities across different regions of the world due to the effects of global climate change (Rodima-Taylor et al., 2011). In recent years, adaptation has emerged as an integral component of both climate policy and research endeavors, as a result of the increasing recognition of the irreversible nature of climate change (Smith et al., 2011). There are significant repercussions on both agricultural output and the economy of the country of India due to variations in annual monsoon rainfall, which range from droughts to floods (Kumar et al., 2004). A rise in minimum temperature is expected to increase the maintenance respiration needs of crops, which will, in turn, reduce the net growth and productivity of crops (Aggarwal, 2003). The increase in temperature has also been associated with an increase in the number of outbreaks and epidemics of diseases as a result of this temperature increase (Barford, 2013).

11.4 CLIMATE CHANGE ADAPTATION AND MITIGATION

In response to climate change, adaptation means making changes in the human or natural ecosystem in order to cope with the changing climate. Adaptation involves some moderate disruptions and sometimes even damages to existing opportunities and systems in the ecosystem (IPCC, 2007a). Essentially, mitigation is any change that is made to a system that reduces the release of greenhouse gases (GHG) into the atmosphere per unit of output by any possible means (IPCC, 2007b). As part of mitigation and adaptation strategies, it is necessary to adjust the sowing date, breed plants that are resistant to climate change, and improve agronomic practices so that they are more resilient. The successful adaptation of a country is not just a function of the government, but of various stakeholders (UN Climate Change, 2023a). Figure 11.1 shows the process of adaptation and the necessary support for its success (UN Climate Change, 2023b).

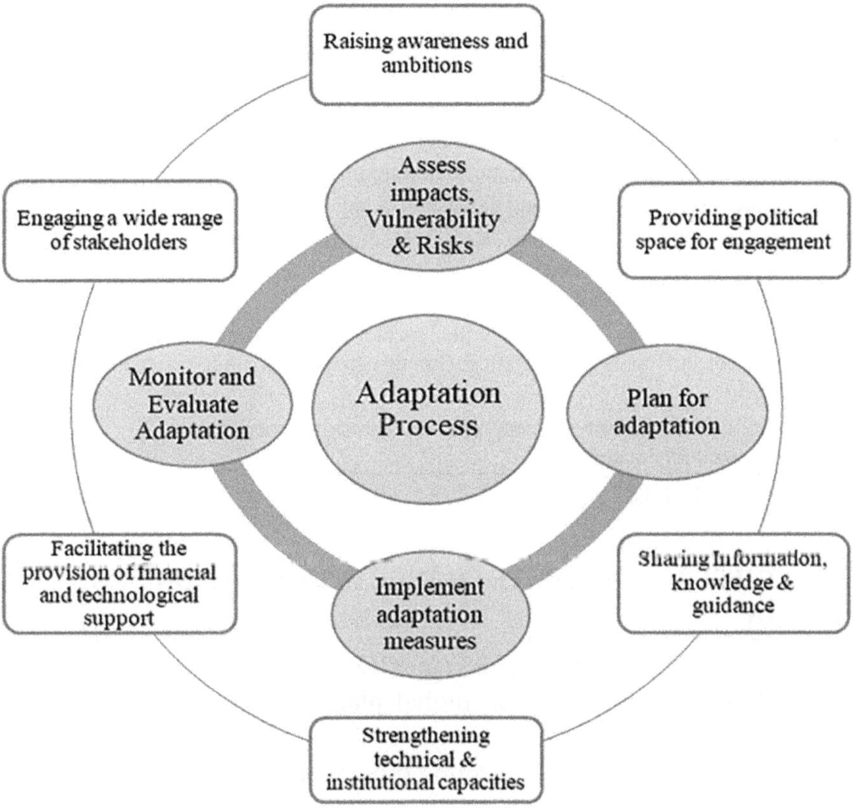

FIGURE 11.1 Adaptation process apropos climate change.

By recognizing the issues of climate change and the importance of innovative adaptation strategies to overcome climate change problems, the Government of India is also approaching adaptation programs through various innovative ways listed in Table 11.3.

TABLE 11.3 Various National Mission on Climate Change

SL. No.	NMCC	Year
1.	National Mission on Enhanced Energy Efficiency: Fostering sustainable business models and fostering conducive regulations and policies will strengthen our energy efficiency market	2009
2.	National Solar Mission: Setting up a solar technology enabling environment	2010
3.	National Mission on Sustainable Habitat: Mitigating and adapting to climate change (CC) through a multi-pronged approach to make urban areas more climate friendly	2010
4.	National Water Mission: Reduces water wastage, conserves water, and ensures equitable distribution of water	2010
5.	National Mission for Sustainable Agriculture: Makes agriculture more productive, high-paying, and climate resilient	2010
6.	National Mission on Strategic Knowledge for Climate Change: Promotes an eco-sustainability knowledge system that informs and supports national action	2010
7.	National Mission for Sustaining the Himalayan Ecosystem: Conserve and protect Himalayan glaciers, mountain ecosystems, biodiversity, and wildlife.	2014
8.	National Mission for a Green India: Biodiversity conservation and increased forest area.	2014

Source: DST CCP (2008).

11.5 DIFFERENT CLIMATE CHANGE INTERVENTION PROJECT IN INDIA

- Farmer field schools (FFS) in NICRA project;
- Mobile technology and digital platforms: HARITA-PRITA and e-Arik;
- Community-based adaptation program.

11.5.1 FARMER FIELD SCHOOLS (FFS) IN NICRA PROJECT

11.5.1.1 FARMER FIELD SCHOOL

"Farmer Field School," or FFS, is an innovative, participative educational model aimed at empowering farmers. In these schools, farmers are taught agriculture in a hands-on, field-based manner (Mbatha et al., 2022) in a risk-free and safe environment (FAO, 2019). In order to fulfill its objectives, FFS trains farmers to understand and have a deep understanding of sustainable agricultural practices. By engaging farmers in practical, experience-based activities, the model fosters learning, and farmers are able to make informed decisions, adapt to changing climatic conditions, which shows the adaptive capacity of farmers (Tomlinson et al., 2018), and effectively manage agricultural challenges with its focus on participatory engagement. Also, as a result of the FFS approach, local communities gain skills in observation and critical analysis, as well as improved decision-making by practicing hands-on learning (FAO, 2011).

11.5.1.2 ROLE OF FARMERS' FIELD SCHOOLS (FFS) IN CLIMATE ADAPTATION

FFS play a vital role in climate adaptation by equipping farmers with the knowledge and skills they need to mitigate the challenges posed by climate change in an effective manner (Osumba et al., 2021). During these schools, farmers come together to learn, experiment, and exchange ideas through participatory and experiential learning methods in order to learn from one another.

- Develops and enhances knowledge and capacity in the area of climate information and climate resilience.
- Supports the development of farmer-based institutions, such as groups and cooperatives.
- Ensures that institutions working on climate resilience are able to collaborate more effectively.
- Establishes a forum for dialogue concerning climate change adaptation practices and creates a platform for dialogue.
- Adopts new practices for adaptation to climate change in order to promote climate change mitigation.
- Increases social cohesion within a community by strengthening relationships between members.
- Promotes a sense of ownership and sustainability for initiatives related to climate change adaptation.

11.5.1.3 NICRA PROJECT

National Innovations on Climate Resilient Agriculture (NICRA) is a network project of the ICAR launched in February 2011 to create a culture of climate resilience in agriculture. Through strategic research and technology demonstrations, the project aims to strengthen the resilience of Indian agriculture to climate change as well as climate vulnerability. The NICRA project is composed of four components, as shown in Figure 11.2 (NICRA, 2011).

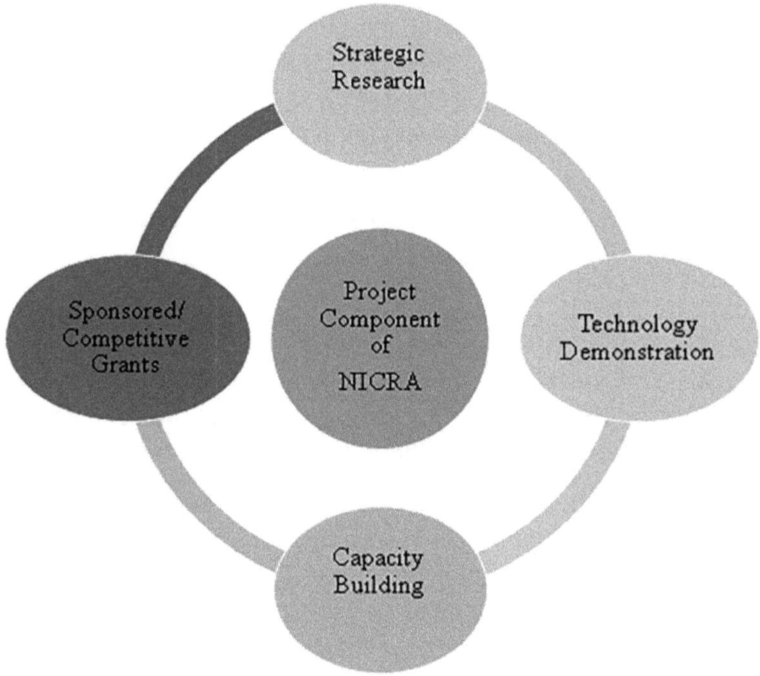

FIGURE 11.2 Project components of NICRA.

11.5.1.3.1 Objectives

- With the development and application of improved production and risk management technologies, Indian agriculture covering crops, livestock, and fisheries will enhance its resilience to climate change and climatic variability in the future.
- Demonstration of technology packages that are tailored to a specific site on farmers' fields in order to help them adapt to current climate risks
- Research and application of climate resilient agriculture will be enhanced by building the capacity of scientists and other stakeholders participating in the program (NICRA, 2011).

11.5.1.3.2 Project components of NICRA

1. **Strategic Research:** A strategic research initiative of ICAR involves its top research institutes in a variety of agricultural areas, including crops, livestock, and natural resource management. By linking weather-based agro-advisories with contingency planning, it assesses impacts, develops climate-tolerant varieties, and develops adaptation and mitigation plans.

2. **Technology Demonstration:** In order to address various climate vulnerabilities around the country, 121 Krishi Vigyan Kendras (KVKs) and the technology transfer division of core institutes have been established under this component to demonstrate climate resilient technologies in eight zones through technology transfer (Balasubramani and Dixit, 2016). Across the country, there are 130 village panchayats that provide services to about one lakh farmers. There are four modules in the Technology Demonstration Module:

    - **Module I (Natural Resources):** Enhances sustainable resource management through *in situ* water conservation, water harvesting, enhanced drainage in flood-prone areas, and conservation tillage;

    - **Module II (Crop Production):** In areas undergoing terminal heat stress, adjusting planting schedules for *rabi* crops, implementing water-saving cultivation methods, frost management in horticulture, and introducing community-based strategies for dealing with monsoon delays;

    - **Module III (Livestock and Fisheries):** In order to reduce livestock heat stress, improve fodder/feed storage methods, prevent disease spread, protect lands from droughts and floods, and manage fish ponds/tanks during water shortages and excesses; and

    - **Module IV (Institutional Interventions):** Establishing seed and fodder banks, community-based resource groups, custom hiring centers, collective marketing efforts, introducing weather index-based insurance, and promoting climate literacy.

3. **Capacity Building:** Training programs are being conducted in India to enhance scientists' simulation modeling, phenotyping,

and greenhouse gas measurements skills. Additionally, sponsored scientists receive international training. Over 100 training initiatives have been conducted nationwide, educating 50,000 farmers about climate change (NICRA, 2011).

4. **Sponsored/Competitive Grants:** Grants in this component address critical issues viz., the impact of climate change on plant pollinators, fisheries in estuarine habitats, hill and mountain ecosystems, hailstorm management, small ruminants, and socioeconomics.

11.5.1.4 IMPLEMENTATION OF FARMER FIELD SCHOOL (FFS) IN NICRA PROJECT

FFS is a popular approach used in agricultural development, such as the NICRA project in India. As part of the NICRA project, FFS is being used to enhance the resilience of agriculture to climate change in India. A typical FFS implementation in NICRA looks like this:

1. **Identifying the Target Areas for the Project:** The NICRA project identifies districts or regions in India that are vulnerable to climate change. As a result, these areas may experience changes in rainfall patterns, increased temperatures, and extreme weather events, making them prime candidates for climate adaptation strategies.
2. **Participant Selection:** The FFS selects farmers based on specific criteria, including their location within the target area, their willingness to participate, and their farming practices. To ensure that the benefits of the project reach a broad cross-section of farmers, efforts are made to include women, smallholders, and marginalized groups.
3. **A Training Program for Facilitators:** As part of the NICRA project, facilitators are trained who are often agricultural extension workers and experts in climate-resilient agriculture. In order to facilitate the Farmer Field School session, these facilitators are responsible for leading the sessions and guiding the participants through the learning process.
4. **Formation of Farmers' Field Schools Groups:** The farmers are organized into small groups, usually consisting of 20–25

individuals. There is a facilitator assigned to each group who will conduct regular sessions and visits to the field on a regular basis.

5. **A Participatory Approach to Learning:** Farmers are actively involved in the FFS sessions, taking part in practical exercises and experiments in an effort to learn. Through hands-on learning, participants gain a deeper understanding of climate-smart agricultural practices, as well as the ability to apply them in their own fields.
6. **Developing Field-Based Skills:** The core of the FFS approach is the learning process that takes place in the field. As part of their learning process during the sessions, the farmers visit their own fields to apply the knowledge and techniques they have acquired. It allows them to experiment with new practices and monitor their results in real time as they see the results in action.
7. **NICRA Monitoring and Evaluation:** As part of the NICRA project, an evaluation and monitoring component will be developed to assess the impact of the FFS on farmers' agricultural practices and their ability to adapt to climate change. Program success is measured by crop yields, resource efficiency, and other metrics.
8. **FFS Programs and Extension Services:** Farmers participating in FFS programs have access to additional resources, information, and government support for climate-resilient agriculture through formal extension services.
9. **A Scaling-Up Approach:** As a result of the success of the FFS approach, the NICRA project may scale up its impact on climate adaptation and resilience in Indian agriculture.

With the implementation of FFS, NICRA intends to provide farmers with information, skills, and practices that will enhance their capability to adapt to climate change, thereby enhancing agriculture's resilience and sustainability in the face of climate change.

11.5.2 MOBILE TECHNOLOGY AND DIGITAL PLATFORMS

The concept of a mobile digital platform implies the use of mobile devices, namely smartphones and tablets, for accessing a variety of digital services, applications, and interactions. There is a wide variety of activities and functions that can be facilitated by these platforms, which are the foundation for mobile computing and connectivity in the modern world. It is possible for digital platforms to serve both digital markets and physical markets. From start-ups to large enterprises, they fulfill a variety of business functions, primarily customer interactions (Patrizio, 2023).

11.5.2.1 IMPORTANCE OF MOBILE TECHNOLOGY AND DIGITAL PLATFORMS IN CLIMATE-SMART AGRICULTURE (CSA)

In CSA, mobile technology and digital platforms play a crucial role by providing farmers and agricultural stakeholders with tools and resources for adapting to changing climate conditions (Zougmoré et al., 2021). Figure 11.3 shows a few of the benefits of mobile technology and digital platforms in climate-smart agriculture. The benefits are:

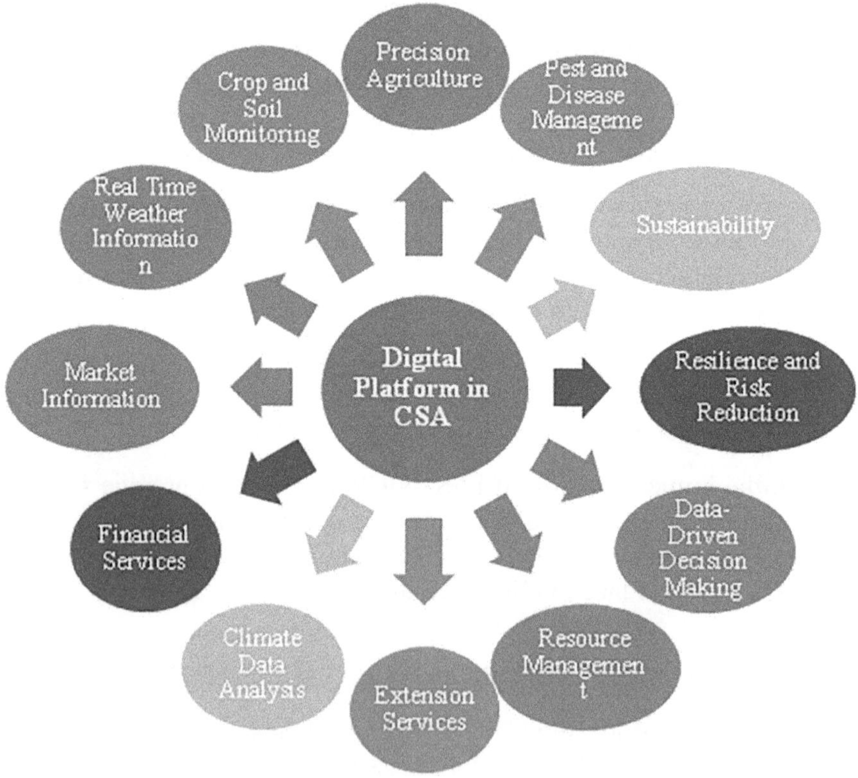

FIGURE 11.3 Benefits of mobile technology and digital platforms in climate-smart agriculture.

1. **Real-Time Weather Information:** Farmers can access real-time weather data and forecasts through mobile apps (Caine et al., 2015) and digital platforms. Farmers can use this information to make informed decisions about planting, irrigation, pest, and disease management, taking into account changing weather patterns.
2. **Crop and Soil Monitoring:** Farmers can monitor the health and condition of their crops and soils using digital tools. With the help of sensors and mobile apps, farmers can monitor soil moisture, nutrient levels, and crop growth in order to optimize resource use and adapt to environmental changes (Mohamed et al., 2021).
3. **Precision Agriculture:** With mobile technology, farmers can tailor their actions to specific fields using precision farming practices. In addition, this reduces resource waste, such as water and fertilizers, and helps improve crop yields.
4. **Pest and Disease Management:** Digital platforms provide information on pest and disease identification and management (Nitin et al., 2020). Using mobile apps, farmers can detect outbreaks early and receive mitigation recommendations.
5. **Market Information:** Farmers can use digital platforms to learn the market price (Deichmann, 2016) of their produce and decide when and where to sell it. As a result, they can maximize their profits and adapt to fluctuations in the market caused by climate-related events.
6. **Financial Services:** Mobile technology facilitates access to financial services, such as digital payments, crop insurance, and loans (Baumüller, 2012). Farmer risk management and climate resilience are aided by these services.
7. **Climate Data Analysis:** Climate data analysis through mobile technology and digital platforms is effective for climate adaptation and climate risk management (Adekola et al., 2023), which can assist farmers with long-term planning decisions, such as crop selection and planting and harvesting schedules.
8. **Knowledge Sharing and Extension Services:** Apps and platforms can provide farmers with information and training on climate-smart farming. Digital communities and discussion forums allow farmers to connect with experts and other farmers (Gandhi et al., 2007).

9. **Resource Management:** Mobile technology facilitates efficient resource management (Papavassiliou et al., 2002). A climate-change-driven drought and water scarcity may pose additional challenges for irrigation scheduling and water conservation, which can be addressed by this technology. It can also minimize energy consumption and reduce water waste.
10. **Data-Driven Decision-Making:** Farmers can use digital platforms to collect and analyze data. It is crucial to adapt to climate change and optimize agricultural practices in order to face the challenges of the future.
11. **Resilience and Risk Reduction:** Digital platforms can help farmers and agricultural stakeholders mitigate climate-related risks by integrating climate data and predictive analytics (Sarker et al., 2019).
12. **Sustainability:** Mobile and digital solutions support climate-smart agriculture practices that are sustainable and environmentally friendly.

Overall, the use of mobile technology and digital platforms in climate-smart agriculture can facilitate farmers in making informed decisions, improving productivity, reducing resource waste, and building resilience regarding climate change problems.

11.5.2.2 HARITA-PRIYA PROJECT

"HARITA-PRIYA" stands for Harmonized Information of Agriculture, Revenue, and Irrigation for a Transformation Agenda – Precision Technology for Agriculture. It is a pilot project implemented by the Center for Development of Advanced Computing (C-DAC), Hyderabad, in collaboration with the Government of Andhra Pradesh (ITU, 2023).

11.5.2.2.1 Objective

- To create a replicable model that allows farmers to collect micro-climate data in real time from their fields using wireless sensor networks (WSNs).
- By using this technology, the Agriculture Department will be able to provide farmers with personalized advice regarding their farming needs.

11.5.2.2.2 Conception

In the Anantapur District, Andhra Pradesh, a team from C-DAC deployed 74 WSN nodes in five villages, covering roughly 450 acres of groundnut crops. Through a field gateway with internet access, the WSN nodes continuously collect microclimate data (Mudda, 2018) at the crop canopy level and transmit this data to a remote server, which deals with crop-centric "Decision Support Models." The HARITA-PRIYA model has revolutionized agricultural advisory services, providing farmers with an average of 18 personalized advisories during the crop season, enhancing the accessibility and impact of weather data, and demonstrating the effectiveness of personalized SMS advisories in the regional language over traditional methods. HARITA-PRIYA is a project that offers farmers access to personalized, real-time advisory services to improve efficiency and sustainability, reduce input costs, and enhance their ability to withstand climate change.

11.5.2.3 E-ARIK PROJECT

e-Arik (e-agriculture) was launched in 2007 with the aim of disseminating "climate-smart farming practices" to ensure food security and environmental sustainability (Raj, 2008). In Yagrung village of Arunachal Pradesh State, India, a computer, scanner, TV, internet facility, printer, phone, and project facilitators have been installed to assist farmers in accessing farm information using information and communication technologies. To facilitate sustainable agricultural practices, e-Arik researchers conducted field visits to observe crop conditions, diagnose diseases and pests, and identify nutrient deficiencies and physiological problems. They then digitally documented crop conditions and communicated them to experts from the CAU, Imphal, for further evaluation (Raj, 2013). As part of the project portal, it provides the latest information on crop cultivation aspects, agricultural, and rural development departments, daily market information, and weather conditions on a daily basis. A number of innovative approaches, such as farmer-to-farmer communication, were implemented as part of the project.

11.5.2.3.1 Features of e-Arik services

- To provide a description of crop cultivation and other agricultural practices.
- To provide information on the government's welfare programs.

- To provide up-to-date daily market information and weather forecasts.
- To provide statistics from relevant government departments dealing with agriculture and rural development (Saravanan and Change, 2011).

11.5.3 COMMUNITY-BASED ADAPTATION PROGRAMS

The deliberate release of greenhouse gases into the atmosphere, mostly by developed nations, is the cause of climate change. Paradoxically, those living in underdeveloped and impoverished nations bear the brunt of the repercussions of greenhouse gas emissions despite having contributed the least to their production. The majority of adaptation initiatives in the recent past have been top-down, with little focus on how communities are adjusting to climate change and how they are coping with their changing surroundings. In order to mitigate the effects of climate change while empowering individuals and enhancing community resilience, community-based initiatives are becoming more and more popular. These approaches are participatory, community-based, and build on the priorities, knowledge, and capacities of locals (Kirkby et al., 2017). The term "community-based adaptation" (CBA) refers to a strengths-based, bottom-up strategy that targets disadvantaged communities in order to increase the adaptive ability of the community. According to Huq and Reid (2007), on page 1, there are those who view it as a "crucial approach to the threat climate change poses to the poor." Nevertheless, in spite of CBA's expanding use and appeal, as well as its rising professional community (Gundel et al., 2013), according to German sociologist Ferdinand Tönnies, a "community" is "a cohesive social entity" that is united by "unity of will" (New World Encyclopaedia, 2020). In this sense, an NGO, a cooperative, a close-knit ethnic group, a hamlet, or even a "society" established by law with particular goals could spearhead a community-based adaptation initiative. The IPCC acknowledged that community-based efforts to adapt to climate change in developing nations offer a variety of insights, but they also have their limitations (IPCC, 2014a). A wide range of climate adaptation programs, such as emergency preparedness, flood/drought protection, resilient livelihood solutions, water resource management, disaster risk reduction (DRR), sustainable agriculture, and food security, have been implemented in climate-vulnerable communities using community-based adaptation approaches (Shammin et al., 2022).

Communities can analyze the root causes and impacts of climate change, integrate community and scientific knowledge of the issue, and design adaptation measures with the use of CBA, an innovative participatory method. Smallholder farmers have a great deal of experience adjusting to their intricate, varied, and hazardous surroundings. The timing of rainy seasons and the pattern of rain within them, however, is becoming increasingly unpredictable, making farming riskier and more difficult. As a result, it is more difficult to cultivate, sow, and harvest crops during dry spells, which increases the need for resources. Climate change looks to be making significant issues like heat stress, pests, and diseases, and a shortage of water at critical periods worse. In order to reduce poverty, improve livelihoods, and lessen susceptibility to natural catastrophes and climate change, community-based adaptation (CBA) must begin with the articulated needs and views of the community. It could also come after working with locals to deal with a calamity like heavy flooding. International development NGOs and supporting donors often leverage local partners, like NGOs or community organizations already embedded and trusted within the local community, for implementing CBA initiatives. Information on climate change and its effects must be incorporated into planning procedures as part of CBA work (Shammin et al., 2021, 2022).

11.5.3.1 CATEGORIES OF CBA ACTIVITIES

Shammin et al. (2022) examined various adaptation strategies promoted as Community-Based Adaptation activities and classified them into six broad categories:

1. **Livelihood Diversification:** Seasonality and other long-term climatic variables constitute the basis for a large number of livelihood activities. This status quo, which sustains millions of people's livelihoods, is altered by climate change, and it is quite evident that tropical agriculture will produce less as a result of climate change. People's livelihoods and means of subsistence will be profoundly impacted by this everywhere in the world, but especially in tropical regions (IPCC, 2014a).

2. **Capacity Building:** Because they are frequently found in isolated locations, vulnerable groups lack access to contemporary technologies and knowledge bases. Therefore, many community-based adaptation projects aim to get the help of organizations, agencies, and funders. These programs create community-based awareness and/or sensitization campaigns in order to introduce new products and technologies.
3. **Ecosystem Integrity:** These objectives can be achieved through reforestation, restoring tidal patterns in coastal rivers, securing riverbanks, and building erosion control structures. At-risk communities can restore nature's ability to handle severe weather through these measures.
4. **Infrastructure:** A number of natural infrastructure projects have been built in South Asia to alleviate climate vulnerability, such as flood prevention, reducing saline intrusion into crop fields, and controlling erosion. For example, in Nepal, roofs are constructed using bamboo, and rainwater can be stored for irrigation, while in other countries like Bangladesh, riverbank erosion is prevented using bamboo structures.
5. **Microfinance and Insurance:** Microfinance is a type of lending that provides small loans to people in local communities. It is a common strategy used by many South Asian nations to diversify their income sources.
6. **Resource Management:** There are a number of community-based organizations engaged in revitalizing deteriorated ecosystems. However, some of these groups are simultaneously safeguarding lakes, rivers, and forests to protect them from the risks associated with climate change.

11.5.3.2 ROLE OF COMMUNITIES IN ADAPTING TO CLIMATE CHANGE

Communities play a vital role in adapting to climate change, such as:

1. **Local Knowledge and Expertise:** Communities possess valuable traditional and local knowledge about their environment,

ecosystems, and weather patterns. This knowledge is crucial for understanding the specific vulnerabilities and risks associated with climate change in their region.

2. **Identifying Vulnerabilities:** Communities can identify their unique vulnerabilities to climate change, such as extreme weather events, sea-level rise, or changing precipitation patterns. Their insights help prioritize adaptation efforts and tailor strategies to address local challenges.

3. **Sustainable Resource Management:** Communities can manage their local resources, such as water, forests, and agriculture, in a sustainable manner. Sustainable resource management enhances resilience to climate change and helps prevent environmental degradation.

4. **Infrastructure and Ecosystem-Based Solutions:** Communities can invest in resilient infrastructure, such as flood defenses or climate-resilient agriculture. They can also implement ecosystem-based adaptation approaches, like reforestation and wetland restoration, which enhance natural resilience.

11.5.3.3 CHARACTERISTICS OF COMMUNITY-BASED ADAPTATION PROGRAMS

The principles of CBA to climate change are overarching guidelines that underpin the approach and philosophy of empowering local communities to address climate change impacts effectively. These principles guide the development and implementation of CBA initiatives. Here are the key principles:

1. **Community-Centered Approach:** Communities are at the center of the adaptation process, and their needs, priorities, and aspirations guide decision-making and actions.

2. **Participation:** Active engagement of all community members, including marginalized and vulnerable groups, in the planning, implementation, and decision-making processes.

3. **Local Knowledge and Practices:** Recognition and incorporation of traditional knowledge and practices that communities have developed over time to adapt to local environmental conditions.

4. **Vulnerability and Risk Assessment:** Thorough assessment of the community's specific vulnerabilities to climate change impacts and understanding the risks they face.
5. **Context-Specific Solutions:** Development and implementation of adaptation strategies tailored to the unique needs and circumstances of the community, taking into account local ecological, social, and economic conditions.
6. **Capacity Building:** Building the capacity of community members through education, training, and skill development to enable them to plan, implement, and manage adaptation initiatives effectively.
7. **Sustainability and Ecosystem-Based Approaches:** Promotion of adaptation strategies that are environmentally sustainable and do not harm local ecosystems.
8. **Local Ownership:** Communities have ownership of the adaptation process, fostering a sense of agency and responsibility for their own well-being.

11.5.3.4 HIGHLIGHT OF SUCCESSFUL EXAMPLES OF COMMUNITY-BASED ADAPTATION PROGRAMS IN INDIA

11.5.3.4.1 Case 1: Community adaptation to flood: Kalana, Odisha

The Kalna community in Odisha, India, has cleverly adapted to flood hazards through diverse measures. This illustrates the institutionalization of community-based adaptation practices, demonstrating local innovation and sustainability (Dash, 2015). Kalna, a panchayat in Jaipur District, covers 600 hectares and is home to six villages with more than 1500 households. A community located between an embankment and the Bramhani River suffers from annual monsoon floods that engulf their homes and fields. The houses in Kalna are notable for being constructed on elevated bases that are dug out of nearby sites that are then turned into ponds. The ponds act as buffers, filling up first during floods, and will also serve as future fish farming sites. There are still many traditional thatched houses with bamboo and earth walls, which are more resistant to floodwaters, even though some have shifted to concrete. A number of storage provisions are included on roofs, and farming and small businesses are the primary activities in the community. As a result of the annual floods, their agricultural practices have evolved from paddy cultivation to winter vegetables. In spite of river erosion challenges, the community has adapted by cultivating peanuts and establishing businesses related to them. On flood embankments, villagers use temporary tents and communal kitchens to stay safe. Accessing submerged houses is done by boat. As a result of the 2001 embankment breach, the community reinforced riverbanks with stone packing to reduce flood durations. A disaster risk assessment must take into account not only potential losses but also livelihoods, housing, and community networks. As a result of recurring flood hazards, adaptation measures have become a vital part of the community's lives.

11.5.3.4.2 Case 2: Community mobilization and disaster mitigation

Case Study of South Odisha Since its founding 40 years ago, Trocaire has been working in Asia, Latin America, and Africa to empower local communities, eradicate poverty, and provide dignity during times of peace and disasters. Trocaire utilizes a Rights-Based Approach to empower local communities (Datta and Furlong, 2015). In 2003, Trocaire initiated long-term development projects in five impoverished districts in South Odisha, following the devastating super cyclone of 1999. In the last eight years, they have mobilized marginalized communities, strengthened Community-Based Organizations (CBOs), and promoted people-centric decentralized planning. Millions were affected, and substantial damage was caused by Cyclone Phailin in 2013. The state government's response was commendable, but Trocaire's remote Kandamal and Gajapati districts suffered as well. In these areas, vulnerable tribal and Dalit communities suffered crop and shelter destruction. Within 36 hours, 2735 households received emergency dry food, while 2000 households received temporary shelter materials, preventing destitution. Their rehabilitation efforts focused on the most vulnerable groups, including female-headed households, child-headed households, and people with disabilities. Beneficiaries were selected using a community-led process involving CBOs and villagers, with rigorous documentation and verification. Many beneficiaries preferred to rebuild their homes instead of receiving shelter repairs from Trocaire. Village Self-Help Groups (SHGs), CBO members, and relatives mobilized resources, including money, materials, and labor. Workers from the local community were engaged, promoting the value of "Srama Dan" (labor donation). Oversight and random checks ensured quality during the repair and reconstruction of the shelters in accordance with community preferences and available materials. It was the CBOs that made the project successful, as they efficiently identified and prioritized vulnerable members, distributed resources fairly, and mobilized additional support. As a result of the crisis, strong CBOs demonstrated their leadership and improved their relationship with the community, emphasizing the critical role that community mobilization and solidarity play in disaster response and recovery. As a result of this experience, we see the potential for local empowerment and collective action in addressing disaster-related challenges.

11.5.3.4.3 Case 3: Water availability spatiotemporal variation and its customary management

Ice reservoirs in Ladakh are deeply influenced by the region's traditional water management systems and historical experiences with water scarcity. Approximately 12.5% of Ladakh's land area is covered by glaciers (7715), which are found in the Upper Indus Basin. As a result of its unique geographical characteristics, Ladakh faces water scarcity despite this abundance. The region's topography, surrounded by towering mountains, makes it difficult to access water from low-lying rivers, which is why settlements are strategically located along glacial streams (Kumar and Saizen, 2023). Water availability varies according to the season, with glacial streams flowing strongly during summer and decreasing during winter. Water scarcity is most acute in spring (April-May), affecting agricultural activities. A short agricultural season further complicates the situation in Ladakh. As a result, Ladakh's inhabitants have developed traditional irrigation water management practices rooted in indigenous knowledge in order to address these natural constraints. The practice involves diverting, storing, and distributing water using canals, channels, and small reservoirs. The Tehsil office also records irrigation customs called "*riwaz-e-abpashi,*" or irrigation rules. The norms dictate that water is distributed on a rotational basis through "Chures" and encourage social bonds among households during their watering turns. As a result of Ladakh's unique geographical challenges, its ice reservoirs incorporate a complex system of water management that integrates indigenous knowledge, historical water rights, and social cooperation in order to utilize the available water resources effectively in an environment where water is scarce.

11.5.3.5 LIMITATIONS TO CAB PROGRAM

An adaptation gap is an important concept that is closely related to the concept of adaptation limits. There will always be a limit to the extent to which adaptation efforts can protect existing goals and values from risks to the extent they cannot provide an acceptable level of protection. As suggested by Dow et al. (2013), adaptation efforts can be regarded as incremental efforts that are aimed at keeping adverse impacts within a range that can be tolerated. For the purpose of preventing the escalation of loss and damage, an actor must either accept or change their behavior dramatically by reaching a threshold for intolerable risks. For example, when an adverse impact occurs with a given frequency and intensity, transformational measures, such as relocation, are adopted to prevent such escalation. As stated by the authors, "such discontinuities in risk or behavior are evidence of a limit to adaptation."

In developing countries, adaptation constraints are caused by a lack of access to credit, land, water, technology, markets, and information (IPCC, 2014b). It has been demonstrated in recent years, however, that CBA initiatives can be developed with low costs by using domestic resources and existing international assistance that is already available in the form of international assistance (Shammin et al., 2021).

11.6 CONCLUSION

In conclusion, the text underscores the critical importance of climate change adaptation in the agricultural sector, particularly in India. The initiatives discussed, such as Farmer Field Schools in the NICRA project, mobile technology, and digital platforms, showcase innovative approaches that empower farmers with knowledge and tools to adapt to changing climate conditions. These technologies provide real-time information, support resource management, and enable data-driven decision-making, ultimately enhancing agricultural resilience. Moreover, projects like HARITA-PRIYA and e-Arik exemplify the potential of advanced technologies in delivering personalized agricultural advice and fostering climate-smart farming practices. These initiatives bridge the gap between traditional farming methods and modern information and communication technologies, aiding farmers in managing climate-related risks and optimizing their yields. Community-based adaptation emerges as a pivotal strategy in addressing climate change, as it leverages local knowledge and empowers communities to identify vulnerabilities and implement context-specific solutions. The guiding principles of community-centered approaches, active participation, and sustainability underscore the importance of placing communities at the heart of climate adaptation efforts. In essence, the text highlights the multi-faceted and collaborative nature of climate change adaptation, showcasing how technology, local wisdom, and community involvement collectively contribute to building resilience in the face of climate challenges. As the world grapples with the ongoing impacts of climate change, these initiatives and strategies offer a promising path forward for safeguarding agricultural livelihoods and ensuring food security in the years to come.

KEYWORDS

- **agricultural productivity**
- **climate adaptation**
- **climate resilience**
- **climate-smart agriculture**
- **crop yields**
- **food security**
- **sustainable agriculture**
- **sustainable livelihood**

REFERENCES

Adekola, O., Lamond, J., Adelekan, I., Bhattacharya-Mis, N., Ekinya, M., Bassey Eze, E., & Ujoh, F. (2023). Towards adoption of mobile data collection for effective adaptation and climate risk management in Africa. *Geoscience Data Journal, 10*(2), 276–290.

Aggarwal, P. K. (2003). Impact of climate change on Indian agriculture. *Journal of Plant Biology, 30*(2), 189–198.

AIDMI. (2015). Successful community-based adaptation in India. All India Disaster Mitigation Institute. Retrieved from: https://reliefweb.int/report/india/southasiadisastersnet-issue-no-124-january-2015-successful-community-based-adaption (accessed on 16 January 2025).

Balasubramani, N., & Dixit, S. (2016). Innovative extension models in promoting climate change adaptation in agriculture. *Journal of Agricultural Extension, 17*(2).

Barford, E. (2013). Crop pests advancing with global warming. *Nature, 10*, 13644.

Baumüller, H. (2012). Facilitating agricultural technology adoption among the poor: The role of service delivery through mobile phones. ZEF Working Paper Series, 93.

Bodansky, D. (1993). The United Nations framework convention on climate change: A commentary. *Yale Journal of International Law, 18*, 451.

Bureau, D. Q. I. (2011). The e-Arik initiative: Weaving ICT into farming. *DATAQUEST*. Retrieved from: https://www.dqindia.com/the-e-arik-initiative-weaving-ict-into-farming/ (accessed on 16 January 2025).

Caine, A., Dorward, P., Clarkson, G., Evans, N., Canales, C., Stern, D., & Stern, R. (2015). Mobile applications for weather and climate information: Their use and potential for smallholder farmers. *CCAFS Working Paper*. https://cgspace.cgiar.org/rest/bitstreams/64569/retrieve (accessed on 16 January 2025).

Dash, B. (2015). Community adaptation to flood: Kalana, Odisha. In *Successful Community-Based Adaptation in India* (Issue No. 124). All India Disaster Mitigation Institute.

Datta, D., & Furlong, M. (2015). Community mobilization and disaster recovery: A case study from South Odisha. In *Successful Community-Based Adaptation in India* (Issue No. 124). All India Disaster Mitigation Institute.

Deichmann, U., Goyal, A., & Mishra, D. (2016). Will digital technologies transform agriculture in developing countries? *Agricultural Economics, 47*(S1), 21–33. https://doi.org/10.1111/agec.12231.

Dow, K., Berkhout, F., Preston, B., Klein, R. J. T., Midgley, G., & Shaw, M. R. (2013). Limits to adaptation. *Nature Climate Change, 3*, 305–307. https://doi.org/10.1038/nclimate1758.

DST CCP. (2008). National action plan on climate change. Climate Change Programme, Department of Science & Technology, GOI. Retrieved from: https://dst.gov.in/climate-change-programme (accessed on 16 January 2025).

Dube, A., & Jaiswal, S. (2023). Climate smart in agriculture: A ray of hope. In *Agricultural Transformation in the Modern Era* (pp. 1–6). Elite Publishing House.

FAO. (2011). Introduction to the Farmer Field School approach. *FAO e-learning Academy, Food and Agriculture Organization*. Retrieved from: https://elearning.fao.org/course/view.php?id=724 (accessed on 16 January 2025).

FAO. (2019). Farmer Field School. *Food and Agriculture Organization*. Retrieved from: https://www.fao.org/3/ca3627en/ca3627en.pdf (accessed on 16 January 2025).

Gandhi, R., Veeraraghavan, R., Toyama, K., & Ramprasad, V. (2007). Digital green: Participatory video for agricultural extension. *International Conference on Information and Communication Technologies and Development* (pp. 1–10). IEEE.

Gundel, S., Anderson, S., Nanki, K., & Schoch, C. (2013). Assessing the CBA community of practice. *International Institute for Environment and Development*. Retrieved from: https://www.iied.org/17152iied (accessed on 16 January 2025).

Huq, S., & Reid, H. (2007). Community-based adaptation: A vital approach to the threat climate change poses to the poor. *International Institute for Environment and Development*. Retrieved from: https://www.iied.org/17005iied (accessed on 16 January 2025).

IPCC. (2007a). *Impacts, Adaptation and Vulnerability: Summary for Policy Makers*. Intergovernmental Panel on Climate Change. Cambridge University Press. https://www.ipcc.ch/site/assets/uploads/2018/03/ar4_wg2_full_report.pdf (accessed on 16 January 2025).

IPCC. (2007b). *Intergovernmental Panel on Climate Change, Fourth Assessment Report*. United Nations. https://www.ipcc.ch/site/assets/uploads/2018/02/ar4_syr_full_report.pdf (accessed on 16 January 2025).

IPCC. (2014a). Technical summary. In *Climate Change 2014: Impacts, Adaptation, and Vulnerability. Part A: Global and Sectoral Aspects*. Contribution of Working Group II to the Fifth Assessment Report of the Intergovernmental Panel on Climate Change (pp. 1–32). Cambridge University Press. Retrieved from: https://www.ipcc.ch/site/assets/uploads/2018/02/WGIIAR5-FrontMatterA_FINAL.pdf (accessed on 16 January 2025).

IPCC. (2014b). Summary for policymakers. In *Climate Change 2014: Impacts, Adaptation, and Vulnerability. Part A: Global and Sectoral Aspects*. Contribution of Working Group II to the Fifth Assessment Report of the Intergovernmental Panel on Climate Change (pp. 1–32). Cambridge University Press.

IPCC. (2018). *IPCC: Global Warming of 1.5°C* (pp. 541–562). Cambridge University Press.

ITU. (2023). Harmonized information of agriculture, revenue, and irrigation for a transformation agenda—Precision technology for agriculture. Center for Development of Advanced Computing. https://www.itu.int/net4/wsis/archive/stocktaking/Project/Details?projectId=1449123327 (accessed on 16 January 2025).

Kapri, A., & Raj, S. (2020). Role of extension services in adaptation to climate change in Himachal Pradesh. *Discussion Paper 12*, MANAGE- Center for Agricultural Extension Innovations, Reforms and Agripreneurship, National Institute for Agricultural Extension Management.

Kirkby, P., Williams, C., & Huq, S. (2017). Community-based adaptation (CBA): Adding conceptual clarity to the approach, and establishing its principles and challenges. *Climate and Development, 10*(7), 577–589.

Kumar, K. K., Kumar, K. R., Ashrit, R. G., Deshpande, N. R., & Hansen, J. W. (2004). Climate impacts on Indian agriculture. *International Journal of Climatology, 24*(11), 1375–1393.

Kumar, T., & Saizen, I. (2023). Social innovation perspective of community-based climate change adaptation: A framework-based study of Ladakh, India. *Water, 15*(7), 1424.

Malhi, G. S., Kaur, M., & Kaushik, P. (2021). Impact of climate change on agriculture and its mitigation strategies: A review. *Sustainability, 13*(3), 1318.

Mbatha, A., Atingi, A., Mengesha, A., Duveskog, D., Kipkogei, O., & Grey, S. (2022). Climate-informed farmer field schools in Eastern Africa: The Agricultural Climate Resilience Enhancement Initiative (ACREI). World Meteorological Organization. https://public.wmo.int/en/climate-informed-farmer-field-schools-eastern-africa-agricultural-climate-resilience-enhancement.

Mohamed, E. S., Belal, A. A., Abd-Elmabod, S. K., El-Shirbeny, M. A., Gad, A., & Zahran, M. B. (2021). Smart farming for improving agricultural management. *The Egyptian Journal of Remote Sensing and Space Science, 24*(3), 971–981.

Mudda, S. K. (2018). GIS/GPS-based precision agriculture model in India: A case study. *Agribusiness and Information Management, 10*(2), 1–7.

National Geographic. (2023). Climate change education. *Encyclopedic Entry*. https://education.nationalgeographic.org/resource/climate-change/ (accessed on 16 January 2025).

New World Encyclopaedia. (2020). Community. *New World Encyclopaedia*. https://www.newworldencyclopedia.org/entry/Community# (accessed on 16 January 2025).

NICRA. (2011). NICRA Revised. *National Innovations on Climate Resilient Agriculture*. http://nicra-icar.in/nicrarevised/index.php/home1 (accessed on 16 January 2025).

Nitin, K. S., Loc, H. C., & Chakravarthy, A. K. (2020). Use of mobile apps and software systems for retrieving and disseminating information on pest and disease management. In *Innovative Pest Management Approaches for the 21st Century: Harnessing Automated Unmanned Technologies* (pp. 103–117).

NOAA. (2021). Climate change impacts. *Climate Resources*. National Oceanic and Atmospheric Administration U.S. Department of Commerce. https://www.noaa.gov/education/resource-collections/climate/climate-change-impacts (accessed on 16 January 2025).

NRDC. (2021). What is climate change? *National Research Development Corporation*. https://www.nrdc.org/stories/what-climate-change#whatis (accessed on 16 January 2025).

Osumba, J. J., Recha, J. W., & Oroma, G. W. (2021). Transforming agricultural extension service delivery through innovative bottom-up climate-resilient agribusiness farmer field schools. *Sustainability, 13*(7), 3938.

Papavassiliou, S., Puliafito, A., Tomarchio, O., & Ye, J. (2002). Mobile agent-based approach for efficient network management and resource allocation: Framework and applications. *IEEE Journal on Selected Areas in Communications, 20*(4), 858–872.

Patrizio, A. (2023). What is a digital platform? *Digital Transformation*. TechTarget. https://www.techtarget.com/searchcio/definition/digital-platform (accessed on 16 January 2025).

Raj, S. (2008). e-Arik: ICTs for agricultural extension services to the tribal farmers. IAALD AFITA WCCA. https://www.cabi.org/gara/mobile/FullTextPDF/2008/20083298081.pdf (accessed on 16 January 2025).

Raj, S. (2013). e-Agriculture prototype for knowledge facilitation among tribal farmers of North-East India: Innovations, impact, and lessons. *The Journal of Agricultural Education and Extension, 19*(2), 113–131.

Rodima-Taylor, D., Olwig, M. F., & Chhetri, N. (2012). Adaptation as innovation, innovation as adaptation: An institutional approach to climate change. *Applied Geography, 33*, 107–111.

Saravanan, R., & Change, C. (2011). eArik: Using ICTs to facilitate "climate-smart agriculture" among tribal farmers of North East India. *Climate Change, Innovation and ICTs Project Case Studies Center for Development Informatics, University of Manchester, UK*. https://g4aw.spaceoffice.nl/blobs/G4AW/overige%20docs/NICCD_AgricAdapt_Case_Study_eArik.pdf (accessed on 16 January 2025).

Sarker, M. N. I., Wu, M., Chanthamith, B., Yusufzada, S., Li, D., & Zhang, J. (2019). Big data-driven smart agriculture: Pathway for sustainable development. In *2nd International Conference on Artificial Intelligence and Big Data* (pp. 60–65). IEEE. https://doi.org/10.1109/ICAIBD.2019.00021.

Shammin, M. R., Firoz, R., & Hasan, R. (2021). Frameworks, stories and learnings from disaster management in Bangladesh. In *Climate Change and Community Resilience: Insights from South Asia* (pp. 239–256). Springer Nature.

Shammin, M. R., Firoz, R., & Hasan, R. (2022). Frameworks, stories and learnings from disaster management in Bangladesh. In A. K. E. Haque, P. Mukhopadhyay, M. Nepal, & M. R. Shammin (Eds.), *Climate Change and Community Resilience* (pp. 239–256). Springer, Singapore.

Shukla, M., Jangid, B. L., Khandelwal, V., Keerthika, A., & Shukla, A. K. (2023). Climate change and agriculture: An Indian perspective: A review. *Agricultural Reviews, 44*(2), 223–230.

Smith, M. S., Horrocks, L., Harvey, A., & Hamilton, C. (2011). Rethinking adaptation for a 4°C world. *Philosophical Transactions of the Royal Society A: Mathematical, Physical and Engineering Sciences, 369*, 196–216.

Srivastava, D., Kumari, A., & Lal, S. P. (2022). Factors influencing adoption of climate-friendly oxo-biodegradable Jan Ausadhi Suvidha sanitary napkins among women in India. *International Journal of Environment and Climate Change, 12*(12), 1754–1760.

Tomlinson, J., & Rhiney, K. (2018). Assessing the role of farmer field schools in promoting pro-adaptive behaviour towards climate change among Jamaican farmers. *Journal of Environmental Studies and Sciences, 8*, 86–98.

UN Climate Change. (2023a). Adaptation and resilience. *UNFCCC*. https://unfccc.int/topics/adaptation-and-resilience/the-big-picture/introduction (accessed on 16 January 2025).

UN Climate Change. (2023b). Adaptation and resilience. *UNFCCC*. https://unfccc.int/topics/adaptation-and-resilience/the-big-picture/introduction (accessed on 16 January 2025).

UN. (2023). What is climate change? *Climate Action*. United Nations. https://www.un.org/en/climatechange/what-is-climate-change (accessed on 16 January 2025).

UNDP. (2023). The climate dictionary: An everyday guide to climate change. *UNDP: Climate Promise*. https://climatepromise.undp.org/news-and-stories/climate-dictionary-everyday-guide-climate-change (accessed on 16 January 2025).

USGS. (2023). What is the difference between global warming and climate change? *USGS*. https://www.usgs.gov/faqs/what-difference-between-global-warming-and-climate-change (accessed on 16 January 2025).

WMO. (2022). What is climate change? *FAQs: Climate*. World Meteorological Organization. https://public.wmo.int/en/about-us/frequently-asked-questions/climate (accessed on 16 January 2025).

World Bank. (2023). Climate change knowledge portal. *World Bank*. https://climateknowledgeportal.worldbank.org/overview (accessed on 16 January 2025).

Zougmoré, R. B., Läderach, P., & Campbell, B. M. (2021). Transforming food systems in Africa under climate change pressure: Role of climate-smart agriculture. *Sustainability, 13*(8), 4305.

Index

D

E

F

T

For Product Safety Concerns and Information please contact our EU
representative GPSR@taylorandfrancis.com
Taylor & Francis Verlag GmbH, Kaufingerstraße 24, 80331 München, Germany

www.ingramcontent.com/pod-product-compliance
Ingram Content Group UK Ltd.
Pitfield, Milton Keynes, MK11 3LW, UK
UKHW021512250926
475702UK00043B/112
* 9 7 8 1 9 9 8 5 1 1 1 4 3 *